CompTIA® Network+ N10-009 Exam Cram

Companion Website and Pearson Test Prep Access Code

Access interactive study tools on this book's companion website, including practice test software, Key Term flash card application, a Cram Sheet, and more!

To access the companion website, simply follow these steps:

1. Go to www.pearsonITcertification.com/register.

2. Enter the **print book ISBN**: 9780135340837.

3. Answer the security question to validate your purchase.

4. Go to your account page.

5. Click on the **Registered Products** tab.

6. Under the book listing, click on the **Access Bonus Content** link.

When you register your book, your Pearson Test Prep practice test access code will automatically be populated with the book listing under the Registered Products tab. You will need this code to access the practice test that comes with this book. You can redeem the code at **PearsonTestPrep.com**. Simply choose Pearson IT Certification as your product group and log into the site with the same credentials you used to register your book. Click the **Activate New Product** button and enter the access code. More detailed instructions on how to redeem your access code for both the online and desktop versions can be found on the companion website.

If you have any issues accessing the companion website or obtaining your Pearson Test Prep practice test access code, you can contact our support team by going to **pearsonitp.echelp.org**.

CompTIA®
Network+
N10-009
Exam Cram

Emmett Dulaney

Pearson

CompTIA® Network+ N10-009 Exam Cram

28 2025

GM K12, Early Career and Professional Learning
Soo Kang

Director, ITP Product Management
Brett Bartow

Executive Editor
Nancy Davis

Development Editor
Ellie Bru

Managing Editor
Sandra Schroeder

Senior Project Editor
Tonya Simpson

Copy Editor
Chuck Hutchinson

Indexer
Timothy Wright

Proofreader
Barbara Mack

Technical Editor
Chris Crayton

Publishing Coordinator
Cindy Teeters

Cover Designer
Chuti Prasertsith

Compositor
codeMantra

Contents at a Glance

Online

Glossary

Table of Contents

About the Author

Emmett Dulaney (CompTIA Network+, Cloud+, Security+, A+, and others) has been the author of several books on certifications and operating systems over the past 25 years. He is a columnist for *Certification Magazine* and a professor at a small university in Indiana. He is currently the editor of a journal devoted to business education (and the business of education).

Dedication

For Harrison, Teresea, Wolfgang, and Elijah: never stop networking

—Emmett Dulaney

Acknowledgments

Thanks are due to Eleanor (Ellie) Bru for working on this title once more and making it as strong as it can be. An enormous amount of credit for this book always goes to Chris Crayton, who goes above and beyond what is expected and without whom the resulting text would only be a shadow of what it is. It is an honor to work with him. Thanks continue to be due to Mike Harwood, who wrote the first few editions, and to the team of talented individuals at Pearson who work behind the scenes and make each title the best it can be.

About the Technical Reviewer

Chris Crayton, MCSE, CISSP, CASP+, PenTest+, Project+, CySA+, Cloud+, S+, N+, A+, ITF+ is a technical consultant, trainer, author, and industry-leading technical editor. He has worked as a computer technology and networking instructor, information security director, network administrator, network engineer, and PC specialist. Chris has served as technical editor and content contributor on numerous technical titles for several of the leading publishing companies. He has also been recognized with many professional and teaching awards, and has served as a state-level SkillsUSA final competition judge.

We Want to Hear from You!

As the reader of this book, *you* are our most important critic and commentator. We value your opinion and want to know what we're doing right, what we could do better, what areas you'd like to see us publish in, and any other words of wisdom you're willing to pass our way.

We welcome your comments. You can email or write to let us know what you did or didn't like about this book—as well as what we can do to make our books better.

Please note that we cannot help you with technical problems related to the topic of this book.

When you write, please be sure to include this book's title and author as well as your name and email address. We will carefully review your comments and share them with the author and editors who worked on the book.

Email: community@informit.com

Reader Services

Register your copy of *CompTIA Network+ N10-009 Exam Cram* for convenient access to downloads, updates, and corrections as they become available. To start the registration process, go to www.pearsonitcertification.com/register and log in or create an account*. Enter the product ISBN 9780135340837 and click Submit. When the process is complete, you will find any available bonus content under Registered Products.

*Be sure to check the box that you would like to hear from us to receive exclusive discounts on future editions of this product.

Introduction

Welcome to *CompTIA Network+ N10-009 Exam Cram*. This book is designed to prepare you to take—and pass—the CompTIA Network+ exam. The Network+ exam has become the leading introductory-level network certification available today. It is recognized by both employers and industry giants as providing candidates with a solid foundation of networking concepts, terminology, and skills. The Network+ exam covers a broad range of networking concepts to prepare candidates for the technologies they are likely to work with in today's network environments.

About Network+ Exam Cram

Exam Crams are designed to give you the information you need to know to prepare for a certification exam. They cut through the extra information, focusing on the areas you need to get through the exam. With this in mind, the elements within the Exam Cram titles are aimed at providing the exam information you need in the most succinct and accessible manner.

In this light, this book is organized to closely follow the actual CompTIA objectives for exam N10-009. As such, it is easy to find the information required for each of the specified CompTIA Network+ objectives. The objective focus design used by this Exam Cram is an important feature because the information you need to know is easily identifiable and accessible. To see what we mean, compare the CompTIA objectives to the book's layout, and you can see that the facts are right where you would expect them to be.

Within the chapters, potential exam hotspots are clearly highlighted with Exam Alerts. They have been carefully placed to let you know that the surrounding discussion is an important area for the exam. To further help you prepare for the exam, a Cram Sheet is included that you can use in the final stages of test preparation. Be sure to pay close attention to the bulleted points on the Cram Sheet because they pinpoint the technologies and facts you probably will encounter on the test.

Finally, great effort has gone into the questions that appear throughout the chapter and the practice tests to ensure that they accurately represent the look and feel of the ones you will see on the real Network+ exam. Be sure, before taking the exam, that you are comfortable with both the format and content of the questions provided in this book.

About the Network+ Exam

The Network+ (N10-009 Edition) exam is the newest iteration of several versions of the exam. The new Network+ objectives are aimed toward those who have at least 9 to 12 months of experience in the IT networking field. While it is helpful if Network+ candidates have A+ certification (or its equivalent), it is not required, and this should not discourage those who do not.

You will have a maximum of 90 minutes to answer the 90 questions on the exam. The allotted time is quite generous, so when you finish, you probably will have time to double-check a few of the answers you were unsure of.

By the time the dust settles, you need a minimum score of 720 to pass the Network+ exam. This is on a scale of 100 to 900. For more information on the specifics of the Network+ exam, refer to CompTIA's main website at http://certification.comptia.org/.

CompTIA Network+ Exam Topics

Table I-1 lists general exam topics (that is, objectives) and specific topics under each general topic (that is, subobjectives) for the CompTIA Network+ N10-009 exam. This table also lists the chapter in which each exam topic is covered.

TABLE I-1   **CompTIA Network+ Exam Topics**

Chapter	N10-009 Exam Objective	N10-009 Exam Subobjective
1 (Networking Models, Ports, Protocols, and Services)	1.0 Networking Concepts	1.1 Explain concepts related to the Open Systems Interconnection (OSI) reference model.
		1.4. Explain common networking ports, protocols, services, and traffic types.
2 (Network Topologies, Architectures, and Types)	1.0 Networking Concepts	1.6 Compare and contrast network topologies, architectures, and types.
3 (Network Addressing, Routing, and Switching)	1.0 Networking Concepts	1.7 Given a scenario, use appropriate IPv4 network addressing.
	2.0 Network Implementation	2.1 Explain characteristics of routing technologies.
	3.0 Network Operations	2.2 Given a scenario, configure switching technologies and features.
		3.4 Given a scenario, implement IPv4 and IPv6 network services.

Chapter	N10-009 Exam Objective	N10-009 Exam Subobjective
4 (Network Implementations)	1.0 Networking Concepts	1.2 Compare and contrast networking appliances, applications, and functions.
		1.8 Summarize evolving use cases for modern network environments.
5 (Cabling Solutions and Issues)	1.0 Networking Concepts	1.5 Compare and contrast transmission media and transceivers.
	5.0 Network Troubleshooting	5.2 Given a scenario, troubleshoot common cabling and physical interface issues.
6 (Wireless Solutions)	2.0 Network Implementation	2.3 Given a scenario, select and configure wireless devices and technologies.
7 (Cloud Computing Concepts and Options)	1.0 Networking Concepts	1.3 Summarize cloud concepts and connectivity options.
8 (Network Operations)	2.0 Network Implementation	2.4 Explain important factors of physical installations.
	3.0 Network Operations	3.1 Explain the purpose of organizational processes and procedures.
		3.2 Given a scenario, use network monitoring technologies.
		3.3 Explain disaster recovery (DR) concepts.
		3.5 Compare and contrast network access and management methods.
9 (Network Security)	4.0 Network Security	4.1 Explain the importance of basic network security concepts.
		4.2 Summarize various types of attacks and their impact to the network.
		4.3 Given a scenario, apply network security features, defense techniques, and solutions.
10 (Network Troubleshooting)	5.0 Network Troubleshooting	5.1 Explain the troubleshooting methodology.
		5.3 Given a scenario, troubleshoot common issues with network services.
		5.4 Given a scenario, troubleshoot common performance issues.
		5.5 Given a scenario, use the appropriate tool or protocol to solve networking issues.

Booking and Taking the Network+ Certification Exam

Unfortunately, testing is not free. You're charged for each test you take, whether you pass or fail. In the United States and Canada, tests are administered by Pearson VUE testing services. To access the VUE contact information and book an exam, refer to the website at http://www.pearsonvue.com or call 1-877-551-7587. When booking an exam, you need to provide the following information:

▶ Your name as you would like it to appear on your certificate.

▶ Your Social Security or Social Insurance number.

▶ Contact phone numbers (to be called in case of a problem).

▶ Mailing address, which identifies the address to which you want your certificate mailed.

▶ Exam number and title.

▶ Email address for contact purposes. This often is the fastest and most effective means to contact you. Test vendors require it for registration.

▶ Credit card information so that you can pay online. You can redeem vouchers by calling the respective testing center.

What to Expect from the Exam

If you haven't taken a certification test, the process can be a little unnerving. Even if you've taken numerous tests, it is not much better. Mastering the inner mental game often can be as much of a battle as knowing the material. Knowing what to expect before heading in can make the process a little more comfortable.

Certification tests are administered on a computer system at a VUE authorized testing center. The format of the exams is straightforward: each question has several possible answers to choose from. The questions in this book provide a good example of the types of questions you can expect on the exam. If you are comfortable with them, the test should hold few surprises. Many of the questions vary in length. Some of them are longer scenario questions, whereas others are short and to the point. Carefully read the questions; the longer questions often have a key point that will lead you to the correct answer.

Most of the questions on the Network+ exam require you to choose a single correct answer, but a few require multiple answers. When there are multiple correct answers, a message at the bottom of the screen prompts you to "Choose all that apply." Be sure to read these messages.

A Few Exam-Day Details

It is recommended that you arrive at the examination room at least 15 minutes early, although a few minutes earlier certainly would not hurt. This will give you time to prepare and will give the test administrator time to answer any questions you might have before the test begins. Many people suggest that you review the most critical information about the test you're taking just before the test. (Exam Cram books provide a reference—the Cram Sheet, located inside the front of this book—that lists the essential information from the book in distilled form.) Arriving a few minutes early will give you some time to compose yourself and mentally review this critical information.

You will be asked to provide two forms of ID, one of which must be a photo ID. Both of the identifications you choose should have a signature. You also might need to sign in when you arrive and sign out when you leave.

Be warned: The rules are clear about what you can and cannot take into the examination room. Books, laptops, note sheets, and so on are not allowed in the examination room. The test administrator will hold these items, to be returned after you complete the exam. You might receive either a wipe board or a pen and a single piece of paper for making notes during the exam. The test administrator will ensure that no paper is removed from the examination room.

After the Test

Whether you want it or not, as soon as you finish your test, your score displays on the computer screen. In addition to the results appearing on the computer screen, a hard copy of the report prints for you. Like the onscreen report, the hard copy displays the results of your exam and provides a summary of how you did on each section and on each technology. If you were unsuccessful, this summary can help you determine the areas you need to brush up on.

Last-Minute Exam Tips

Studying for a certification exam is no different than studying for any other exam, but a few hints and tips can give you the edge on exam day:

▶ **Read all the material:** CompTIA has been known to include material not expressly specified in the objectives. This book has included additional information not reflected in the objectives to give you the best possible preparation for the examination.

▶ **Watch for the Exam Tips and Notes:** The Network+ objectives include a wide range of technologies. ExamAlerts, Tips, and Notes found throughout each chapter are designed to pull out exam-related hotspots. These can be your best friends when preparing for the exam.

▶ **Use the questions to assess your knowledge:** Don't just read the chapter content; use the exam questions to find out what you know and what you don't. If you struggle, study some more, review, and then assess your knowledge again.

▶ **Review the exam objectives:** Develop your own questions and examples for each topic listed. If you can develop and answer several questions for each topic, you should not find it difficult to pass the exam.

Good luck!

Companion Website

Register this book to get access to the Pearson Test Prep practice test software and other study materials plus additional bonus content. Check this site regularly for new and updated postings written by the author that provide further insight into the more troublesome topics on the exams. Be sure to check the box that you would like to hear from us to receive updates and exclusive discounts on future editions of this product or related products.

To access this companion website, follow these steps:

1. Go to www.pearsonITcertification.com/register and log in or create a new account.

2. Enter the ISBN: 9780135340837.

3. Answer the challenge question as proof of purchase.

4. Click the **Access Bonus Content** link in the Registered Products section of your account page, to be taken to the page where your downloadable content is available.

Please note that many of our companion content files can be very large, especially image and video files.

If you are unable to locate the files for this title by following these steps, please visit www.pearsonITcertification.com/contact and select the Site Problems/Comments option. Our customer service representatives will assist you.

How to Access the Pearson Test Prep (PTP) App

You have two options for installing and using the Pearson Test Prep application: a web app and a desktop app. To use the Pearson Test Prep application, start by accessing the registration code that comes with the book. You can access the code in these ways:

▶ You can get your access code by registering the print ISBN 9780135340837 on https://www.pearsonitcertification.com/register. Make sure to use the print book ISBN, regardless of whether you purchased an eBook or the print book. After you register the book, your access code will be populated on your account page under the Registered Products tab. Instructions for how to redeem the code are available on the book's companion website by clicking the **Access Bonus Content** link.

▶ If you purchase the Premium Edition eBook and Practice Test directly from the Pearson IT Certification website, the code will be populated on your account page after purchase. Just log in at https://www.pearsonitcertification.com, click **Account** to see details of your account, and click the **Digital Purchases** tab.

> **Note**
>
> After you register your book, your code can always be found in your account under the Registered Products tab.

Once you have the access code, to find instructions about both the PTP web app and the desktop app, follow these steps:

Step 1. Open this book's companion website as shown on the first page of the book.

Step 2. Click the Practice Exams button.

Step 3. Follow the instructions listed there for both installing the desktop app and using the web app.

Note that if you want to use the web app only at this point, just navigate to https://www.pearsontestprep.com, log in using the same credentials used to register your book or purchase the Premium Edition, and register for this book's practice tests using the registration code you just found. The process should take only a couple of minutes.

Customizing Your Exams

After you are in the exam settings screen, you can choose to take exams in one of three modes:

▶ Study Mode: Enables you to fully customize your exams and review answers as you are taking the exam. This is typically the mode you would use first to assess your knowledge and identify information gaps.

▶ Practice Exam Mode: Locks certain customization options because it is presenting a realistic exam experience. Use this mode when you are preparing to test your exam readiness.

▶ Flash Card Mode: Strips out the answers and presents you with only the question stem. This mode is great for late-stage preparation when you really want to challenge yourself to provide answers without the benefit of seeing multiple-choice options. This mode will not provide the detailed score reports that the other two modes will, so it should not be used if you are trying to identify knowledge gaps.

In addition to these three modes, you will be able to select the source of your questions. You can choose to take exams that cover all the chapters, or you can narrow your selection to a single chapter or the chapters that make up specific parts in the book. All chapters are selected by default. If you want to narrow your focus to individual chapters, first deselect all the chapters; then select only those on which you want to focus in the Objectives area.

You can also select the exam banks on which to focus. Each exam bank comes complete with a full exam of questions that cover topics in every chapter. The two exams printed in the book are available to you as well as two additional exams of unique questions. You can have the test engine serve up exams from all four banks or just from one individual bank by selecting the desired banks in the exam bank area.

You can make several other customizations to your exam from the exam settings screen, such as the time of the exam, the number of questions, whether to randomize questions and answers, whether to show the number of correct answers for multiple answer questions, or whether to serve up only specific types of questions. You can also create custom test banks by selecting only questions that you have marked or questions on which you have added notes.

Updating Your Exams

If you are using the online version of the Pearson Test Prep software, you should always have access to the latest version of the software as well as the exam data. If you are using the Windows desktop version, every time you launch the software, it will check to see if there are any updates to your exam data and automatically download any changes that were made since the last time you used the software. This requires that you are connected to the Internet at the time you launch the software.

Sometimes, due to many factors, the exam data may not fully download when you activate your exam. If you find that figures or exhibits are missing, you may need to manually update your exams.

To update a particular exam you have already activated and downloaded, select the **Tools** tab and then click the **Update Products** button. Again, this is an issue only with the desktop Windows application.

If you want to check for updates to the Pearson Test Prep exam engine software, Windows desktop version, select the **Tools** tab and click the **Update Application** button. This will ensure that you are running the latest version of the software engine.

Assessing Exam Readiness

Exam candidates never really know whether they are adequately prepared for the exam until they have completed about 30 percent of the questions. At that point, if you are not prepared, it is too late. The best way to determine your readiness is to work through the CramSaver quizzes at the beginning of each chapter and review the exam objectives and ExamAlerts presented in each chapter. It is best to work your way through the entire book unless you can complete each subject without having to do any research or look up any answers.

Figure Credits

Figures 1.3, 3.5, 3.9, 6.2, 6.4, 6.5, 8.13, 9.3, 10.2, 10.3: Linksys Holdings, Inc

Figures 3.2–3.4, 4.3, 4.4, 5.15, 8.9–8.12, 10.4: Microsoft Corporation

Chapter 10, Cram Quiz under "Troubleshooting Common Networking Issues," screenshot of Internet Protocol Version 4 (TCP/IPv4) Properties: Microsoft Corporation

CHAPTER 1

Networking Models, Ports, Protocols, and Services

This chapter covers the following official Network+ objectives:

▶ 1.1 Explain concepts related to the Open Systems Interconnection (OSI) reference model.

▶ 1.4 Explain common networking ports, protocols, services, and traffic types.

This chapter covers CompTIA Network+ objectives 1.1, and 1.4. For more information on the official Network+ exam topics, see the "About the Network+ Exam" section in the Introduction.

One of the most important networking concepts to understand is the *Open Systems Interconnection (OSI)* reference model. This conceptual model, created by the *International Organization for Standardization (ISO)* in 1978 and revised in 1984, describes a network architecture that enables data to be passed between computer systems.

This chapter looks at the OSI model and describes how it relates to real-world networking. It also examines how common network devices relate to the OSI model. Even though the OSI model is conceptual, an appreciation of its purpose and function can help you better understand how protocol suites and network architectures work in practical applications.

> **Note**
>
> The TCP/IP model, which performs the same functions as the OSI model, except in four layers instead of seven, is not a standalone Network+ objective, but this is the protocol suite predominantly in use today, and you should have a solid understanding of it since so many other objectives are based on it. Since it is still important to know it to understand the underlying principles of networking, we refer to it where it is appropriate to do so.

The OSI Networking Model

▶ **1.1 Explain concepts related to the Open Systems Interconnection (OSI) reference model.**

CramSaver

If you can correctly answer these questions before going through this section, save time by skimming the ExamAlerts in this section and then completing the Cram Quiz at the end of the section.

1. Which layer of the OSI model converts data from the application layer into a format that can be sent over the network?

2. True or false: Transport protocols, such as UDP, map to the transport layer of the OSI model and are responsible for transporting data across the network.

3. At what layer of the OSI model do HTTP and SSH map?

Answers

1. The presentation layer converts data from the application layer into a format that can be sent over the network. It also converts data from the session layer into a format the application layer can understand.

2. True. Transport protocols map to the transport layer of the OSI model and are responsible for transporting data across the network. UDP is a transport protocol.

3. HTTP and SSH (along with many other protocols) map to the application layer of the OSI model.

For networking, two models commonly are referenced: the OSI model and the TCP/IP model. Both offer a framework, theoretical and actual, for how networking is implemented. Objective 1.1 of the Network+ exam focuses only on the OSI model. A thorough discussion of it follows with a brief discussion of the TCP/IP model tossed in for further understanding.

The OSI Seven-Layer Model

As shown in Figure 1.1, the OSI reference model is built, bottom to top, in the following order: physical, data link, network, transport, session, presentation, and application. The physical layer is classified as Layer 1, and the top layer of the model, the application layer, is Layer 7.

> **ExamAlert**
>
> The OSI model can be used as a bottom-to-top troubleshooting tool. For example, troubleshooting a network interface card (NIC) or network wiring would begin at Layer 1, the physical layer where electrical functions support physical connections. If the problem is not found there, then the next step would be to run a loopback test on the NIC (moving up to Layer 2), and so on.

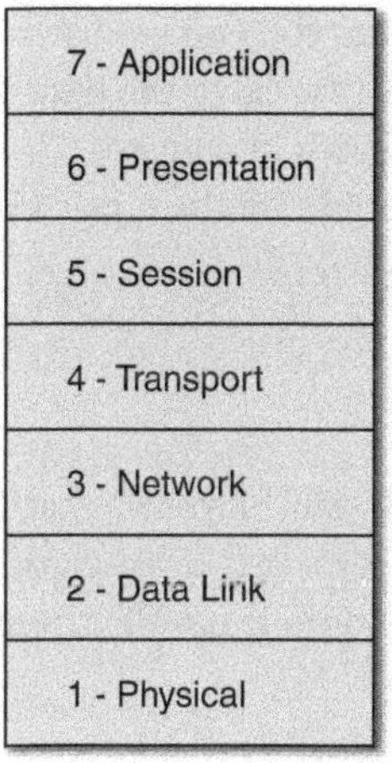

7 - Application
6 - Presentation
5 - Session
4 - Transport
3 - Network
2 - Data Link
1 - Physical

FIGURE 1.1   **The OSI Seven-Layer Model**

> **ExamAlert**
>
> On the Network+ exam, you might see an OSI layer referenced either by its name, such as network layer, or by its layer number. For instance, you might find that a router is referred to as a Layer 3 device. An easy mnemonic that you can use to remember the layers from top to bottom is: All People Seem To Need Data Processing.

Each layer of the OSI model has a specific function. The following sections describe the function of each layer, starting with the physical layer and working up the model.

Physical Layer (Layer 1)

The physical layer of the OSI model identifies the network's physical character-istics, including the following specifications:

▶ **Hardware:** The type of media used on the network, such as type of cable, type of connector, and pinout format for cables.

▶ **Topology:** The physical layer identifies the topology to be used in the network. Common topologies today include mesh, star/hub and spoke, spine and leaf, point-to-point and hybrid, and these are discussed in Chapter 2.

Protocols and technologies such as USB, Ethernet, Bluetooth, DSL, ISDN, T-carrier links (T1 and T3), GSM, and Integrated Services Digital Networks (ISDN) operate at the physical layer.

In addition to these characteristics, the physical layer defines the voltage used on a given medium and the frequency at which the signals that carry the data operate. These characteristics dictate the speed and bandwidth of a given medium, as well as the maximum distance over which a certain media type can be used.

Data Link Layer (Layer 2)

The data link layer is responsible for getting data to the physical layer so that it can transmit over the network. The data link layer is also responsible for error detection, error correction, and hardware addressing. The term *frame* describes the logical grouping of data at the data link layer.

The data link layer has two distinct sublayers:

▶ **Media Access Control (MAC) layer:** The MAC address is defined at this layer. The MAC address is the physical or hardware address burned into each NIC. The MAC sublayer also controls access to network media. The MAC layer specification is included in the IEEE 802.1 standard.

▶ **Logical Link Control (LLC) layer:** The LLC layer is responsible for the error and flow-control mechanisms of the data link layer. The LLC layer is specified in the IEEE 802.2 standard.

Protocols and technologies such as *High-Level Data Link Control (HDLC)*, *Layer 2 Tunneling Protocol (L2TP)*, *Point-to-Point Protocol (PPP)*, *Point-to-Point Tunneling Protocol (PPTP)*, *Spanning Tree Protocol (STP)*, and *virtual LANs (VLANs)* operate at the data link layer.

Network Layer (Layer 3)

The primary responsibility of the network layer is *routing*—providing mechanisms by which data can be passed from one network system to another. The network layer does not specify how the data is passed but rather provides the mechanisms to do so. Functionality at the network layer is provided through routing protocols, which are software components.

Protocols at the network layer are also responsible for *route selection*, which refers to determining the best path for the data to take throughout the network. The network layer contains Internet Protocol (IP) headers. In contrast to the data link layer, which uses MAC addresses to communicate on the LAN, network layer protocols use software-configured addresses and special routing protocols to communicate on the network. The term *packet* describes the logical grouping of data at the network layer.

When you're working with networks, routes can be configured in two ways: *statically* or *dynamically*. In a static routing environment, routes are manually added to the routing tables. In a dynamic routing environment, routing protocols such as *Routing Information Protocol (RIP)* and *Open Shortest Path First (OSPF)* are used. These protocols communicate routing information between networked devices on the network. Other important network layer protocols include *Internet Protocol (IP)*, *Address Resolution Protocol (ARP)*, *Reverse Address Resolution Protocol (RARP)*, *Asynchronous Transfer Mode (ATM)*, *Intermediate System-to-Intermediate System (IS-IS)*, *IP Security (IPsec)*, *Internet Control Message Protocol (ICMP)*, and *Multiprotocol Label Switching (MPLS)*.

Transport Layer (Layer 4)

The basic function of the transport layer is to provide mechanisms to transport data between network devices. Primarily, it does this in three ways:

▶ **Error checking:** Protocols at the transport layer ensure that data is correctly sent or received.

▶ **Service addressing:** A number of protocols support many network services. The transport layer ensures that data is passed to the right service at the upper layers of the OSI model.

▶ **Segmentation:** To traverse the network, blocks of data need to be broken into packets of a manageable size for the lower layers to handle. This process, called *segmentation*, is the responsibility of the transport layer.

Protocols that operate at the transport layer can either be connectionless, such as *User Datagram Protocol (UDP)*, or connection oriented, such as *Transmission Control Protocol (TCP)*.

The transport layer is also responsible for *data flow control*, which refers to how the receiving device can accept data transmissions. Two common methods of flow control are used:

▶ **Buffering:** When buffering flow control is used, data is temporarily stored and waits for the destination device to become available. Buffering can cause a problem if the sending device transmits data much faster than the receiving device can manage.

▶ **Windowing:** In a windowing environment, data is sent in groups of segments that require only one acknowledgment. The size of the window (that is, how many segments fit into one acknowledgment) is defined when the session between the two devices is established. As you can imagine, the need to have only one acknowledgment for every five segments, for instance, can greatly reduce overhead.

Session Layer (Layer 5)

The session layer is responsible for managing and controlling the synchronization of data between applications on two devices. It does this by establishing, maintaining, and breaking sessions. Whereas the transport layer is responsible for setting up and maintaining the connection between the two nodes (devices), the session layer performs the same function on behalf of the application. In other words, conversations between applications are established, coordinated, and terminated at the session layer.

Protocols that operate at the session layer include NetBIOS, *Network File System (NFS)*, and *Server Message Block (SMB)*.

Presentation Layer (Layer 6)

The presentation layer's basic function is to convert the data intended for or received from the application layer into another format. Such conversion is necessary because of how data is formatted so that it can be transported across the network. Applications cannot necessarily read this conversion. Some common data formats handled by the presentation layer include the following:

▶ **Graphics files:** JPEG, TIFF, GIF, and so on are graphics file formats that require the data to be formatted in a certain way.

▶ **Text and data:** The presentation layer can translate data into different formats, such as *American Standard Code for Information Interchange (ASCII)* and *Extended Binary Coded Decimal Interchange Code (EBCDIC)*.

▶ **Sound/video:** MPEG, MP3, and MIDI files all have their own data formats to and from which data must be converted.

Another important function of the presentation layer is *encryption*, which is the scrambling of data so that it can't be read by anyone other than the intended recipient. Given the basic role of the presentation layer—that of data-format translator—it is the obvious place for encryption and decryption to take place. For example, the cryptographic protocol *Transport Layer Security (TLS)* operates at the presentation layer.

Application Layer (Layer 7)

In simple terms, the function of the application layer is to take requests and data from the users and pass them to the lower layers of the OSI model. Incoming information is passed to the application layer, which then displays the information to the users. Some of the most basic application layer services include file and print capabilities.

The most common misconception about the application layer is that it represents applications used on a system, such as a web browser, word processor, or spreadsheet. Instead, the application layer defines the processes that enable applications to use network services. For example, if an application needs to open a file from a network drive, the functionality is provided by components that reside at the application layer. Protocols defined at the application layer include *Secure Shell (SSH)*, *Border Gateway Protocol (BGP)*, *Dynamic Host Configuration Protocol (DHCP)*, *Domain Name System (DNS)*, *Network Time Protocol (NTP)*, *Real-time Transport Protocol (RTP)*, *Session Initiation Protocol (SIP)*, *Simple Mail Transfer Protocol (SMTP)*, *Server Message Block (SMB)*, *File Transfer Protocol (FTP)*, *Hypertext Transfer Protocol (HTTP)*, *Hypertext Transfer Protocol Secure (HTTPS)*, *Internet Message Access Protocol (IMAP)*, and *Post Office Protocol version 3 (POP3)*.

> **ExamAlert**
>
> Be sure you understand the OSI model and its purpose. You will almost certainly be asked questions on it during the exam. Know Table 1.1 well!

OSI Model Summary

Table 1.1 summarizes the seven layers of the OSI model and describes some of the most significant points of each layer.

TABLE 1.1 **OSI Model Summary**

OSI Layer	Major Function
Physical (Layer 1)	Defines the physical structure of the network and the topology.
Data link (Layer 2)	Provides error detection and correction. Uses two distinct sublayers: the Media Access Control (MAC) and Logical Link Control (LLC) layers. Identifies the method by which media are accessed. Defines hardware addressing through the MAC sublayer.
Network (Layer 3)	Handles the discovery of destination systems and addressing. Provides the mechanism by which data can be passed and routed from one network system to another.
Transport (Layer 4)	Provides connection services between the sending and receiving devices and ensures reliable data delivery. Manages flow control through buffering or windowing. Provides segmentation, error checking, and service identification.
Session (Layer 5)	Synchronizes the data exchange between applications on separate devices.
Presentation (Layer 6)	Translates data from the format used by applications into one that can be transmitted across the network. Handles encryption and decryption of data. Provides compression and decompression functionality. Formats data from the application layer into a format that can be sent over the network.
Application (Layer 7)	Provides access to the network for applications.

Comparing OSI to the Four-Layer TCP/IP Model

The OSI model does a fantastic job outlining how networking should occur and the responsibility of each layer. However, TCP/IP also has a reference model and has to perform the same functionality with only four layers. Figure 1.2 shows how these four layers line up with the seven layers of the OSI model.

TCP/IP Model	OSI Model
Application Layer	Application Layer Presentation Layer Session Layer
Transport Layer	Transport Layer
Internet Layer	Network Layer
Network Interface Layer	Data Link Layer Physical Layer

FIGURE 1.2 The TCP/IP Model Compared to the OSI Model

The network interface layer in the TCP/IP model is sometimes referred to as the network access or link layer, and this is where Ethernet, wireless networks, network interface cards (NICs), or any other physical technology can run. The Internet layer is where IP runs (along with ICMP and others). The transport layer is where TCP and its counterpart UDP operate. The application layer enables any number of protocols to be plugged in, such as HTTP, SMTP, *Simple Network Management Protocol (SNMP)*, DNS, and many others.

Identifying the OSI Layers at Which Various Network Components Operate

When you understand the OSI model, you can relate network connectivity devices to the appropriate layer of the OSI model. Knowing at which OSI layer a device operates enables you to better understand how it functions on the network. Table 1.2 identifies various network devices and maps them to the OSI model.

> **ExamAlert**
>
> For the Network+ exam, you are expected to identify at which layer of the OSI model certain network devices operate.

TABLE 1.2 **Mapping Network Devices to the OSI Model**

Device	OSI Layer
Hub	Physical (Layer 1)
Repeater	Physical (Layer 1)
NIC	Data link (Layer 2)
Access point (AP)	Data link (Layer 2)
Bridge	Data link (Layer 2)
Switch	Data link (Layer 2) or Network (Layer 3)
Router	Network (Layer 3)
Firewall	Network (Layer 3) or Transport (Layer 4)
IDS/IPS	Network (Layer 3)
Load Balancer	Transport (Layer 4) or Application (Layer 7)
Proxy Server	Application (Layer 7)

Data Encapsulation/Decapsulation and OSI

As data moves down the model (and through the devices on that host), it is encapsulated with a header added to the beginning and a trailer to the end. Once the data arrives at the receiving host, it moves up the model (and through the devices) and is decapsulated in that the header and trailer are stripped off as it moves up.

> **Note**
>
> There are a great many topics beneath exam objective 1.1. In the interest of our discussion building in a logical way, the focus here is still on the networking model in order to complete the discussion of it. Later in this chapter, we visit headers again and some of the other topics the objectives include but that do not fit well with the dialogue yet.

> **ExamAlert**
>
> Adding protocol information to data as it passes through layers is known as encapsulation. Removing protocol information to data as it passes through layers is known as decapsulation.

In the encapsulation/decapsulation process, each layer on the receiving host does the opposite of what was done at that layer on the sending host: the receiving host's network layer, for example, strips off what was added by the network layer on the sending host. Table 1.3 shows what encapsulation/decapsulation occurs at each of the layers of the OSI model.

TABLE 1.3 **OSI Model Encapsulation/Decapsulation**

OSI Layer	Encapsulation/Decapsulation Function	Representation
Application (Layer 7) Presentation (Layer 6) Session (Layer 5)	The data is created in the application(s) and passed to/from the transport layer.	DATA
Transport (Layer 4)	A segment header is added to, or removed from, the data.	SEGMENT HEADER | DATA
Network (Layer 3)	A packet header is added to, or removed from, the data.	PACKET HEADER | SEGMENT HEADER | DATA

OSI Layer	Encapsulation/Decapsulation Function	Representation
Data link (Layer 2)	A frame header is added to, or removed from, the data. A frame trailer is added to, or removed from, the data.	FRAME HEADER | PACKET HEADER | SEGMENT HEADER | DATA | FRAME TRAILER

It should be noted that the physical layer (Layer 1) does not appear in Table 1.3 because it does not add or remove anything, but sends what it has (on the sending host) and receives what comes to it (on the receiving host).

It should also be noted that the unit of data worked with at each layer of the model (such as a frame at Layer 2 or a packet at Layer 3) is called a *protocol data unit (PDU)*.

Cram Quiz

1. At which OSI layer does an AP operate?

   ○ **A.** Network

   ○ **B.** Physical

   ○ **C.** Data link

   ○ **D.** Session

2. Which of the following are sublayers of the data link layer? (Choose two.)

   ○ **A.** MAC

   ○ **B.** LCL

   ○ **C.** Session

   ○ **D.** LLC

3. At which OSI layers can a switch operate? (Choose two.)

   ○ **A.** Layer 1

   ○ **B.** Layer 2

   ○ **C.** Layer 3

   ○ **D.** Layer 4

4. Which of the following OSI layers is responsible for establishing connections between two devices?

   ○ **A.** Network

   ○ **B.** Transport

   ○ **C.** Session

   ○ **D.** Data link

5. What happens to data as it moves from the upper to the lower layers of the OSI model on a host system?

   ○ **A.** The header and trailer are stripped off through decapsulation.

   ○ **B.** The data is sent in groups of segments that require two acknowledgments.

   ○ **C.** The data moves from the physical layer to application layer.

   ○ **D.** It is encapsulated with a header at the beginning and a trailer at the end.

Cram Quiz Answers

1. **C.** A wireless access point (AP) operates at the data link layer of the OSI model. An example of a network layer device is a router. An example of a physical layer device is a hub. Session layer components normally are software, not hardware.

2. **A and D.** The data link layer is broken into two distinct sublayers: Media Access Control (MAC) and Logical Link Control (LLC). LCL is not a valid term. Session is another of the OSI model layers.

3. **B and C.** A switch uses the MAC addresses of connected devices to make its forwarding decisions. Therefore, it is called a data link, or Layer 2, network device. It can also operate at Layer 3 or be called a multilayer switch or Layer 3 switch. Devices or components that operate at Layer 1 typically are media based, such as cables or connectors. Layer 4 components typically are software based, not hardware based.

4. **B.** The transport layer is responsible for establishing a connection between networked devices. The network layer is most commonly associated with route discovery and datagram delivery. Protocols at the session layer synchronize the data exchange between applications on separate devices. Protocols at the data link layer perform error detection and handling for the transmitted signals and define the method by which the medium is accessed.

5. **D.** As data moves down the model (and through the devices on that host), it is encapsulated with a header added to the beginning and a trailer to the end. Once the data arrives at the receiving host, it moves up the model (and through the devices) and is decapsulated in that the header and trailer are stripped off as it moves up. In a windowing environment, data is sent in groups of segments that require only one acknowledgment. On the sending host system, data moves from the application layer down to the physical layer. On the receiving system, data moves from the physical layer upwards to the application layer.

Ports, Protocols, Services, and Traffic Types

▶ **1.4 Explain common networking ports, protocols, services, and traffic types.**

CramSaver

If you can correctly answer these questions before going through this section, save time by skimming the ExamAlerts in this section and then completing the Cram Quiz at the end of the section.

1. With TCP, a data session is established through a three-step process. This is known as a three-way _____.

2. The SSH protocol is a more secure alternative to what protocol?

3. What ports do the HTTPS, RDP, and DHCP protocols use?

Answers

1. This is known as a three-way handshake.

2. SSH is a more secure alternative to Telnet, which transmits in cleartext.

3. HTTPS uses port 443, RDP uses port 3389, and DHCP uses ports 67 and 68.

When computers were restricted to standalone systems, there was little need for mechanisms to communicate between them. However, it wasn't long before the need to connect computers for the purpose of sharing files and printers became a necessity. Establishing communication between network devices required more than a length of cabling; a method or a set of rules was needed to establish how systems would communicate. Protocols provide that method.

It would be nice if a single protocol facilitated communication between all devices, but this is not the case. You can use a number of protocols on a network, each of which has its own features, advantages, and disadvantages. What protocol you choose can have a significant impact on the network's functioning and performance. This section explores some of the more common protocols you can expect to work with as a network administrator.

Connection-Oriented Protocols Versus Connectionless Protocols

Before getting into the characteristics of the various network protocols and protocol suites, you must first identify the difference between connection-oriented and connectionless protocols.

In a *connection-oriented* communication, data delivery is guaranteed. The sending device resends any packet that the destination system does not receive. Communication between the sending and receiving devices continues until the transmission has been verified. Because of this, connection-oriented protocols have a higher overhead and place greater demands on bandwidth.

> **ExamAlert**
>
> Connection-oriented protocols such as TCP can accommodate lost or dropped packets by asking the sending device to retransmit them. They can do this because they wait for all the packets in a message to be received before considering the transmission complete. On the sending end, connection-oriented protocols also assume that a lack of acknowledgment is sufficient reason to retransmit.

In contrast to connection-oriented communication, *connectionless* protocols such as User Datagram Protocol (UDP) offer only a best-effort delivery mechanism. Basically, the information is just sent; there is no confirmation that the data has been received. If an error occurs in the transmission, there is no mechanism to resend the data, so transmissions made with connectionless protocols are not guaranteed. Connectionless communication requires far less overhead than connection-oriented communication, so it is popular in applications such as streaming audio and video, where a small number of dropped packets might not represent a significant problem.

> **ExamAlert**
>
> As you work through the various protocols, keep an eye out for those that are connectionless and those that are connection oriented. Also, look for protocols such as TCP that guarantee delivery of data and those such as UDP that are a fire-and-forget or best-delivery method.

Internet Protocol

Internet Protocol (IP), which is defined in RFC 791, is the protocol used to transport data from one node on a network to another. IP is connectionless, which means that it doesn't guarantee the delivery of data; it simply makes its best effort to do so. To ensure that transmissions sent via IP are completed, a higher-level protocol such as TCP is required.

> **Note**
>
> In this chapter and throughout the book, the term *Request For Comments (RFC)* is used. RFCs are standards published by the *Internet Engineering Task Force (IETF)* and describe methods, behaviors, research, or innovations applicable to the operation of the Internet and Internet-connected systems. Each new RFC has an associated reference number. Looking up this number gives you information on the specific technology. For more information on RFCs, look for the Internet Engineering Task Force online.

> **ExamAlert**
>
> IP operates at the network layer of the OSI model.

In addition to providing best-effort delivery, IP also performs fragmentation and reassembly tasks for network transmissions. Fragmentation is necessary because the *maximum transmission unit (MTU)* size is limited in IP. In other words, network transmissions that are too big to traverse the network in a single packet must be broken into smaller chunks and reassembled at the other end. Another function of IP is addressing. IP addressing is a complex subject. Refer to Chapter 3, "Network Addressing, Routing, and Switching," for a complete discussion of IP addressing.

Transmission Control Protocol

Transmission Control Protocol (TCP), which is defined in RFC 793, is a connection-oriented transport layer protocol. Being connection-oriented means that TCP establishes a mutually acknowledged session between two hosts before communication takes place. TCP provides reliability to IP communications. Specifically, TCP adds features such as flow control, sequencing, and error detection and correction. For this reason, higher-level applications that need guaranteed delivery use TCP rather than its lightweight and connectionless brother, UDP.

How TCP Works

When TCP wants to open a connection with another host, it follows this procedure:

1. It sends a message called a SYN to the target host.

2. The target host opens a connection for the request and sends back an acknowledgment message called an ACK (or SYN ACK).

3. The host that originated the request sends back another acknowledgment, saying that it has received the ACK message and that the session is ready to be used to transfer data.

When the data session is completed, a similar process is used to close the session. This three-step session establishment and acknowledgment process is called the *TCP three-way handshake*.

> **ExamAlert**
>
> TCP operates at the transport layer of the OSI model.

TCP is a reliable protocol because it has mechanisms that can accommodate and handle errors. These mechanisms include time-outs, which cause the sending host to automatically retransmit data if its receipt is not acknowledged within a given time period.

User Datagram Protocol

User Datagram Protocol (UDP), which is defined in RFC 768, is the brother of TCP. Like TCP, UDP is a transport protocol, but the big difference is that UDP does not guarantee delivery like TCP does. In a sense, UDP is a "fire-and-forget" protocol; it assumes that the data sent will reach its destination intact. The checking of whether data is delivered is left to upper-layer protocols. UDP operates at the transport layer of the OSI model.

Unlike TCP, with UDP no session is established between the sending and receiving hosts, which is why UDP is called a connectionless protocol. The upshot of this is that UDP has much lower overhead than TCP. A TCP packet header has 14 fields, whereas a UDP packet header has only 4 fields. Therefore, UDP is much more efficient than TCP. In applications that don't need the added features of TCP, UDP is much more economical in terms of bandwidth and processing effort.

ExamAlert

Remember that TCP is a connection-oriented protocol and UDP is a connectionless protocol.

Internet Control Message Protocol

Internet Control Message Protocol (ICMP), which is defined in RFC 792, is a protocol that works with the IP layer to provide error checking and reporting functionality. In effect, ICMP is a tool that IP uses in its quest to provide best-effort delivery.

ICMP can be used for a number of functions. Its most common function is probably the widely used and incredibly useful ping utility, which can send a stream of ICMP echo requests to a remote host. If the host can respond, it does so by sending echo reply messages back to the sending host. In that one simple process, ICMP enables the verification of the protocol suite configuration of both the sending and receiving nodes and any intermediate networking devices.

However, ICMP's functionality is not limited to the use of the ping utility. ICMP also can return error messages such as "Destination unreachable" and "Time exceeded." (The former message is reported when a destination cannot be contacted and the latter when the *time to live [TTL]* of a datagram has been exceeded.)

Note

The traceroute command also uses TTL. Traceroute uses IP packet time-to-live time-outs to discover the path a packet takes as it traverses an internetwork.

In addition to these and other functions, ICMP performs source quench. In a source quench scenario, the receiving host cannot handle the influx of data at the same rate as the data is sent. To slow down the sending host, the receiving host sends ICMP source quench messages, telling the sender to slow down. This action prevents packets from dropping and having to be re-sent.

ICMP is a useful protocol. Although ICMP operates largely in the background, the ping utility makes it one of the most valuable of the protocols discussed in this chapter.

IPsec

The *Internet Protocol Security (IPsec)* protocol is designed to provide secure communications between systems. This includes system-to-system communication in the same network, as well as communication to systems on external networks. IPsec is an IP layer security protocol that can both encrypt and authenticate network transmissions. In a nutshell, IPsec is composed of two separate protocols: *Authentication Header (AH)* and *Encapsulating Security Payload (ESP)*. AH provides the authentication and integrity checking for data packets, and ESP provides encryption services.

> ### ExamAlert
>
> IPsec relies on two underlying protocols: AH and ESP. AH provides authentication services, and ESP provides encryption services.

Using both AH and ESP, data traveling between systems can be secured, ensuring that transmissions cannot be viewed, accessed, or modified by those who should not have access to them. It might seem that protection on an internal network is less necessary than on an external network; however, much of the data you send across networks has little or no protection, allowing unwanted eyes to see it.

> ### Note
>
> The Internet Engineering Task Force (IETF) created IPsec, which you can use on both IPv4 and IPv6 networks.

IPsec provides three key security services:

▶ **Data verification:** Verifies that the data received is from the intended source

▶ **Protection from data tampering:** Ensures that the data has not been tampered with or changed between the sending and receiving devices

▶ **Private transactions:** Ensures that the data sent between the sending and receiving devices is unreadable by any other devices

Internet Key Exchange (IKE) is a key management protocol used in conjunction with IPsec to establish secure communication channels and negotiate cryptographic keys for encrypted VPN connections. IKE is responsible for establishing Security Associations (SAs) between communicating devices, which define the parameters for secure communication, including encryption algorithms, integrity algorithms, and authentication methods. During the IKE negotiation process, the devices exchange IKE messages to agree on the parameters for the SA, and the negotiation begins with Phase 1, during which the devices establish a secure channel to exchange keying material and negotiate the parameters for the IKE SA. IKE is also responsible for key management, including generating, distributing, and refreshing cryptographic keys used by IPsec for encryption and authentication. IKE periodically refreshes the keys and rekeys the SAs to maintain security and prevent cryptographic attacks.

IPsec operates at the network layer of the Open Systems Interconnection (OSI) reference model and provides security for protocols that operate at the higher layers. Thus, by using IPsec, you can secure practically all TCP/IP-related communications.

> **Note**
>
> For Network+ study, it is important to note that next-generation firewalls (NGFWs) can provide content filtering and threat protections and are very often used to manage multiple IPsec site-to-site connections.

Generic Routing Encapsulation

Generic Routing Encapsulation (GRE) is a Cisco-created tunneling protocol. It is an encapsulating protocol used to wrap data and securely send it across VPNs and Point-to-Point (or point-to-multipoint) links. GRE provides a mechanism for creating a virtual point-to-point connection between two endpoints, allowing data packets to be transmitted transparently across intermediate IP networks, and it works as follows:

▶ **Encapsulation:** In a GRE tunnel, data packets from a source network are encapsulated within GRE headers before being transmitted over the IP network. The original packet, including its header and payload, is encapsulated within a GRE header, which includes a GRE header and a new IP header.

▶ **GRE Header:** The GRE header contains fields such as the protocol type (indicating the type of encapsulated payload), a key field (optional for security purposes), and a sequence number field (optional for packet ordering).

▶ **New IP Header:** After encapsulation, a new IP header is added to the GRE-encapsulated packet. This new IP header specifies the source and destination IP addresses of the tunnel endpoints, as well as other IP header fields such as the protocol type (usually set to protocol number 47, indicating GRE) and the time-to-live (TTL) field.

▶ **Transmission:** The GRE-encapsulated packets are then transmitted over the IP network like any other IP packet. The intermediate routers in the network route the packets based on the destination IP address specified in the new IP header.

▶ **Decapsulation:** Upon reaching the destination endpoint of the GRE tunnel, the GRE-encapsulated packets are decapsulated to extract the original packets. The GRE header and the new IP header are removed, leaving the original packet intact. The decapsulated packets are then forwarded to the destination network.

GRE tunnels are often used in scenarios where direct Layer 2 connectivity between two endpoints is not possible or practical, such as connecting remote sites over a public IP network or creating VPN tunnels between network devices. GRE tunnels are protocol-independent, meaning they can transport any network layer protocol, including IPv4, IPv6, and even non-IP protocols such as Ethernet frames.

File Transfer Protocol (FTP)

As its name suggests, *File Transfer Protocol (FTP)* provides for the uploading and downloading of files from a remote host running FTP server software. As well as uploading and downloading files, FTP enables you to view the contents of folders on an FTP server and rename and delete files and directories if you have the necessary permissions. FTP, which is defined in RFC 959, uses TCP as a transport protocol to guarantee delivery of packets.

FTP has weak security mechanisms used to authenticate users. However, rather than create a user account for every user, you can configure FTP server software to accept anonymous logons. When you do this, the username is anonymous, and the password normally is the user's email address. Most FTP servers that offer files to the general public operate in this way. Even when logins are used, FTP is still considered insecure in today's environment. SFTP/SSH should be used in its place in almost every scenario.

In addition to being popular as a mechanism for distributing files to the general public over networks such as the Internet, FTP can also be used by organizations that need to frequently exchange large files with other people or organizations. Such a system can be used when the files being exchanged are

larger than can be easily accommodated using email. A number of apps/
programs are available that simplify the process. For example, FileZilla is a
cross-platform graphical FTP, SFTP, and FTPS file management tool for
Windows, Linux, macOS, and more (more information on FileZilla can be
found at https://sourceforge.net/projects/filezilla/).

> **ExamAlert**
>
> Remember that FTP is an application layer protocol. FTP uses ports 20 and 21 and
> sends information unencrypted, making it unsecure.

All the common network operating systems offer FTP server capabilities;
however, whether you use them depends on whether you need FTP services. All
popular workstation operating systems offer FTP client functionality, although
it is common to use third-party utilities such as FileZilla (mentioned earlier),
CuteFTP, or SmartFTP instead. By default, FTP operates on ports 20 and 21
and sends messages in cleartext. It is a client/server-based protocol, meaning
that the two computers involved communicate with each other in a request-
response pattern. The client sends a request to the server, and the server
responds with the requested data (an architecture known as client/server).

FTP assumes that files uploaded or downloaded are straight text (that is,
ASCII) files. If the files are not text, which is likely, the transfer mode must
be changed to binary. With sophisticated FTP clients, such as CuteFTP, the
transition between transfer modes is automatic. With more basic utilities, you
must manually perform the mode switch.

Secure File Transfer Protocol (SFTP)

One of the big problems associated with FTP is that it is considered unsecure.
Even though simple authentication methods are associated with FTP, it is still
susceptible to relatively simple hacking approaches. In addition, FTP transmits
data between sender and receiver in an unencrypted format. By using a packet
sniffer, a hacker could easily copy packets from the network and read the
contents. In today's high-security computing environments, you need a more
robust solution.

That solution is the *Secure File Transfer Protocol (SFTP)*, which, based on Secure
Shell (SSH) technology, provides robust authentication between sender and
receiver. It also provides encryption capabilities, which means that even if
packets are copied from the network, their contents remain hidden from
prying eyes.

SFTP is implemented through client and server software available for all commonly used computing platforms. SFTP uses port 22 (the same port SSH uses) for secure file transfers.

Secure Shell (SSH)

Created by students at the Helsinki University of Technology, *Secure Shell (SSH)* is a secure alternative to Telnet. SSH provides security by encrypting data as it travels between systems. This makes it difficult for hackers using packet sniffers and other traffic-detection systems. It also provides more robust authentication systems than Telnet.

Two versions of SSH are available: SSH1 and SSH2. Of the two, SSH2 is considered more secure. The two versions are incompatible. If you use an SSH client program, the server implementation of SSH that you connect to must be the same version. By default, SSH operates on port 22.

Although SSH, like Telnet, is associated primarily with UNIX and Linux systems, implementations of SSH are available for all commonly used computing platforms, including Windows and macOS.

> **ExamAlert**
>
> Remember that SSH uses port 22 and is a more secure alternative to Telnet.

Telnet

Telnet, which is defined in RFC 854, is a virtual terminal protocol. It enables sessions to be opened on a remote host, and then commands can be executed on that remote host. For many years, Telnet was the method by which clients accessed multiuser systems such as mainframes and minicomputers. It also was the connection method of choice for UNIX systems. Today, Telnet is still used to access routers and other managed network devices. By default, Telnet operates on port 23.

One of the problems with Telnet is that it is not secure. As a result, remote session functionality is now almost always achieved by using alternatives such as the previously discussed SSH protocol.

> **ExamAlert**
>
> Telnet is used to access UNIX and Linux systems. Telnet uses port 23 and is insecure. SSH is considered the secure replacement for Telnet.

Simple Mail Transfer Protocol (SMTP)

Simple Mail Transfer Protocol (SMTP), which is defined in RFC 821, is a protocol that defines how mail messages are sent between hosts. SMTP uses TCP connections to guarantee error-free delivery of messages. SMTP is not overly sophisticated and requires that the destination host always be available. For this reason, mail systems spool incoming mail so that users can read it later. How the user then reads the mail depends on how the client accesses the SMTP server. The default port used by SMTP is 25.

> **Note**
>
> SMTP can be used to both send and receive mail. Post Office Protocol version 3 (POP3) and Internet Message Access Protocol version 4 (IMAP4) can be used only to receive mail.

Domain Name System (DNS)

Domain Name System (DNS)—also known as *Domain Name Service*—resolves hostnames, such as www.pearsonitcertification.com, to IP addresses, such as 168.146.67.180. By default, DNS operates on port 53, and it constitutes one of the few network services that CompTIA wants you to know quite a bit about. As such, it is discussed in more detail in Chapter 3.

Dynamic Host Configuration Protocol (DHCP)

Dynamic Host Configuration Protocol (DHCP) is defined in RFC 2131. It enables ranges of IP addresses, known as *scopes*, or predefined groups of addresses within *address pools* to be defined on a system running a DHCP server application. When another system configured as a DHCP client is initialized, it asks the server for an address and is leased one. By default, DHCP uses ports 67 and 68. Figure 1.3 shows an example of a configuration interface for DHCP on a SOHO router.

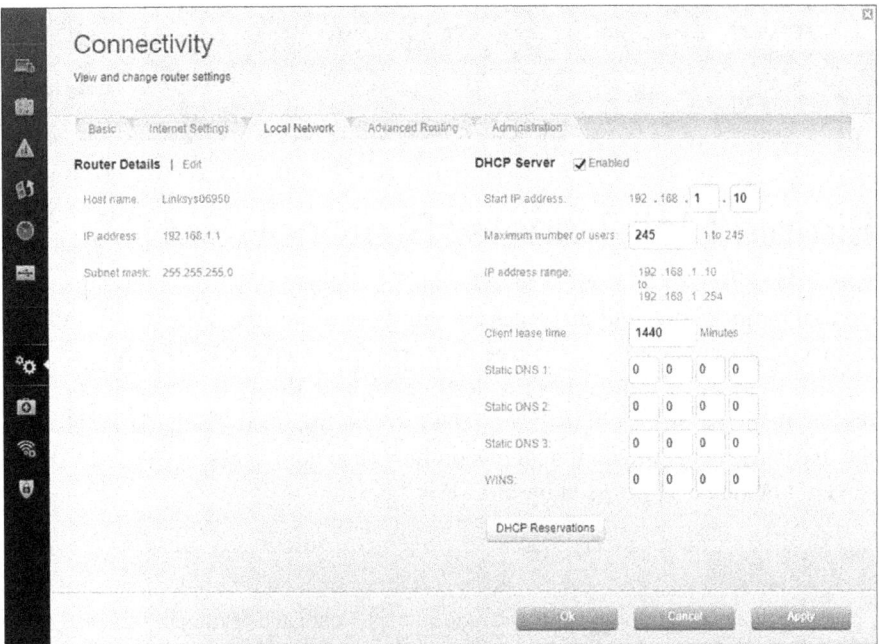

FIGURE 1.3   **Configuring DHCP on a SOHO Router**

> **Note**
>
> DNS, DHCP, and NTP appear in the exam objectives for this section, but also appear in much more depth in the next set of objectives. To avoid overlap with the discussion of the objectives, all three are discussed in more depth later in this chapter.

Trivial File Transfer Protocol (TFTP)

A variation on FTP is *Trivial File Transfer Protocol (TFTP)*, which is also a file transfer mechanism. However, TFTP does not have the security capability or the level of functionality that FTP has. TFTP, which is defined in RFC 1350, is most often associated with simple downloads, such as those associated with transferring firmware to a device such as a router and booting diskless workstations.

Another feature that TFTP does not offer is directory navigation. Whereas in FTP, commands can be executed to navigate and manage the file system, TFTP offers no such capability. TFTP requires that you request not only exactly what you want but also the particular location. Unlike FTP, which uses TCP as its transport protocol to guarantee delivery, TFTP uses UDP. By default, TFTP operates on port 69.

> **ExamAlert**
>
> TFTP is an application layer protocol that uses UDP, which is a connectionless transport layer protocol. For this reason, TFTP is called a *connectionless file transfer method*.

Hypertext Transfer Protocol (HTTP)

Hypertext Transfer Protocol (HTTP), which is defined in RFC 2068, is the protocol that enables text, graphics, multimedia, and other material to be downloaded from an HTTP server. HTTP defines what actions can be requested by clients and how servers should answer those requests.

In a practical implementation, HTTP clients (that is, web browsers) make requests on port 80 in an HTTP format to servers running HTTP server applications (that is, web servers). Files created in a special language such as *Hypertext Markup Language (HTML)* are returned to the client, and the connection is closed.

> **ExamAlert**
>
> Make sure that you understand that HTTP is a connection-oriented protocol that uses TCP as a transport protocol. By default, it operates at port 80. HTTP is insecure and has been replaced for the most part by HTTPS everywhere.

Today, HTTP over port 80 is considered deprecated and insecure and often replaced by HTTPS (over port 443). Both HTTP and HTTPS use a *uniform resource locator (URL)* to determine what page should be downloaded from the remote server. The URL contains the type of request (for example, http:// or https://), the name of the server contacted (for example, www.microsoft.com or just microsoft.com since the web portion is the default), and optionally the page requested (for example, /support). The result is the syntax that Internet-savvy people are familiar with: https://support.microsoft.com/.

Network Time Protocol (NTP)

Network Time Protocol (NTP), which is defined in RFC 958, is the part of the TCP/IP protocol suite that facilitates the communication and synchronization of time between systems. NTP operates over UDP port 123.

Simple Network Management Protocol (SNMP)

The *Simple Network Management Protocol (SNMP)* uses port 161 to send data and port 162 to receive it. It enables network devices to communicate information about their state to a central system. It also enables the central system to pass configuration parameters to the devices.

> **ExamAlert**
>
> SNMP uses ports 161 and 162. It is a protocol that facilitates network management functionality. It is not, in itself, a *network management system (NMS)*, simply the protocol that makes NMS possible.

Components of SNMP

In an SNMP configuration, a central system known as a *manager* acts as the central communication point for all the SNMP-enabled devices on the network. On each device to be managed and monitored via SNMP, software called an SNMP agent is set up and configured with the manager's IP address. Depending on the configuration, the SNMP manager then communicates with and retrieves information from the devices running the SNMP agent software. In addition, the agent can communicate the occurrence of certain events to the SNMP manager as they happen. These messages are known as *traps*. Figure 1.4 shows how an SNMP system works.

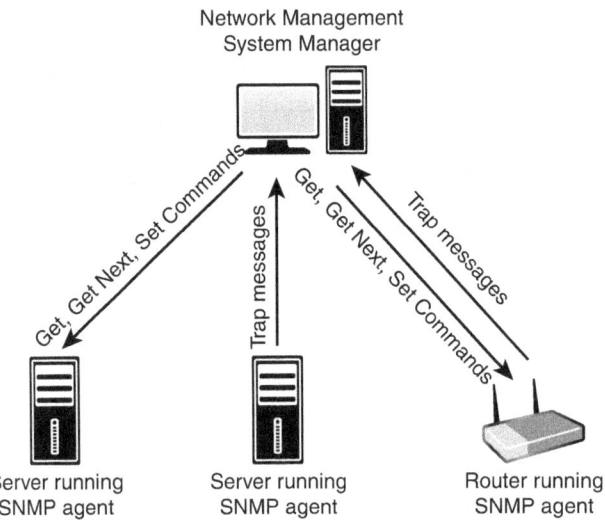

FIGURE 1.4 **How SNMP Works**

As Figure 1.4 illustrates, there are a number of components to SNMP. The following discussion looks at the management system, the agents, the management information base, and communities.

SNMP Management Systems

An SNMP management system is a computer running a special piece of software called a *network management system (NMS)*. These software applications can be free, or they can cost thousands of dollars. The difference between the free applications and those that cost a great deal of money normally boils down to functionality and support. All NMS applications, regardless of cost, offer the same basic functionality. Today, most NMS applications use graphical maps of the network to locate a device and then query it. The queries are built in to the application and are triggered by pointing and clicking. You can issue SNMP requests from a command-line utility, but with so many tools available, this is unnecessary.

> **Note**
>
> Some people call SNMP managers or NMSs *trap managers*. This reference is misleading, however, because an NMS can do more than just accept trap messages from agents.

Using SNMP and an NMS, you can monitor all the devices on a network, including switches, hubs, routers, servers, and printers, as well as any device that supports SNMP, from a single location. Using SNMP, you can see the amount of free disk space on a server in Jakarta or reset the interface on a router in Helsinki—all from the comfort of your desk in San Jose. Such power, though, brings with it some considerations. For example, because an NMS enables you to reconfigure network devices, or at least get information from them, it is common practice to implement an NMS on a secure workstation platform such as a Linux or Windows server and to place the NMS PC in a secure location.

SNMP Agents

Although the SNMP manager resides on a PC or server, each device that is part of the SNMP structure also needs to have SNMP functionality enabled. This is performed through a software component called an *agent*.

An SNMP agent can be any device that can run a small software component that facilitates communication with an SNMP manager. SNMP agent functionality is supported by almost any device designed to be connected to a network.

In addition to providing a mechanism for managers to communicate with them, agents can tell SNMP managers when a threshold is surpassed. When this happens, on a device running an SNMP agent, a trap is sent to the NMS, and the NMS then performs an action, depending on the configuration. Basic NMS systems might sound an alarm or flash a message on the screen. Other more advanced products might dial a cell phone or send an email message.

Management Information Bases

Although the SNMP trap system might be the most commonly used aspect of SNMP, manager-to-agent communication is not a one-way street. In addition to reading information from a device using the SNMP commands **Get** and **Get Next**, SNMP managers can issue the **Set** command. If you have a large sequence of **Get Next** commands to perform, you can use the **Walk** command to automatically move through them. The purpose of this command is to save a manager's time: you issue one command on the root node of a subtree, and the command "walks" through, getting the value of every node in the subtree.

To demonstrate how SNMP commands work, imagine that you and a friend each have a list on which the following four words are written: *four*, *book*, *sky*, and *table*. If you, as the manager, ask your friend for the first value, they, acting as the agent, can reply "four." This is analogous to an SNMP **Get** command. Now, if you ask for the next value, they would reply "book." This is analogous to an SNMP **Get Next** command. If you then say "set green," and your friend changes the word *book* to *green*, you have performed the equivalent of an SNMP **Set** command. Sound simplistic? If you can imagine expanding the list to include 100 values, you can see how you could navigate and set any parameter in the list, using just those commands. The key, though, is to make sure that you and your friend have exactly the same list—which is where *Management Information Bases (MIBs)* come in.

SNMP uses databases of information called Management Information Bases (MIBs) to define what parameters are accessible, which of the parameters are read-only, and which can be set. MIBs are available for thousands of devices and services, covering every imaginable need.

To ensure that SNMP systems offer cross-platform compatibility, MIB creation is controlled by the *International Organization for Standardization (ISO)*. An organization that wants to create a MIB can apply to the ISO. The ISO then assigns the organization an ID under which it can create MIBs as it sees fit. The assignment of numbers is structured within a conceptual model called the *hierarchical name tree*.

SNMP Communities

Another feature of SNMP that enables manageability is communities. *SNMP communities* are logical groupings of systems. When a system is configured as part of a community, it communicates only with other devices that have the same community name. In addition, it accepts **Get**, **Get Next**, or **Set** commands only from an SNMP manager with a community name it recognizes. Typically, two communities are defined by default: a public community, intended for read-only use, and a private community, intended for read-and-write operations.

> **ExamAlert**
>
> For the exam, you should understand the SNMP concepts of **Get**, **Trap**, **Walk**, and MIBs.

Whether you use SNMP depends on how many devices you have and how distributed your network infrastructure is. Even in environments that have just a few devices, SNMP can be useful because it can act as your eyes and ears, notifying you if a problem on the network occurs.

SNMPv2c

One iteration of SNMPv2 is v2c—a sub-version—in which the *c* stands for *community*. SNMPv2c primarily differs in its reliance on a community-based security model and the absence of advanced security features. While SNMPv2c provides enhancements in protocol operations and error handling compared to SNMPv1, it lacks the robust security mechanisms introduced in SNMPv2.

Devices that communicate using SNMPv2c must share a community string, acting as a form of password. These community strings serve as a simple form of authentication. SNMPv2c maintains backward compatibility with SNMPv1, allowing devices supporting SNMPv2c to communicate with devices using SNMPv1.

SNMPv3

SNMP, which runs by default on port 161, is now on its third version (v2c was a sub-iteration), and this version has some significant differences. One of the most noticeable changes is that, unlike earlier versions, SNMPv3 supports authentication and encryption:

▶ **Authentication:** Authentication protocols ensure that the message is from a valid source.

▶ **Encryption:** Encryption protocols ensure that data cannot be read by unintended sources.

> **ExamAlert**
>
> You might be asked to know the differences between SNMPv2, SNMPv2c, and SNMPv3. Remember, SNMPv3 supports authentication and encryption.

> **Note**
>
> There are many ways in which SNMP can help make network administration easier. For example, consider a company that needs to monitor computer room temperatures in several remote locations. This can be implemented with SNMP using a collection server and setting up SNMP agents on each sensor, then sending the data to a collector server, which in turn monitors and automates temperatures automatically without the need for manual intervention.

> **ExamAlert**
>
> Anytime you need to collect data outside the norm, consider SNMP. For example, one instance where it can help would be to configure SNMP traps on a network to monitor a VoIP network that is experiencing jitter and high latency.

Lightweight Directory Access Protocol (LDAP)

Lightweight Directory Access Protocol (LDAP) is a protocol that provides a mechanism to access and query directory services systems. LDAP uses port 389 by default (a more secure version, LDAPS, runs on port 636 by default). In the context of the Network+ exam, these directory services systems are most likely to be UNIX/Linux based or Microsoft Active Directory (AD) based. Although LDAP supports command-line queries executed directly against the directory database, most LDAP interactions are via utilities such as an authentication program (network logon) or locating a resource in the directory through a search utility.

Hypertext Transfer Protocol Secure (HTTPS)

As noted earlier in the chapter, one of the downsides of using HTTP is that HTTP requests are sent in cleartext. For some applications, such as e-commerce, this method to exchange information is unsuitable—a more secure method is needed. The solution is *Hypertext Transfer Protocol Secure (HTTPS)*, which encrypts the information sent between the client and host (changing the port from 80 to 443). The data is encrypted using Transport Layer Security (TLS) or, formerly, Secure Sockets Layer (SSL). The protocol is therefore also referred to as HTTP over TLS, or HTTP over SSL.

For HTTPS to be used, both the client and server must support it. All popular browsers now support HTTPS, as do web server products, such as Microsoft *Internet Information Services (IIS)*, Apache, and almost all other web server applications that provide sensitive applications. When you access an application that uses HTTPS, the URL starts with https rather than http—for example, https://www.mybankonline.com.

Server Message Block (SMB)

Server Message Block (SMB) is used on a network for providing access to resources such as files, printers, ports, and so on that are running on Windows. If you were wanting to connect Linux-based hosts to Windows-shared printers, for example, you would need to implement support for SMB; it runs, by default, on port 445.

One of the most common ways of implementing SMB support is by running Samba, which is an open-source implementation of the SMB/CIFS protocol that allows files and printers to be shared easily across Windows and Linux systems.

Syslog

Most UNIX/Linux-based systems include the capability to write messages (either directly or through applications) to log files via *syslog*. This can be done for security or management reasons and provides a central means by which devices that otherwise could not write to a central repository can easily do so (often by using the logger utility). The default port for syslog is 514.

> **ExamAlert**
>
> Know that syslog is a standard for message logging. It is available on most network devices (e.g., routers, switches, and firewalls), as well as printers, and UNIX/Linux–based systems. Over a network, a syslog server listens for and then logs data messages coming from the syslog client. Syslog server messages are often used for troubleshooting and determining how systems may have been compromised after an attack. Remember that syslog uses port 514.

Simple Mail Transfer Protocol Secure (SMTPS)

SMTP TLS, more commonly known as *SMTPS (Simple Mail Transfer Protocol Secure)* uses Transport Layer Security (TLS) to provide authentication of the communication partners along with data integrity and confidentiality by wrapping SMTP data in TLS. This is similar to how HTTPS wraps HTTP data inside TLS. The default port for SMTPS is 587.

> **Note**
>
> Some implementations of SMTP with security use port 465. This port was proposed for SMTP with SSL and was never officially approved. It is good practice to avoid using this port and to use 587 instead.

Lightweight Directory Access Protocol over SSL (LDAPS)

Lightweight Directory Access Protocol over SSL (LDAPS), also known as Secure LDAP, adds an additional layer of security to LDAP. It operates at port 636 and differs from LDAP in two ways: (1) upon connection, the client and server establish a TLS session before any LDAP messages are transferred (without a start operation), and (2) the LDAPS connection must be closed if TLS closes.

> **ExamAlert**
>
> Remember that LDAP uses port 389, and LDAPS (secure LDAP) uses port 636.

Structured Query Language (SQL) Server

The SQL database server uses port 1433 by default, while Oracle's SQLnet uses port 1521 and the default port for MySQL is 3306. The most common language used to speak to databases is *Structured Query Language (SQL)*. SQL allows queries to be configured in real time and passed to database servers. This flexibility causes a major vulnerability when it isn't implemented securely.

> **Note**
>
> Most commercial relational database management systems (Oracle, Microsoft SQL Server, MySQL, PostgreSQL, and so forth) use SQL. A NoSQL database is a relatively new phenomenon: it is a relational database that does not use SQL. These databases are less common than relational databases but often used where scaling is important.

Remote Desktop Protocol (RDP)

Remote Desktop Protocol (RDP) is used in a Windows environment for remote connections. It operates, by default, on port 3389. *Remote Desktop Services (RDS*, formerly known as Terminal Services) provides a way for a client system to connect to a server, such as Windows Server, and, by using RDP, operate on the server as if it were a local client application. Such a configuration is known as *thin client computing*, whereby client systems use the resources of the server instead of their local processing power.

Windows Server products and recent Windows client systems have built-in support for remote connections using the Windows program Remote Desktop Connection. The underlying protocol used to manage the connection is RDP. RDP is a low-bandwidth protocol used to send mouse movements, keystrokes, and bitmap images of the screen on the server to the client computer. RDP does not actually send data over the connection—only screenshots and client keystrokes.

> **ExamAlert**
>
> Remember that Remote Desktop Protocol (RDP) can be used to remotely manage and control Windows systems and runs over port 3389.

Session Initiation Protocol (SIP)

Long-distance calls are expensive, in part because it is costly to maintain phone lines and employ technicians to keep those phones ringing. *Voice over IP (VoIP)* provides a cheaper alternative for phone service. VoIP technology enables regular voice conversations to occur by traveling through IP packets and via the Internet. VoIP avoids the high cost of regular phone calls by using the existing infrastructure of the Internet. No monthly bills or expensive long-distance charges are required. But how does it work?

Like every other type of network communication, VoIP requires protocols to make the magic happen. For VoIP, one such protocol is *Session Initiation Protocol (SIP)*, which is an application layer protocol designed to establish and maintain multimedia sessions, such as Internet telephony calls. This means that SIP can create communication sessions for such features as audio/videoconferencing, online gaming, and person-to-person conversations over the Internet. SIP does not operate alone; it uses TCP or UDP as a transport protocol. Remember, TCP enables guaranteed delivery of data packets, whereas UDP is a fire-and-forget transfer protocol. The default ports for SIP are 5060 and 5061.

> **ExamAlert**
>
> SIP operates at the application layer of the OSI model and is used to maintain a multimedia session. SIP uses ports 5060 and 5061.

> **Tip**
>
> SIP also includes a suite of security services, which include denial-of-service (DoS) prevention, authentication (both user-to-user and proxy-to-user), integrity protection, and encryption and privacy services.

Understanding Port Functions

As protocols were mentioned in this chapter, the default ports were also given. Each TCP/IP or application has at least one default port associated with it. When a communication is received, the target port number is checked to determine which protocol or service it is destined for. The request is then forwarded to that protocol or service. For example, consider HTTPS, whose assigned port number is 443. When a web browser forms a request for a secure web page, that request is sent to port 443 on the target system. When the target system receives the request, it examines the port number. When it sees that the port is 443, it forwards the request to the web server application.

TCP/IP has 65,535 ports available, with 0 to 1023 labeled as the well-known ports. Although a detailed understanding of the 65,535 ports is not necessary for the Network+ exam, you need to understand the numbers of some well-known ports. Network administration often requires you to specify port assignments when you work with applications and configure services. Table 1.4 shows some of the most common port assignments.

ExamAlert

You should concentrate on the information provided in Table 1.4 and apply it to any port-related questions you might receive on the exam. For example, the exam may present you with a situation in which you can't access a particular service; you may have to determine whether a port is open or closed on a firewall. For example, let's say you are troubleshooting a remote Windows system with Remote Desktop Protocol (RDP), but you cannot connect to the system. We know that RDP uses port 3389. So, one of the first things to ensure is that port 3389 is open on the firewall.

Tip

Default RDP port is 3389 (this example may have to be changed according to the configured RDP port). Open a command prompt, type in **telnet**, and press Enter. For example, we would type **telnet 192.168. 8.1 3389**. If a blank screen appears, then the port is open, and the test is successful.

TABLE 1.4 **TCP/UDP Port Assignments for Commonly Used Protocols**

Protocol	Port Assignment
TCP Ports	
FTP	20/21
SSH/SFTP	22
Telnet	23
SMTP	25
DNS	53
HTTP	80
LDAP	389
HTTPS	443

Protocol	Port Assignment
SMB	445
SMTPS	587
LDAPS	636
SQL Server	1433
RDP	3389
SIP	5060/5061
UDP Ports	
DNS	53
DHCP (and BOOTP server)	67
DHCP (and BOOTP client)	68
TFTP	69
NTP	123
SNMP	161/162
Syslog	514
RDP	3389
SIP	5060/5061

ExamAlert

The term *well-known ports* identifies the ports ranging from 0 to 1023. If/when an exam question refers to "well-known ports," this is what it refers to. For the N+ exam, know Table 1.4 well!

Note

You might have noticed in Table 1.4 that two ports are associated with FTP (and some other protocols as well). With FTP, port 20 is considered the data port, and port 21 is considered the control port. In practical use, FTP connections use port 21. Port 20 is rarely used in modern implementations.

Traffic Types

Different types of data may need to be sent through the network in different ways. Because of this, four different types of networking traffic are used to deliver data packets to one or more destinations in a network: unicast, multicast, anycast, and broadcast. You need to distinguish among these four types.

Unicast

With *unicast*, traffic is one-to-one communication between a single sender and a single receiver, and this is the most common type of traffic in IP networks. It is used for point-to-point communication between devices, such as client/server interactions and most Internet browsing. Each packet in a unicast transmission is addressed specifically to a unique destination IP address. Data sent with unicast addressing is delivered to a specific node identified by the address. It is a point-to-point, one-to-one, address link.

Multicast

Multicasting is a mechanism by which groups of network devices can send and receive data between the members of the group at one time (one-to-many), instead of separately sending messages to each device in the group. The multicast grouping is established by configuring each device with the same multicast IP address. Each packet in a multicast transmission is addressed to a multicast group, identified by a multicast IP address. Devices interested in receiving multicast traffic join specific multicast groups and receive packets destined for those groups. Multicasting is commonly used for applications such as IPTV, video streaming, online gaming, and real-time data distribution.

Anycast

Anycast traffic is one-to-nearest communication from a sender to the nearest of multiple receivers. Multiple devices share the same anycast IP address, but packets are routed to the nearest (or best) destination based on routing metrics such as shortest path or lowest latency. Anycast is often used for services that benefit from geographic or topological proximity, such as content delivery networks (CDNs), DNS servers, and load balancers.

Broadcast

A *broadcast* is at the opposite end of the spectrum from a unicast. Broadcast traffic is one-to-all communication from a single sender to all devices within a specific network segment. Each packet in a broadcast transmission is addressed to a special broadcast IP address (e.g., 255.255.255.255 in IPv4) or a Layer 2 broadcast address (e.g., MAC address FF:FF:FF:FF:FF:FF). Broadcast is used for sending messages or data to all devices within a local network segment, such as ARP requests, DHCP requests, and some network discovery protocols. Broadcast traffic is typically limited to the local network segment and does not propagate across routers.

Cram Quiz

1. TCP is an example of what kind of transport protocol?

   ○ **A.** Connection oriented

   ○ **B.** Connection reliant

   ○ **C.** Connection dependent

   ○ **D.** Connectionless

2. Which of the following are considered transport protocols? (Choose the two best answers.)

   ○ **A.** TCP

   ○ **B.** IP

   ○ **C.** UDP

   ○ **D.** THC

3. What is the function of NTP?

   ○ **A.** It provides a mechanism for the sharing of authentication information.

   ○ **B.** It is used to access shared folders on a Linux system.

   ○ **C.** It is used to communicate utilization information to a central manager.

   ○ **D.** It is used to communicate time synchronization information between systems.

4. Which of the following protocols offers guaranteed delivery?

   ○ **A.** FTP

   ○ **B.** POP

   ○ **C.** IP

   ○ **D.** TCP

5. By default, which protocol uses port 68?

   ○ **A.** DHCP

   ○ **B.** DNS

   ○ **C.** SMB

   ○ **D.** SMTP

6. What are SNMP databases called?

   ○ **A.** HOSTS

   ○ **B.** MIBs

   ○ **C.** WINS

   ○ **D.** Agents

7. What are logical groupings of SNMP systems known as?

   ○ **A.** Communities

   ○ **B.** Pairs

   ○ **C.** Mirrors

   ○ **D.** Nodes

8. What are two features supported in SNMPv3 and not previous versions?

   ○ **A.** Authentication

   ○ **B.** Dynamic mapping

   ○ **C.** Platform independence

   ○ **D.** Encryption

9. Which of the following is a key management protocol used with IPsec to establish secure communication channels and negotiate cryptographic keys for encrypted VPN connections?

   ○ **A.** GRE

   ○ **B.** SMB

   ○ **C.** SIP

   ○ **D.** IKE

10. Which is a standard for message logging that uses port 514 and is available on most network devices (e.g., routers, switches, and firewalls), as well as printers, and UNIX/Linux–based systems?

   ○ **A.** SNMP

   ○ **B.** Syslog

   ○ **C.** SMTP TLS

   ○ **D.** SSH

11. Which network traffic type traffic is a one-to-one communication between a single sender and a single receiver and is the most common type of traffic in IP networks?

   ○ **A.** Multicast

   ○ **B.** Anycast

   ○ **C.** Broadcast

   ○ **D.** Unicast

Cram Quiz Answers

1. **A.** TCP is an example of a connection-oriented transport protocol. UDP is an example of a connectionless protocol. *Connection reliant* and *connection dependent* are not terms commonly associated with protocols.

2. **A and C.** Both TCP and UDP are transport protocols. IP is a network protocol, and THC is not a valid protocol.

3. **D.** NTP is used to communicate time-synchronization information between devices. Network File System (NFS) is a protocol typically associated with accessing shared folders on a Linux system. Utilization information is communicated to a central management system most commonly by using SNMP.

4. **D.** TCP is a connection-oriented protocol that guarantees delivery of data. FTP is a protocol used to transfer large blocks of data. POP stands for Post Office Protocol and is not the correct choice. IP is a network layer protocol responsible for tasks such as addressing and route discovery.

5. **A.** DHCP uses port 68 by default (along with 67). DNS uses port 53, SMB uses 445, and SMTP uses port 25.

6. **B.** SNMP uses databases of information called MIBs to define what parameters are accessible, which of the parameters are read-only, and which can be set.

7. **A.** SNMP communities are logical groupings of systems. When a system is configured as part of a community, it communicates only with other devices that have the same community name.

8. **A and D.** SNMPv3 supports authentication and encryption.

9. **D.** Internet Key Exchange (IKE) is a key management protocol used with IPsec to establish secure communication channels and negotiate cryptographic keys for encrypted VPN connections. IKE is responsible for establishing Security Associations (SAs) between communicating devices, which define the parameters for secure communication, including encryption algorithms, integrity algorithms, and authentication methods. Generic Routing Encapsulation (GRE) is a Cisco-created tunneling protocol. It is an encapsulating protocol used to wrap data and securely send it across VPNs and Point-to-Point (or point-to-multipoint) links. Server Message Block (SMB) is used on a network for providing access to resources such as files, printers, ports, and so on that are running on Windows. If you were wanting to connect Linux-based hosts to Windows-shared printers, for example, you would need to implement support for SMB; it runs, by default, on port 445. Session Initiation Protocol (SIP) is an application layer protocol designed to establish and maintain multimedia sessions, such as Internet telephony calls. This means that SIP can create communication sessions for such features as audio/videoconferencing, online gaming, and person-to-person conversations over the Internet.

10. **B.** Syslog is a standard for message logging. It is available on most network devices (e.g., routers, switches, and firewalls), as well as printers, and UNIX/Linux–based systems. Over a network, a syslog server listens for and then logs data messages coming from the syslog client. Syslog server messages are often used for troubleshooting and determining how systems may have been compromised after an attack. The Simple Network Management Protocol (SNMP) uses

port 161 to send data and port 162 to receive it. It enables network devices to communicate information about their state to a central system. It also enables the central system to pass configuration parameters to the devices. SMTP TLS, more commonly known as SMTPS (Simple Mail Transfer Protocol Secure), uses Transport Layer Security (TLS) to provide authentication of the communication partners along with data integrity and confidentiality by wrapping SMTP data in TLS. This is similar to how HTTPS wraps HTTP data inside TLS. The default port for SMTPS is 587. Secure Shell (SSH) is a secure alternative to Telnet. SSH provides security by encrypting data as it travels between systems. This makes it difficult for hackers using packet sniffers and other traffic-detection systems. It also provides more robust authentication systems than Telnet. By default, SSH operates on port 22.

11. **D.** Unicast network traffic is one-to-one communication between a single sender and a single receiver, and this is the most common type of traffic in IP networks. It is used for point-to-point communication between devices, such as client/server interactions and most Internet browsing. Each packet in a unicast transmission is addressed specifically to a unique destination IP address. Data sent with unicast addressing is delivered to a specific node identified by the address. It is a point-to-point, one-to-one, address link. Multicasting is a mechanism by which groups of network devices can send and receive data between the members of the group at one time (one-to-many), instead of separately sending messages to each device in the group. The multicast grouping is established by configuring each device with the same multicast IP address. Each packet in a multicast transmission is addressed to a multicast group, identified by a multicast IP address. Anycast traffic is one-to-nearest communication from a sender to the nearest of multiple receivers. Multiple devices share the same anycast IP address, but packets are routed to the nearest (or best) destination based on routing metrics such as shortest path or lowest latency. A broadcast is at the opposite end of the spectrum from a unicast. Broadcast traffic is one-to-all communication from a single sender to all devices within a specific network segment. Each packet in a broadcast transmission is addressed to a special broadcast IP address (e.g., 255.255.255.255 in IPv4) or a Layer 2 broadcast address (e.g., MAC address FF:FF:FF:FF:FF:FF).

What's Next?

This chapter created a foundation upon which Chapter 2, "Network Topologies, Architectures, and Types," builds.

CHAPTER 2

Network Topologies, Architectures, and Types

This chapter covers the following official Network+ objective:

▶ 1.6 Compare and contrast network topologies, architectures, and types.

For more information on the official CompTIA Network+ exam topics, see the "About the Network+ Exam" section in the Introduction.

A variety of physical and logical network layouts are in use today. As a network administrator, you might find yourself working on these different network layouts or topologies. Therefore, you must understand how they are designed to function.

This chapter reviews general network considerations, such as the various topologies used on today's networks, *local-area networks* (LANs), and *wide-area networks* (WANs).

Network Topologies

▶ **1.6 Compare and contrast network topologies, architectures, and types.**

CramSaver

If you can correctly answer these questions before going through this section, save time by skimming the ExamAlerts in this section and then completing the Cram Quiz at the end of the section.

1. Which topology is commonly known as a hub and spoke model?

2. With which topology does every node have a direct connection to every other node?

3. True or false: Traffic flows entering and leaving a datacenter are known as East-West traffic.

4. True or false: In the three-tier hierarchical model, the access/edge layer ensures data is delivered to edge/end devices.

Answers

1. The star topology is commonly known as a hub and spoke model as it utilizes a centralized switch, or hub, and devices extend from it.

2. With a mesh topology, every node has a direct connection to every other node.

3. False. Traffic flows entering and leaving a datacenter are known as North-South traffic. East-West traffic refers to network traffic that flows within a datacenter between servers.

4. True. In the three-tier hierarchical model, the access/edge layer is the place where switches connect to and ensure data is delivered to edge/end devices.

A *topology* refers to a network's physical and logical layout. A network's *physical topology* refers to the actual layout of the computer cables and other network devices. A network's *logical topology* refers to the way in which the network appears to the devices that use it. Network topology diagrams are used to identify network components and how they are physically or logically connected.

Several topologies are in use on networks today. Some of the more common topologies are the star/hub and spoke, mesh, and hybrid models. The following sections provide an overview of each as well as look at some older topologies you may encounter.

Star/Hub and Spoke

In the *star topology*, all computers and other network devices connect to a central device called a *hub* or *switch* and, for that reason, it is also called a *hub and spoke network*. Each connected device requires a single cable to be connected to the hub or switch, creating a point-to-point connection between the device and the hub or switch.

Using a separate cable to connect to the hub or switch allows the network to be expanded without disruption. A break in any single cable does not cause the entire network to fail. Figure 2.1 shows a star/hub and spoke topology.

> **ExamAlert**
>
> Among the network topologies discussed in this chapter, the star topology is the easiest to expand in terms of the number of devices connected to the network.

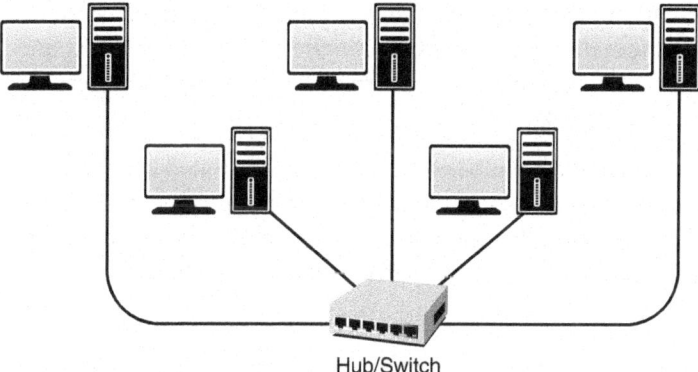

Hub/Switch

FIGURE 2.1   **Star/Hub and Spoke Topology**

The star/hub and spoke topology is the most widely implemented network design in use today, but it is not without shortcomings. Because all devices connect to a centralized hub or switch, this creates a single point of failure for the network. If the hub or switch fails, any device connected to it cannot access the network. Because of the number of cables required and the need for network devices, the cost of a star/hub and spoke network is often higher than other topologies. Table 2.1 summarizes the advantages and disadvantages of the star/hub and spoke topology.

TABLE 2.1 **Advantages and Disadvantages of the Star/Hub and Spoke Topology**

Advantages	Disadvantages
Star/hub and spoke networks are easily expanded without disruption to the network.	This topology requires more cable than most of the other topologies.
Cable failure affects only a single user.	A central connecting device allows for a single point of failure.
It is easy to troubleshoot and implement.	It requires additional networking equipment to create the network layout.

Mesh Topology

When it comes to the *mesh topology*, it is helpful to differentiate between a wired and wireless implementation, so we focus on the former here and the latter in Chapter 6, "Wireless Solutions." Mesh incorporates a unique network design in which each computer on the network connects to every other, creating a point-to-point connection between every device on the network. Since this is often done physically, the term *wired mesh* or *wired mesh topology* is sometimes used. The purpose of the mesh design is to provide a high level of *redundancy*. If one network cable fails, the data always has an alternative path to get to its destination; each node can act as a relay.

The wiring for a mesh network can be complicated, as illustrated by Figure 2.2. Furthermore, the cabling costs associated with the mesh topology can be high, and troubleshooting a failed cable can be tricky. As a result, the mesh topology is not the first choice for many wired networks but is more popular with servers/routers.

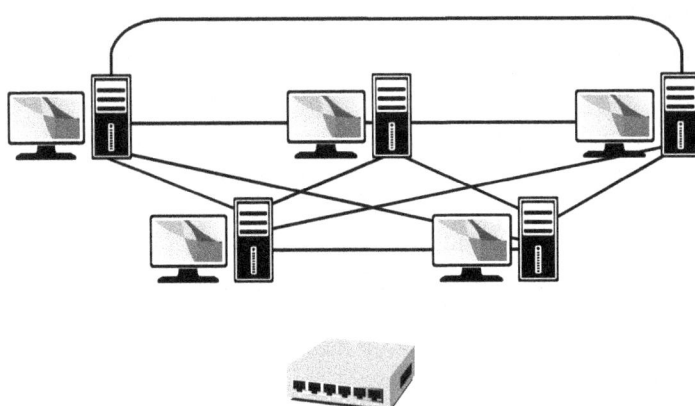

Hub/Switch

FIGURE 2.2 **Mesh Topology**

Table 2.2 summarizes the advantages and disadvantages of the mesh topology.

> **ExamAlert**
>
> Because of the redundant connections, the mesh topology offers better fault tolerance than other topologies.

> **Note**
>
> Fault tolerance is a process or capability of a network or system to continue working after there is a problem. It refers to a network or system's ability to continue operating after a malfunction or failure.

TABLE 2.2 **Advantages and Disadvantages of the Mesh Topology**

Advantages	Disadvantages
Mesh provides redundant paths between LAN topologies.	It requires more cable than the other topologies.
The network can be expanded without disruption to current users.	The implementation is complicated.

Hybrid Topology

A variation on a true mesh topology is the *hybrid* or *hybrid mesh*. It creates a redundant point-to-point network connection between only specific network devices (such as the servers). The hybrid mesh is most often seen in WAN implementations but can be used in any network.

Another way of describing the degree of mesh implementation is by labeling it as either *partial* or *full*. If it is a true mesh network with connections between each device, it can be labeled full mesh, and if it is less than that—a hybrid of any sort—it is called a *partial mesh network*.

Many of the topologies found in large networking environments are a hybrid of physical topologies. An example of a hybrid topology is the star/hub and spoke bus—a combination of the star/hub and spoke topology and the bus topology (explored further later in this chapter). Figure 2.3 shows how this might look in a network implementation.

> **ExamAlert**
>
> Another meaning: The term *hybrid topology* also can refer to the combination of wireless and wired networks. For the Network+ exam, however, the term *hybrid* most likely refers to the combination of physical networks.

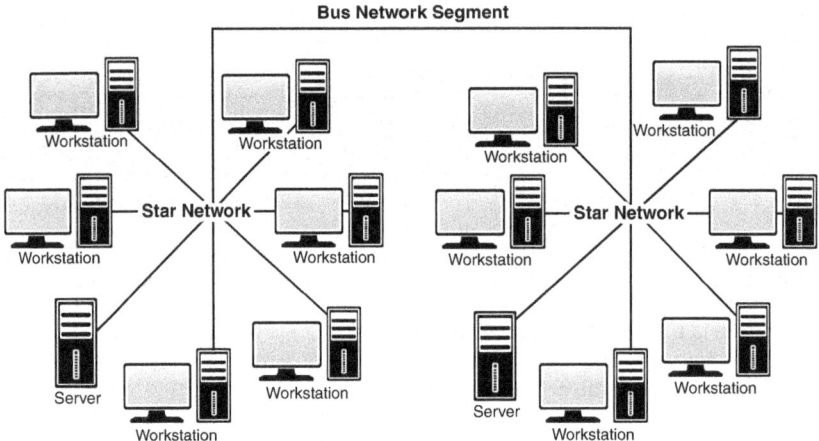

FIGURE 2.3   A Star/Hub and Spoke Bus Topology

Point to Point

A *point to point* (often written with hyphens: *point-to-point*) networking topology is a network configuration where two devices or nodes are directly connected to each other. In this topology, there is a dedicated communication link between the two endpoints, and data flows directly between them without the involvement of any intermediary devices. While the distinction can be nitpicky, mesh and star/hub and spoke networks are used to connect multiple devices to a network, while point-to-point topology is used to connect two devices.

It might not be the ideal choice for larger, more complex networks. In such cases, more scalable and flexible topologies like star/hub and spoke or mesh networks are typically preferred, since they can accommodate multiple devices and provide greater redundancy and fault tolerance.

> **Note**
>
> More information on point to point, as it relates to wireless networking, is provided in Chapter 6.

Spine and Leaf

Tiered models of computer network architecture are a way of organizing and structuring a network infrastructure into distinct layers (or *tiers*), each with specific functions and responsibilities. These models help simplify network design, management, and scalability.

A two-tier model that Cisco promotes for switches is the *spine and leaf* model. In this model, the spine is the *backbone* of the network, just as it would be in a skeleton and is responsible for interconnecting all the leaf switches in a full-mesh topology. Thanks to the mesh, every leaf is connected to every spine, and the path is randomly chosen so that the traffic load is evenly distributed among the top-tier switches. If one of the switches at the top tier were to fail, there would only be a slight degradation in performance throughout the datacenter.

Because of the design of this model, no matter which leaf switch is connected to a server, the traffic always has to cross the same number of devices to get to another server. This keeps latency at a steady level.

> **Note**
>
> Latency is the typical amount of time that it takes for packets of data to travel from one computer or system to the next. The higher the latency, the worse the experience when it comes to real-time video conferencing, webinars, gaming, and so on.

When *top-of-rack (ToR) switching* is incorporated into the network architecture, switches located within the same rack are connected to an in-rack network switch, which is connected to aggregation switches (usually via fiber cabling). The big advantage of this setup is that the switches within each rack can be connected with cheaper copper cabling and the cables to each rack are all that need to be fiber.

> **ExamAlert**
>
> Remember that in a spine and leaf model, the spine is the backbone of the network and is responsible for interconnecting all the leaf switches in a full-mesh topology. Know, as well, that all data flows require the same number of hops, they incorporate a full-mesh switching topology, and they use software-defined networking (SDN) to direct traffic, rather than blocking ports using the Spanning Tree Protocol (STP).

Three-Tier Hierarchical Model

Just as the spine and leaf is a two-tiered model, it is possible to improve system performance, as well as to improve security, by implementing an *n*-tiered model (wherein the *n*- can be one of several different numbers).

If we were looking at a database, for example, with a one-tier model, or single-tier environment, the database and the application exist on a single system. This is common on desktop systems running a standalone database. Early UNIX implementations also worked in this manner; each user would sign on to a terminal and run a dedicated application that accessed the data. With two-tier architecture, the client workstation or system runs an application that communicates with the database that is running on a different server. This common implementation works well for many applications. With *three-tiered architecture*, otherwise known as a three-tier hierarchical model, security is enhanced. In this model, the end user is effectively isolated from the database by the introduction of a middle-tier server. This server accepts requests from clients, evaluates them, and then sends them on to the database server for processing. The database server sends the data back to the middle-tier server, which then sends the data to the client system. Becoming common in business today, this approach adds both capability and complexity.

While the examples are of database tiering, this same approach can be taken with devices such as routers, switches, and other servers. In a three-tiered model of routing and switching, the three tiers would be the core, the distribution/aggregation layer, and the access/edge. We walk through each of the layers present in this scenario.

Core Layer

The *core* layer is the backbone: the place where switching and routing meet (switching ends, routing begins). It provides high-speed, highly redundant forwarding services to move packets between distribution layer devices in different regions of the network. The core switches and routers would be the most powerful in the enterprise (in terms of their raw forwarding power) and would be used to manage the highest-speed connections (such as 100 Gigabit Ethernet). Core switches also incorporate internal firewall capability as part of their features, helping with segmentation and control of traffic moving from one part of the network to another.

Distribution/Aggregation Layer

The *distribution layer*, or *aggregation layer* (sometimes called the workgroup layer), is the layer in which management takes place. This is the place where Quality of Service (QoS) policies are managed, filtering is done, and routing takes place. Distribution layer devices can be used to manage individual branch-office WAN connections, and this is considered to be smart (usually offering a larger feature set than switches used at the access/edge layer). Lower latency and larger MAC address table sizes are important features for switches used at this level because they aggregate traffic from thousands of users rather than hundreds (as access/edge switches do).

Access/Edge Layer

Switches that allow end users and servers to connect to the enterprise are called access switches or edge switches, and the layer where they operate in the three-tiered model is known as the *access layer*, or *edge layer*. Devices at this layer may or may not provide Layer 3 switching services; the traditional focus is on mini-mizing the cost of each provisioned Ethernet port (known as "cost-per-port") and providing high port density. Because the focus is on connecting client nodes, such as workstations to the network, this is sometimes called the desktop layer.

> **Note**
>
> As was discussed in Chapter 1, "Networking Models, Ports, Protocols, and Services," a switch can work at either Layer 2 (the data link layer) or Layer 3 (the network layer) of the OSI model. When it filters traffic based on the MAC address, it is called a Layer 2 switch, since MAC addresses exist at Layer 2 of the OSI model (if it operated only with IP traffic, it would be a Layer 3 switch).

Table 2.3 highlights each of the layers of the three-tier hierarchical model.

TABLE 2.3 Three-tier Hierarchical Model Layers

Layer	Core	Distribution	Access
Description	Backbone of the network, provides high-speed connectivity between distribution layers and serves as a transit for all traffic	Aggregates traffic from access layer devices and distributes it to the appropriate destinations, provides policy enforcement and access control	Interfaces directly with end devices such as computers, printers, and IP phones, provides connectivity to the network
Function	Provides high-speed, low-latency forwarding of packets between distribution layer devices	Aggregates and filters traffic, enforces security policies, implements VLANs, routing protocols, and Quality of Service (QoS)	Delivers network services to end devices, such as Ethernet ports, wireless access points, and VLANs
Scale	Typically has the highest capacity and fastest speeds, often utilizes high-performance networking equipment	Capacity and performance requirements are moderate, often use Layer 3 switches and routers for routing and filtering	Usually consists of a large number of ports to accommodate end devices, employs switches with basic Layer 2 functionality

Layer	Core	Distribution	Access
Redundancy	Redundancy and high availability are critical, often implemented using redundant links and protocols like Virtual Router Redundancy Protocol (VRRP)	Redundancy is important but may not be as critical as in the core, often utilizes redundant uplinks and EtherChannel bundles	Redundancy is essential to ensure connectivity for end devices, typically implemented using redundant switches and network paths
Traffic flow	Handles transit traffic between distribution layer devices, typically high-speed and low-latency	Aggregates and filters traffic from access layer devices before forwarding it to the core or other distribution layer devices	Facilitates traffic between end devices and the rest of the network, including user data, management traffic, and control messages
Examples	High-speed routers, switches with large forwarding tables, MPLS networks	Layer 3 switches, VLANs, Quality of Service (QoS) policies, access control lists (ACLs)	Ethernet switches, wireless access points, Power over Ethernet (PoE) switches

> **ExamAlert**
>
> Remember: The core layer is the backbone of the network (where the fastest routers and switches operate to manage separate networks), whereas the distribution/aggregation layer (between the access/edge and core layers) is the "boundary" layer where ACLs and Layer 3 switches operate to properly manage data between VLANs and subnetworks. The access/edge layer is the place where switches connect to and ensure data is delivered to edge/end devices, such as computers and servers.

Collapsed Core

With a *collapsed core* architecture, the three-tier model becomes a two-tier model as the core and distribution layers are combined. While a three-tier model is necessary for complex installations that require access by multiple sites, devices, and users, the collapsed core approach is commonly used in datacenters and enterprise networks.

By collapsing the core and distribution layers into a single layer, the streamlined design provides both advantages and trade-offs. Advantages include simplicity, cost savings (via a reduced number of network devices and switches), efficient communication (with fewer layers, there are fewer network hops between devices), and scalability (it can scale relatively well for medium-sized networks and datacenters with a moderate number of devices). Disadvantages include limited redundancy (a failure at the collapsed core layer could potentially impact the entire network), scalability constraints (it is not ideal for extremely large networks or datacenters with high traffic demands), and potential bottlenecks (there is a risk of network congestion if not properly designed and managed).

Traffic Flows

Traffic flows within a datacenter typically occur within the framework of one of two models: East-West or North-South. The names may not be the most intuitive, but the East-West traffic model means that data is flowing among devices within a specific datacenter, whereas North-South means that data is flowing into the datacenter (from a system physically outside the datacenter) or out of it (to a system physically outside the datacenter).

The naming convention comes from the way diagrams are drawn: data staying within the datacenter is traditionally drawn on the same horizontal line (East-to-West), while data leaving or entering is typically drawn on a vertical line (North-to-South). With the increase in virtualization being implemented at so many levels, the East-West traffic has increased in recent years. Table 2.4 summarizes the traffic flow possibilities.

> **Note**
>
> Network Functions Virtualization (NFV) is a network architecture concept that uses the proven technologies of IT virtualization. It delivers the network services needed to support an infrastructure independent from hardware by decoupling network functions from proprietary hardware appliances. This is covered in more detail in Chapter 7, "Cloud Computing Concepts and Options."

TABLE 2.4 **Traffic Flow Options**

Traffic Flow	Description	Characteristics	Examples
North-South	Refers to the traffic flow between a client and external resources	Typically involves communication between internal users or devices and external networks or services	Internet browsing, accessing cloud services
East-West	Relates to the traffic flow between internal resources within a network	Occurs within the boundaries of a datacenter or local network, involving communication between servers, virtual machines, or applications	Inter-server communication, database queries

Older Topologies: Bus and Ring

There are two topologies that have been removed from this iteration of the CompTIA Network+ exam objectives that you will very likely encounter in the workplace: bus and ring. For that reason, it is highly suggested that you be aware of them, and coverage of them is included at the end of this chapter rather than in with the exam fodder.

A *bus topology* uses a trunk or backbone to connect all the computers on the network, as shown in Figure 2.4. Systems connect to this backbone using *T connectors* or taps (known as a vampire tap, if you must pierce the wire). To avoid signal reflection, a physical bus topology requires that each end of the physical bus be terminated, with one end also being grounded. Note that a hub or switch is not needed in this installation, and loose or missing terminators from a bus network disrupt data transmissions.

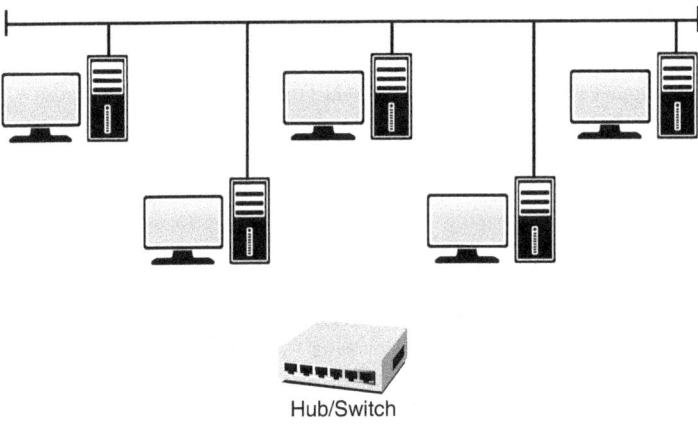

Hub/Switch

FIGURE 2.4 **Physical Bus Topology**

Table 2.5 summarizes the advantages and disadvantages of the bus topology.

TABLE 2.5 **Advantages and Disadvantages of the Bus Topology**

Advantages	Disadvantages
Compared to other topologies, a bus is cheap and easy to implement.	Network disruption might occur when computers are added or removed.
A bus requires less cable than other topologies.	Because all systems on the network connect to a single backbone, a break in the cable prevents all systems from accessing the network.
A bus does not use any specialized network equipment.	It is difficult to troubleshoot.

The *ring topology* is a logical ring, meaning that the data travels in a circular fashion from one computer to another on the network. It is not a physical ring topology. Figure 2.5 shows the logical layout of a ring topology. Note that a hub or switch is not needed in this installation either.

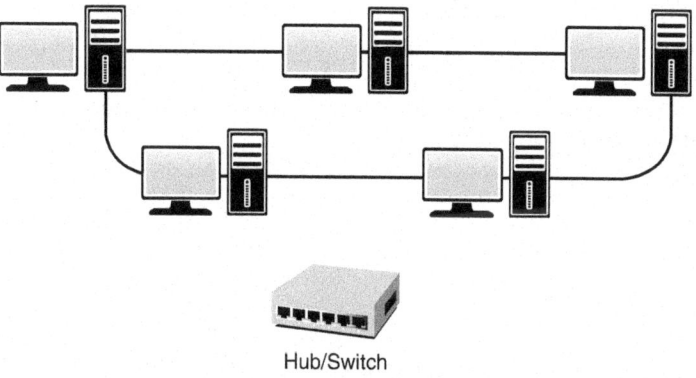

Hub/Switch

FIGURE 2.5 **Logical Design of a Ring Topology**

In a true ring topology, if a single computer or section of cable fails, the signal is interrupted. The entire network becomes inaccessible. Network disruption can also occur when computers are added to or removed from the network, making it an impractical network design in environments where the network changes often.

Ring networks can be set up in a fault-tolerant design, meaning that they have primary and secondary rings. If one ring fails, data can use the second ring to reach its destination. Naturally, the addition of the second ring adds to the cost of the network as well as the complexity. Table 2.6 summarizes the advantages and disadvantages of the ring topology.

TABLE 2.6 **Advantages and Disadvantages of the Ring Topology**

Advantages	Disadvantages
Cable faults are easily located, making troubleshooting easier.	Expansion to the network can cause network disruption.
Ring networks are moderately easy to install.	A single break in the cable can disrupt the entire network.

Cram Quiz

1. You have been asked to install a network that will give the network users the greatest amount of fault tolerance. Which of the following network topologies would you choose?

   ○ **A.** Star/hub and spoke

   ○ **B.** Ring

   ○ **C.** Mesh

   ○ **D.** Bus

2. Which of the following topologies allows for network expansion with the least amount of disruption for the current network users?

   ○ **A.** Bus

   ○ **B.** Ring

   ○ **C.** LAN

   ○ **D.** Star/hub and spoke

3. Which network topology offers the greatest level of redundancy but has the highest implementation cost?

   ○ **A.** Wireless mesh

   ○ **B.** Wired mesh

   ○ **C.** Hybrid star

   ○ **D.** Bus network

4. What traffic pattern refers to data that travels outside the datacenter or enterprise?

   ○ **A.** East-to-West

   ○ **B.** North-to-South

   ○ **C.** On-premises

   ○ **D.** West-to-South

5. What layer in three-tier hierarchical model network architecture is considered the backbone of a network?

   ○ **A.** Core layer

   ○ **B.** Distribution/aggregation layer

   ○ **C.** Access/edge layer

   ○ **D.** Application layer

6. Which topology utilizes a dedicated connection between two endpoints?

   ○ **A.** Mesh

   ○ **B.** Spine and leaf

   ○ **C.** Point to point

   ○ **D.** Star

7. Which topology is commonly used in datacenter environments for high scalability and flexibility?

   ○ **A.** Ad hoc

   ○ **B.** Hybrid

   ○ **C.** Star

   ○ **D.** Spine and leaf

8. This network architecture includes simplicity, cost savings (via a reduced number of network devices and switches), efficient communication (with fewer layers, there are fewer network hops between devices), and scalability (it can scale relatively well for medium-sized networks and datacenters with a moderate number of devices). Which network architecture is being described?

   ○ **A.** Collapsed core

   ○ **B.** NFV

   ○ **C.** ToR switching

   ○ **D.** Four-tier hierarchical model

9. Which layer in the three-tier hierarchical model network architecture facilitates traffic between end devices and the rest of the network?

   ○ **A.** Core layer

   ○ **B.** Distribution layer

   ○ **C.** Point-to-point layer

   ○ **D.** Access layer

Cram Quiz Answers

1. **C.** A mesh network uses a point-to-point connection to every device on the network. This creates multiple points for the data to be transmitted around the network and therefore creates a high degree of redundancy. The star/hub and spoke, ring, and bus topologies do not offer the greatest amount of fault tolerance.

2. **D.** On a star/hub and spoke network, each network device uses a separate cable to make a point-to-point connection to a centralized device, such as a hub or a switch. With such a configuration, a new device can be added to the network by attaching the new device to the hub or switch with its own cable. This process does not disrupt the users who are currently on the network. Answers A and B are incorrect because the addition of new network devices on a ring or bus network can cause a disruption in the network and cause network services to be unavailable during the installation of a new device.

3. **B.** The wired mesh topology requires each computer on the network to be individually connected to every other device. This configuration provides maximum reliability and redundancy for the network. However, it is very costly to implement because of the multiple wiring requirements.

4. **B.** North-South refers to data transfers between the datacenter and outside of the network. East-West traffic is a concept referring to network traffic flow within a datacenter between servers. On-premises can be thought of as the old, traditional approach: the data and the servers are kept in-house. Although West-to-South is a direction, it is not a valid specified data path.

5. **A.** The core layer is the backbone of the network where the fastest routers and switches operate to manage separate networks. The distribution/aggregation layer is between the access/edge and core layers. This is the "boundary" layer where ACLs and Layer 3 switches operate. The access/edge layer is the place where switches connect to and ensure data is delivered to edge/end devices. The application layer is the seventh and top layer of the OSI reference model.

6. **C.** Point-to-point topology involves a direct link between two devices, providing a dedicated connection for communication. It's commonly used in WAN connections, such as leased lines or serial connections.

7. **D.** Spine and leaf topology is prevalent in datacenters due to its scalability and high-performance characteristics. It consists of spine switches interconnected with leaf switches, offering multiple paths and high bandwidth for data traffic.

8. **A.** With a collapsed core architecture, the three-tier model becomes a two-tier model as the core and distribution layers are combined. By collapsing the core and distribution layers into a single layer, the streamlined design provides both advantages and trade-offs. Advantages include simplicity, cost savings (via a reduced number of network devices and switches), efficient communication (with fewer layers, there are fewer network hops between devices), and scalability (it can scale relatively well for medium-sized networks and datacenters with a moderate number of devices). Network Functions Virtualization (NFV) is a technology that allows for the virtualization of a replica of a network's physical topology and the way it behaves without changing the logical topology and the way that devices are managed. NFV allows for the virtualization of network functions such

as routers, firewalls, and switches, resulting in increased flexibility and scalability. When top-of-rack (ToR) switching is incorporated into the network architecture, switches located within the same rack are connected to an in-rack network switch, which is connected to aggregation switches (usually via fiber cabling). There is no such thing as the four-tier hierarchical model.

9. **D.** The access layer in three-tier hierarchical model network architecture facilitates traffic between end devices and the rest of the network, including user data, management traffic, and control messages. The core layer is the backbone of the network, provides high-speed connectivity between distribution layers, and serves as a transit for all traffic. The distribution layer aggregates and filters traffic; enforces security policies; and implements VLANs, routing protocols, and Quality of Service (QoS). Point to point is not a layer in the three-tier hierarchical model. It is a network configuration where two devices or nodes are directly connected to each other. In this topology, there is a dedicated communication link between the two endpoints, and data flows directly between them without the involvement of any intermediary devices.

What's Next?

The TCP/IP suite is the most widely implemented protocol on networks today. As such, it is an important topic on the Network+ exam. Chapter 3, "Network Addressing, Routing, and Switching," starts by discussing one of the more complex facets of TCP/IP: IP addresses.

CHAPTER 3

Network Addressing, Routing, and Switching

This chapter covers the following official Network+ objectives:

▶ 1.7 Given a scenario, use appropriate IPv4 network addressing.

▶ 2.1 Explain characteristics of routing technologies.

▶ 2.2 Given a scenario, configure switching technologies and features.

▶ 3.4 Given a scenario, implement IPv4 and IPv6 network services.

This chapter covers CompTIA Network+ objectives 1.7, 2.1, 2.2, and 3.4. For more information on the official Network+ exam topics, see the "About the Network+ Exam" section in the Introduction.

Without question, TCP/IP is the most widely implemented protocol suite on networks today. As such, it is an important topic on the Network+ exam. To pass the exam, you definitely need to understand the material presented in this chapter.

This chapter deals with the concepts that govern routing and switching. It starts, however, by discussing one of the more complex facets of TCP/IP: addressing.

IPv4 Network Addressing

▶ **1.7 Given a scenario, use appropriate IPv4 network addressing.**

CramSaver

If you can correctly answer these questions before going through this section, save time by skimming the ExamAlerts in this section and then completing the Cram Quiz at the end of the section.

1. How many octets does a Class A address use to represent the network portion?

2. What is the range that Class C addresses span in the first octet?

3. What are the reserved private IPv4 ranges for private networks?

Answers

1. A Class A address uses only the first octet to represent the network portion, a Class B address uses two octets, and a Class C address uses three octets.

2. Class C addresses span from 192 to 223, with a default subnet mask of 255.255.255.0.

3. A private network is any network to which access is restricted. Reserved IP addresses are 10.0.0.0 to 10.255.255.255, 172.16.0.0 to 172.31.0.0, and 192.168.0.0 to 192.168.255.255.

IP addressing is one of the most challenging aspects of TCP/IP. It can leave even the most seasoned network administrators scratching their heads. Fortunately, the Network+ exam requires only a fundamental knowledge of IP addressing. The following sections look at how IP addressing works for IPv4.

To communicate on a network using TCP/IP, each system must be assigned a unique address. The address defines both the number of the network to which the device is attached and the number of the node on that network. In other words, the IP address provides two pieces of information. It's a bit like a street name and house number in a person's home address.

ExamAlert

A *node* or *host* is any device connected to the network. A node might be a client computer, a server computer, a printer, a router, or a gateway.

Each device on a logical network segment must have the same network address as all the other devices on the segment. All the devices on that network segment must then have different node (host) addresses.

In IP addressing, another set of numbers, called a subnet mask, defines which portion of the IP address refers to the network address and which refers to the node (host) address.

IP addressing is different in IPv4 and IPv6. This discussion focuses on IPv4. IPv6 addressing is discussed in Chapter 4, "Network Implementations."

An Overview of IPv4

An IPv4 address is composed of four sets of 8 binary bits, which are called *octets*. The result is that IP addresses contain 32 bits. Each bit in each octet is assigned a decimal value. The far-left bit has a value of 128, followed by 64, 32, 16, 8, 4, 2, and 1, left to right.

Each bit in the octet can be either a 1 or a 0. If the value is 1, it is counted as its decimal value, and if it is 0, it is ignored. If all the bits are 0, the value of the octet is 0. If all the bits in the octet are 1, the value is 255, which is 128 + 64 + 32 + 16 + 8 + 4 + 2 + 1.

By using the set of 8 bits and manipulating the 1s and 0s, you can obtain any value between 0 and 255 for each octet.

Table 3.1 shows some examples of decimal-to-binary value conversions.

TABLE 3.1 **Decimal-to-Binary Value Conversions**

Decimal Value	Binary Value	Decimal Calculation
10	00001010	8 + 2 = 10
192	11000000	128 + 64 = 192
205	11001101	128 + 64 + 8 + 4 + 1 = 205
223	11011111	128 + 64 + 16 + 8 + 4 + 2 + 1 = 223

IP Address Classes

IP addresses are grouped into logical divisions called *classes*. The IPv4 address space has five address classes (A through E); however, only three (A, B, and C) assign addresses to clients. Class D is reserved for multicast addressing, and Class E is reserved for future development and research.

Of the three classes available for address assignments, each uses a fixed-length subnet mask to define the separation between the network and the node (host) address. A Class A address uses only the first octet to represent the network portion; a Class B address uses two octets; and a Class C address uses the first three octets. The upshot of this system is that Class A has a small number of network addresses, but each Class A address has a large number of possible host addresses. Class B has a larger number of networks, but each Class B address has a smaller number of hosts. Class C has an even larger number of networks, but each Class C address has an even smaller number of hosts. The exact numbers are provided in Table 3.2.

Be prepared for exam questions asking you to identify IP class ranges, such as the IP range for a Class A network.

TABLE 3.2 **IPv4 Address Classes and the Number of Available Network/ Host Addresses**

Address Class	Range	Number of Networks	Number of Hosts Per Network	Binary Value of First Octet
A	1 to 126	126	16,777,214	0xxxxxxx
B	128 to 191	16,384	65,534	10xxxxxx
C	192 to 223	2,097,152	254	110xxxxx
D	224 to 239	N/A	N/A	1110xxxx
E	240 to 255	N/A	N/A	1111xxxx

Note

Notice in Table 3.2 that the network number 127 is not included in any of the ranges. The 127.0.0.1 network ID is reserved for the IPv4 local *loopback*. The local loopback, or *localhost*, address is a function of the protocol suite used in the troubleshooting process.

ExamAlert

For the Network+ exam, be prepared to identify into which class a given address falls. Also be prepared to identify the IPv4 loopback address. The loopback address is 127.0.0.1. Similarly, the address used for the IPv6 loopback address is 0:0:0:0:0:0:0:1, which is normally expressed as ::1, and this is discussed a bit later.

Subnet Mask Assignment

Like an IP address, a *subnet mask* is most commonly expressed in 32-bit dotted-decimal format. Unlike an IP address, though, a subnet mask performs just one function: it defines which parts of the IP address refers to the network address and which refers to the node (host) address. Each class of the IP address used for address assignment has a default subnet mask associated with it. Table 3.3 lists the default subnet masks.

TABLE 3.3 **Default Subnet Masks Associated with IP Address Classes**

Address Class	Default Subnet Mask
A	255.0.0.0
B	255.255.0.0
C	255.255.255.0

ExamAlert

You will likely see questions about address classes and the corresponding default subnet mask. Review Table 3.3 before taking the exam.

Subnetting

Now that you have looked at how IPv4 addresses are used, you can learn the process of subnetting. *Subnetting* is a process by which the node (host) portions of an IP address create more networks than you would have if you used the default subnet mask.

To illustrate subnetting, for example, suppose that you have been assigned the Class B address 150.150.0.0. Using this address and the default subnet mask, you could have a single network (150.150) and use the rest of the address as node (host) addresses. This would give you a large number of possible node addresses, which in reality is probably not useful. With subnetting, you use bits from the node portion of the address to create more network addresses. Doing so reduces the number of nodes per network, but you probably will still have more than enough.

Following are two main reasons for subnetting:

▶ It enables you to more effectively use IP address ranges.

▶ It makes IP networking more secure and manageable by providing a mechanism to create multiple networks rather than having just one. Using multiple networks confines traffic to the network that it needs to be on, which reduces overall network traffic levels. Multiple subnets also create more broadcast domains, which in turn reduces network-wide broadcast traffic. A difference exists between *broadcast domains* and *collision domains*: the latter is all the connected nodes, whereas the former is all the logical nodes that can reach each other. As such, collision domains are typically subsets of broadcast domains.

> **ExamAlert**
>
> Subnetting does not increase the number of IP addresses available. It increases the number of network IDs and, as a result, decreases the number of node IDs per network. It also creates more broadcast domains. Broadcasts are not forwarded by routers, so they are limited to the network on which they originate.

With *Variable Length Subnet Masking (VLSM)*, it is possible to use a different subnet mask for the same network number on different subnets. This way, a network administrator can use a long mask on networks with few hosts and a short mask on subnets with many hosts, thus allowing each subnet in a routed system to be correctly sized for the required size. The routing protocol used (EIGRP, OSPF, RIPv2, IS-IS, or BGP) must be able to advertise the mask for each subnet in the routing update, which means that it must be classless. Classless interdomain routing (CIDR) is discussed shortly.

Identifying the Differences Between IPv4 Public and Private Networks

IP addressing involves many considerations, not the least of which are public and private networks:

▶ A public network is a network to which anyone can connect. The best (and perhaps only pure) example of such a network is the Internet.

▶ A private network is any network to which access is restricted. A corporate network and a network in a school are examples.

> **Note**
>
> The Internet Assigned Numbers Authority (IANA) is responsible for assigning IP addresses to public networks. However, because of the workload involved in maintaining the systems and processes to do this, IANA has delegated the assignment process to a number of regional authorities. For more information, visit www.iana.org/numbers.

The main difference between public and private networks, other than access—a private network is tightly controlled and access to a public network is not—is that the addressing of devices on a public network must be carefully considered. Addressing on a private network has a little more latitude.

As already discussed, for hosts on a network to communicate by using TCP/IP, they must have unique addresses. This number defines the logical network that each host belongs to and the host's address on that network. On a private network with, for instance, three logical networks and 100 nodes on each network, addressing is not a difficult task. On a network on the scale of the Internet, however, addressing is complex.

If you connect a system to the Internet, you need to get a valid registered IP address. Most commonly, you obtain this address from your *Internet service provider (ISP)*. Alternatively, if you want a large number of addresses, for example, you could contact the organization responsible for address assignment in your area. You can determine who the regional numbers authority for your area is by visiting the IANA website.

Because of the nature of their business, ISPs have large blocks of IP addresses that they can assign to their clients. If you need a registered IP address, getting one from an ISP is almost certainly a simpler process than going through a regional numbers authority. Some ISP plans include blocks of registered IP addresses, working on the principle that businesses want some kind of permanent presence on the Internet. However, if you discontinue your service with the ISP, you can no longer use the provided IP address.

To provide flexibility in addressing, and to prevent an incorrectly configured network from polluting the Internet, certain address ranges are set aside for private use. These address ranges are called *private ranges* because they are designated for use only on private networks. These addresses are special because Internet routers are configured to ignore any packets they see that use these addresses. This means that if a private network "leaks" onto the Internet, it won't get any farther than the first router it encounters. So a private address cannot be on the Internet because it cannot be routed to public networks.

Three ranges are defined in *RFC 1918*: one each from Classes A, B, and C. You can use whichever range you want; however, the Class A and B address ranges offer more addressing options than Class C. Table 3.4 defines the private address ranges for Class A, B, and C addresses.

TABLE 3.4 **Private Address Ranges**

Class	Address Range	Default Subnet Mask
A	10.0.0.0 to 10.255.255.255	255.0.0.0
B	172.16.0.0 to 172.31.255.255	255.255.0.0
C	192.168.0.0 to 192.168.255.255	255.255.255.0

> **ExamAlert**
>
> You can expect questions on RFC 1918, private IP address ranges, and their corresponding default subnet masks.

Classless Interdomain Routing

Classless interdomain routing (CIDR) is an IPv4 method of assigning addresses outside the standard Class A, B, and C structure. Specifying the number of bits in the subnet mask offers more flexibility than the three standard class definitions.

Using CIDR, addresses are assigned using a value known as the *slash*. The actual value of the slash depends on how many bits of the subnet mask are used to express the network portion of the address. For example, a subnet mask that uses all 8 bits from the first octet and 4 from the second would be described as /12, or "slash 12." A subnet mask that uses all the bits from the first three octets would be called /24. Why the slash? In addressing terms, the CIDR value is expressed after the address, using a slash. So, the address 192.168.2.1/24 means that the node's IP address is 192.168.2.1, and the subnet mask is 255.255.255.0.

> **Note**
>
> You can find a great CIDR calculator that can compute values from ranges at www.subnet-calculator.com/.

> **ExamAlert**
>
> You will likely see IP addresses in their CIDR format on the exam. Be sure that you understand CIDR addressing and IPv4 notation for the exam.

Default Gateways

Default gateways are the means by which a device can access hosts on other networks for which it does not have a specifically configured route. Most workstation configurations default to using default gateways rather than having any static routes configured. This enables workstations to communicate with other network segments or with other networks, such as the Internet.

ExamAlert

You will be expected to identify the purpose and function of a default gateway. You may also be asked to place the IP address of the default gateway (or other specified system) in the correct location within a performance-based question.

When a system wants to communicate with another device, it first determines whether the host is on the local network or a remote network. If the host is on a remote network, the system looks in the routing table to determine whether it has an entry for the network on which the remote host resides. If it does, it uses that route. If it does not, the data is sent to the default gateway.

Note

Although it might seem obvious, it's worth mentioning that the default gateway must be on the same network as the nodes that use it.

In essence, the default gateway is simply the path out of the network for a given device. Figure 3.1 shows how a default gateway fits into a network infrastructure.

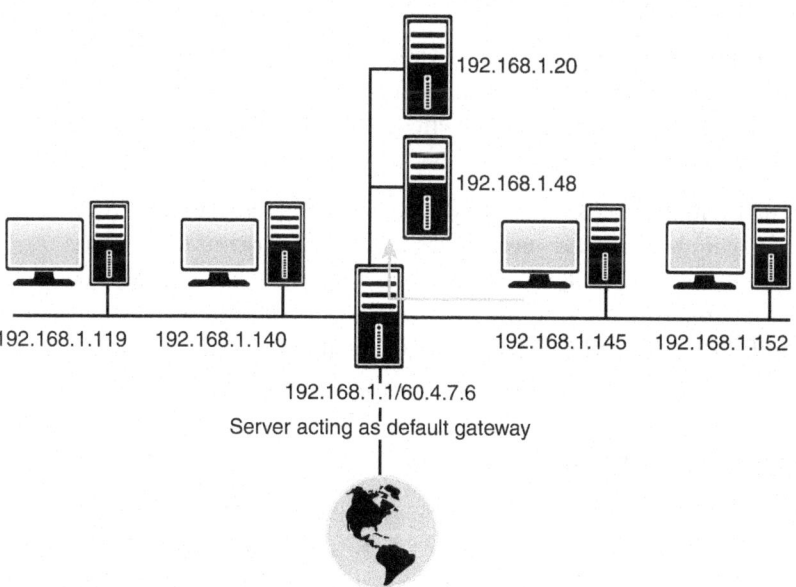

FIGURE 3.1 **The Role of a Default Gateway**

On the network, a default gateway could be a router or a computer with network interfaces (multihomed) for all segments to which it is connected. These interfaces have local IP addresses for the respective segments. If a system is not configured with any static routes or a default gateway, it is limited to operating on its own network segment.

> **ExamAlert**
>
> For the exam, know that any system that does not have a default gateway or any static routes configured is limited to operating on its own network segment.

Assigning IP Addresses

Now that you understand the need for each system on a TCP/IP-based network to have a unique address, the following sections examine how those systems receive their addresses.

Static Addressing

Static addressing refers to the manual assignment of IP addresses to a system. This approach has two main problems:

▶ Statically configuring one system with the correct address is simple, but in the course of configuring, for instance, a few hundred systems, mistakes are likely. If the IP addresses are entered incorrectly, the system probably cannot connect to other systems on the network.

▶ If the IP addressing scheme for the organization changes, each system must again be manually reconfigured. In a large organization with hundreds or thousands of systems, such a reconfiguration could take a considerable amount of time. These drawbacks of static addressing are so significant that nearly all networks use dynamic IP addressing.

Dynamic Addressing

Dynamic addressing refers to the automatic assignment of IP addresses. On modern networks, the mechanism used to do this is *Dynamic Host Configuration Protocol (DHCP)*. DHCP, part of the TCP/IP suite, enables a central system to provide client systems with IP addresses. Automatically assigning addresses with DHCP alleviates the burden of address configuration and reconfiguration that occurs with static IP addressing.

The basic function of the DHCP service is to automatically assign IP addresses to client systems. To do this, ranges of IP addresses, known as *scopes*, are defined on a system running a DHCP server application. When another system configured as a DHCP client is initialized, it asks the server for an address. If all things are as they should be, the server assigns an address to the client for a predetermined amount of time, known as the *lease*, from the scope.

> **ExamAlert**
>
> If you wanted to ensure that a device with a specific MAC address is always assigned the same IP address from a DHCP server, you would need to configure a DHCP reservation. If it is determined that a DHCP scope has become exhausted, an administrator can reduce the DHCP lease time to free up available DHCP addresses. This eliminates the need to create a new DHCP pool.

A DHCP server typically can be configured to assign more than just IP addresses. It often is used to assign the subnet mask, the default gateway, and *Domain Name Service (DNS)* information.

Using DHCP means that administrators do not need to manually configure each client system with a TCP/IP address. This removes the common problems associated with statically assigned addresses, such as human error. The potential problem of assigning duplicate IP addresses is also eliminated. DHCP also removes the need to reconfigure systems if they move from one subnet to another, or if you decide to make a wholesale change in the IP addressing structure.

> **ExamAlert**
>
> Even when a network is configured to use DHCP, several mission-critical network systems continue to use static addressing: DHCP server, DNS server, web server, network printers, and more. They do not have dynamic IP addressing because their IP addresses can never change. If they do, client systems may be unable to access the resources from that server.

Configuring a client for TCP/IP can be relatively complex, or it can be simple. Any complexity involved is related to the possible need to manually configure TCP/IP. The reason for the simplicity is that TCP/IP configuration can occur automatically via DHCP or through Automatic Private IP Addressing (APIPA). At the very least, a system needs an IP address and subnet mask to log on to a network. The default gateway and DNS server IP information is optional, but network functionality is limited without them. The following list briefly explains the IP-related settings used to connect to a TCP/IP network, many of which are shown in Figure 3.2:

▶ **IP address:** This value is the unique address that each system must be assigned so that it can communicate on the network.

▶ **Subnet mask:** This value enables the system to determine what portion of the IP address represents the network address and what portion represents the node address.

▶ **Default gateway:** This value identifies the node on the network that enables the system to communicate on a remote network, without the need for explicit routes to be defined.

▶ **DNS server addresses:** This value identifies the server that is enabling dynamic hostname resolution to be performed. It is common practice to have two DNS server addresses defined so that if one server becomes unavailable, the other can be used.

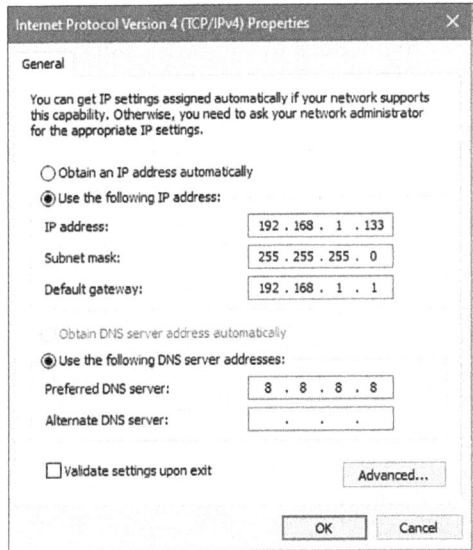

FIGURE 3.2   Configuration Options in Windows for TCP/IP

ExamAlert

At the very minimum, an IP address and subnet mask are required to connect to a TCP/IP network. With this minimum configuration, connectivity is limited to the local segment, and DNS resolution is not possible.

Note

DHCP6 (or, more correctly, DHCPv6) is the IPv6 counterpart to DHCP. It issues the necessary configuration information for clients on IPv6-based networks. When it comes to assigning the addresses, the preferred method of assigning IP addresses in an IPv6 network is to use Stateless Address Auto Configuration (SLAAC). With SLAAC, devices send the router a request for the network prefix, and the device then uses the prefix along with its own MAC address to create an IP address.

BOOT Protocol (BOOTP)

BOOTP was originally created so that diskless workstations could obtain information needed to connect to the network, such as the TCP/IP address, subnet mask, and default gateway. Such a system was necessary because diskless workstations had no way to store the information.

When a system configured to use BOOTP is powered up, it broadcasts for a BOOTP server on the network. If such a server exists, it compares the MAC address of the system issuing the BOOTP request with a database of entries. From this database, it supplies the system with the appropriate information. It can also notify the workstation about a file that it must run on BOOTP.

In the unlikely event that you use BOOTP, you should be aware that, like DHCP, it is a broadcast-based system. Therefore, routers must be configured to forward BOOTP broadcasts.

Automatic Private IP Addressing

Automatic Private IP Addressing (APIPA) was introduced with Windows 98 and has been included in all subsequent Windows versions. The function of APIPA is that a system can give itself an IP address if it is incapable of receiving an address dynamically from a DHCP server. Then APIPA assigns the system an address from the 169.254.0.0 address range (between 169.254.0.1 and 169.254.255.254) and configures an appropriate subnet mask (255.255.0.0). However, it doesn't configure the system with a default gateway address. As a result, communication is limited to the local network. So, if you can connect to other devices on a local network but can't reach the Internet, for example, it is likely that your DHCP server is down and you are currently using an APIPA address.

> **ExamAlert**
>
> If a system that does not support APIPA cannot get an address from a DHCP server, it typically assigns itself an IP address of 0.0.0.0. Keep this in mind when troubleshooting IP addressing problems on non-APIPA platforms.

The idea behind APIPA is that systems on a segment can communicate with each other if DHCP server failure occurs. In reality, the limited usability of APIPA makes it little more than a last resort. For example, imagine that a system is powered on while the DHCP server is operational and receives an IP address of 192.168.100.2. Then the DHCP server fails. Now, if the other systems on the segment are powered on and cannot get an address from the DHCP server because it is down, they would self-assign addresses in the 169.254.0.0 address range via APIPA. The systems with APIPA addresses would talk to each other, but they couldn't talk to a system that received an address from the DHCP server. Likewise, any system that receives an IP address via DHCP cannot talk to systems with APIPA-assigned addresses. This, and the absence of a default gateway, is why APIPA is of limited use in real-world environments.

> **ExamAlert**
>
> Be prepared to answer APIPA questions. Know what it is and how you can tell whether you have been assigned an APIPA address and why.

Identifying MAC Addresses

Many times this book refers to MAC addresses and how certain devices use them. However, it has not yet discussed why MAC addresses exist, how they are assigned, and what they consist of.

> **Note**
>
> A MAC address is sometimes called a physical address because it is physically embedded in the interface (network interface card).

A MAC address is a 6-byte (48-bit) hexadecimal address that enables a NIC to be uniquely identified on the network. The MAC address forms the basis of network communication, regardless of the protocol used to achieve network connection. Because the MAC address is so fundamental to network communication, mechanisms are in place to ensure that duplicate addresses cannot be used.

To combat the possibility of duplicate MAC addresses being assigned, the *Institute of Electrical and Electronics Engineers (IEEE)* took over the assignment of MAC addresses. But rather than be burdened with assigning individual addresses, the IEEE decided to assign each manufacturer an ID and then let the manufacturer further allocate IDs. The result is that in a MAC address, the first 3 bytes define the manufacturer, and the last 3 are assigned by the manufacturer.

For example, consider the MAC address of the computer on which this book is being written: 00:D0:59:09:07:51. The first 3 bytes (00:D0:59) identify the manufacturer of the card. Because only this manufacturer can use this address, it is known as the *organizationally unique identifier (OUI)*. The last 3 bytes (09:07:51) are called the *universal LAN MAC address*: they make this interface unique. You can find a complete listing of organizational MAC address assignments at http://standards-oui.ieee.org/oui.txt.

Because MAC addresses are expressed in hexadecimal, only the numbers 0 through 9 and the letters *A* through *F* can be used in them. If you get an exam question about identifying a MAC address and some of the answers contain letters and numbers other than 0 through 9 and the letters *A* through *F*, you can immediately discount those answers.

You can discover the NIC's MAC address in various ways, depending on what system or platform you work on (several of the ways can be found at https://carleton.ca/its/help-centre/how-to-find-your-mac-address/). Table 3.5 defines various platforms and methods you can use to view an interface's MAC address.

TABLE 3.5 **Methods of Viewing the MAC Addresses of NICs**

Platform	Method
Windows	Enter **ipconfig /all** at a command prompt.
Linux/some UNIX	Enter the **ifconfig -a** command.
Cisco router	Enter the **sh int interface name** command.

ExamAlert

Be sure that you know the commands used to identify the MAC address in various operating system formats.

Just as there was fear that there would not be enough IP addresses for all the devices needed to access the Internet if we stayed with IPv4, there has also been considerable fear that there are not enough MAC addresses to assign. To deal with this, 64-bit addresses are now available. The IEEE refers to 48-bit addresses as *EUI-48* (for *extended unique identifier*) and longer addresses as *EUI-64*. It is projected that there are a sufficient number of 48-bit addresses to last for quite some time, but the IEEE is encouraging the adoption of the 64-bit addressing as soon as possible. EUI-64 is used to automatically configure IPv6 host addresses by using the MAC address of its interface to generate a 64-bit interface ID. The MAC address is split in two, and "FFFE" is inserted into the middle. Then the 7th bit of the interface ID is inverted, and EUI-64 uses hyphens between number sets instead of colons. A good explanation or overview can be found at https://community.cisco.com/t5/networking-documents/understanding-ipv6-eui-64-bit-address/ta-p/3116953.

> **ExamAlert**
>
> Be sure that you know what EUI-64 is for the exam.

Cram Quiz

1. Which of the following is a Class B address?

   - ○ **A.** 129.16.12.200
   - ○ **B.** 126.15.16.122
   - ○ **C.** 211.244.212.5
   - ○ **D.** 193.17.101.27

2. You are the administrator for a network with two Windows Server systems and 65 Windows desktop systems. At 10 a.m., three users call to report that they are experiencing network connectivity problems. Upon investigation, you determine that the DHCP server has failed. How can you tell that the DHCP server failure is the cause of the connectivity problems experienced by the three users?

   - ○ **A.** When you check their systems, they have an IP address of 0.0.0.0.
   - ○ **B.** When you check their systems, they have an IP address in the 192.168.x.x address range.
   - ○ **C.** When you check their systems, they have a default gateway value of 255.255.255.255.
   - ○ **D.** When you check their systems, they have an IP address from the 169.254.x.x range.

3. Which of the following IP addresses is not from a private address range?

   ○ **A.** 192.168.200.117

   ○ **B.** 172.16.3.204

   ○ **C.** 127.45.112.16

   ○ **D.** 10.27.100.143

4. You have been assigned to set up a new network with TCP/IP. For the external interfaces, you decide to obtain registered IP addresses from your ISP, but for the internal network, you choose to configure systems by using one of the private address ranges. Of the following address ranges, which one would you not consider?

   ○ **A.** 192.168.0.0 to 192.168.255.255

   ○ **B.** 131.16.0.0 to 131.16.255.255

   ○ **C.** 10.0.0.0 to 10.255.255.255

   ○ **D.** 172.16.0.0 to 172.31.255.255

5. You ask your ISP to assign a public IP address for the external interface of your Windows server, which is running a proxy server application. In the email message that contains the information, the ISP tells you that you have been assigned the IP address 203.15.226.12/24. When you fill out the subnet mask field on the IP configuration dialog box on your system, what subnet mask should you use?

   ○ **A.** 255.255.255.255

   ○ **B.** 255.255.255.0

   ○ **C.** 255.255.240.0

   ○ **D.** 255.255.255.240

6. Examine the diagram shown here. What is the most likely reason that user Spencer cannot communicate with user Evan?

User: Evan
IP address: 192.168.1.121
Subnet mask: 255.255.255.0
Default gateway: 192.168.1.1

User: Spencer
IP address: 192.168.1.127
Subnet mask: 255.255.248.0
Default gateway: 192.168.1.1

   ○ **A.** The default gateways should have different values.

   ○ **B.** Spencer's IP address is not a loopback address.

   ○ **C.** The subnet values should be the same.

   ○ **D.** There is no problem identifiable by the values given.

Cram Quiz Answers

1. **A.** Class B addresses fall into the range 128 to 191. Answer A is the only address listed that falls into that range. Answer B is a Class A address, and answers C and D are Class C IP addresses.

2. **D.** When a Windows desktop system that is configured to obtain an IP address via DHCP fails to obtain an address, it uses APIPA to assign itself an address from the 169.254.x.x address range. An address of 0.0.0.0 normally results from a system that does not support APIPA. APIPA does not use the 192.168.x.x address range. The IP address 255.255.255.255 is the broadcast address. A DHCP failure would not lead to a system assigning itself this address.

3. **C.** The 127.x.x.x network range is reserved for the loopback function. It is not one of the recognized private address ranges. The private address ranges as defined in RFC 1918 are 10.x.x.x, 172.16.x.x to 172.31.x.x, and 192.168.x.x.

4. **B.** The 131.16 range is from the Class B range and is not one of the recognized private IP address ranges. All the other address ranges are valid private IP address ranges.

5. **B.** In CIDR terminology, the number of bits to be included in the subnet mask is expressed as a slash value. If the slash value is 24, the first three octets form the subnet mask, so the value is 255.255.255.0.

6. **C.** The most likely problem, given the IP values for each user's workstation, is that the subnet value is not correct on Spencer's machine and should be 255.255.255.0.

Routing and Switching Technologies

▶ **2.1 Explain characteristics of routing technologies.**

▶ **2.2 Given a scenario, configure switching technologies and features.**

CramSaver

If you can correctly answer these questions before going through this section, save time by skimming the ExamAlerts in this section and then completing the Cram Quiz at the end of the section.

1. What is a common link-state protocol?

2. What is convergence?

3. What term is used when specific routes are combined into one route?

4. True or false: With the help of FSL, STP avoids or eliminates loops on Layer 2 bridges.

Answers

1. One of the most common link-state protocols is OSPF.

2. Convergence represents the time it takes routers to detect change on the network.

3. The term *route aggregation* applies when specific routes are combined into one route.

4. False. With the help of Spanning Tree Algorithm (STA), STP avoids or eliminates loops on a Layer 2 bridge.

Because today's networks branch out between interconnected offices all over the world, networks may have any number of separate physical network segments connected using routers. Routers are devices that direct data between networks. Essentially, when a router receives data, it must determine the destination for the data and send it there. To accomplish this, the network router uses two key pieces of information: the gateway address and the routing tables.

The Default Gateway

A default gateway is the router's IP address, which is the pathway to any and all remote networks. To get a packet of information from one network to another, the packet is sent to the default gateway, which helps forward the packet to its destination network. Computers that live on the other side of routers are said to be on remote networks. Without default gateways, Internet communication is not possible because your computer does not have a way to send a packet destined for any other network. On the workstation, it is common for the default gateway option to be configured automatically through DHCP configuration.

Routing Tables

Before a data packet is forwarded, a chart is reviewed to determine the best possible path for the data to reach its destination. This chart is the computer's routing table. Maintaining an accurate routing table is essential for effective data delivery. Every computer on a TCP/IP network has a routing table stored locally. Figure 3.3 shows the routing table on a Windows system.

> **Note**
>
> You can use the **route print** command to view the routing table on a client system. To see the configuration of a specific port on a switch, the network technician should use the **show interface** command. This command provides detailed information about the interface, including the current configuration, status, and statistics for the interface.

As shown in Figure 3.3, the information in the routing table includes the following:

▶ **Network Destination:** The host IP address.

▶ **Netmask:** The subnet mask value for the destination parameter.

▶ **Gateway:** Where the IP address is sent. This may be a gateway server, a router, or another system acting as a gateway.

▶ **Interface:** The address of the interface that's used to send the packet to the destination.

▶ **Metric:** A measurement of the directness of a route. The lower the metric, the faster the route. If multiple routes exist for data to travel, the one with the lowest metric is chosen.

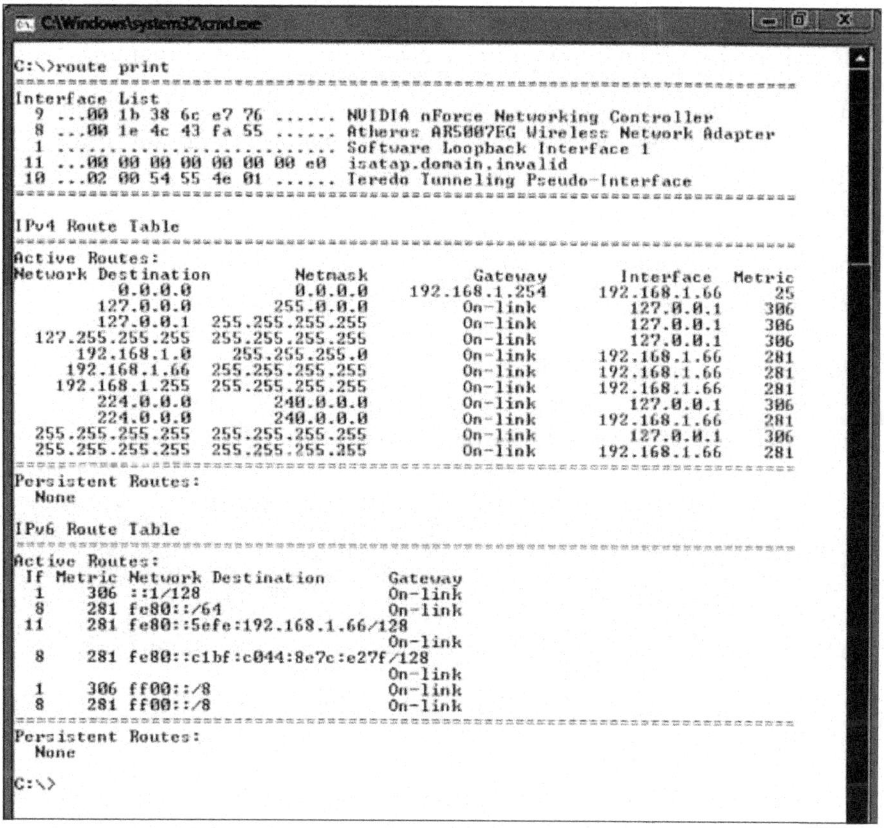

```
C:\Windows\system32\cmd.exe                                      _  □  X

C:\>route print
===========================================================================
Interface List
  9 ...00 1b 38 6c e7 76 ...... NVIDIA nForce Networking Controller
  8 ...00 1e 4c 43 fa 55 ...... Atheros AR5007EG Wireless Network Adapter
  1 .......................... Software Loopback Interface 1
 11 ...00 00 00 00 00 00 00 e0 isatap.domain.invalid
 10 ...02 00 54 55 4e 01 ...... Teredo Tunneling Pseudo-Interface
===========================================================================

IPv4 Route Table
===========================================================================
Active Routes:
Network Destination        Netmask          Gateway       Interface  Metric
          0.0.0.0          0.0.0.0      192.168.1.254   192.168.1.66     25
        127.0.0.0        255.0.0.0         On-link        127.0.0.1    306
        127.0.0.1  255.255.255.255         On-link        127.0.0.1    306
  127.255.255.255  255.255.255.255         On-link        127.0.0.1    306
      192.168.1.0    255.255.255.0         On-link     192.168.1.66    281
     192.168.1.66  255.255.255.255         On-link     192.168.1.66    281
    192.168.1.255  255.255.255.255         On-link     192.168.1.66    281
        224.0.0.0        240.0.0.0         On-link        127.0.0.1    306
        224.0.0.0        240.0.0.0         On-link     192.168.1.66    281
  255.255.255.255  255.255.255.255         On-link        127.0.0.1    306
  255.255.255.255  255.255.255.255         On-link     192.168.1.66    281
===========================================================================
Persistent Routes:
  None

IPv6 Route Table
===========================================================================
Active Routes:
 If Metric Network Destination       Gateway
  1    306 ::1/128                   On-link
  8    281 fe80::/64                 On-link
 11    281 fe80::5efe:192.168.1.66/128
                                     On-link
  8    281 fe80::c1bf:c044:8e7c:e27f/128
                                     On-link
  1    306 ff00::/8                  On-link
  8    281 ff00::/8                  On-link
===========================================================================
Persistent Routes:
  None

C:\>
```

FIGURE 3.3   **The Routing Table on a Windows System**

Routing tables play an important role in the network routing process. They are the means by which the data is directed through the network. For this reason, a routing table needs to be two things. It must be up to date and complete. The router can get the information for the routing table in two ways: through static routing or dynamic routing.

Static Routing

In environments that use *static routing*, routes and route information are manually entered into the routing tables. Not only can this be a time-consuming task, but also errors are more common. In addition, when a change occurs to the network's layout, or topology, statically configured routers must be manually updated with the changes. Again, this is a time-consuming and potentially error-laden task. For these reasons, static routing is suited to only the smallest environments, with perhaps just one or two routers. A far more practical solution, particularly in larger environments, is to use dynamic routing.

You can add a static route to a routing table using the **route add** command. To do this, specify the route, the network mask, and the destination IP address of the network card your router will use to get the packet to its destination network.

The syntax for the **route add** command is as follows:

```
route add 192.168.2.1 mask (255.255.255.0) 192.168.2.4
```

Adding a static address is not permanent; in other words, it will most likely be gone when the system reboots. To make it persistent (the route is still in the routing table on boot), you can use the **-p** switch with the command.

> **ExamAlert**
>
> The **route add** command adds a static route to the routing table. The **route add** command with the **-p** switch makes the static route persistent. You might want to try this on your own before taking the Network+ exam.

Default Route

In environments that use dynamic routing, there is usually one static route defined that is known as the *default route*. The default route, sometimes called the route (or gateway) of last resort, specifies the path to be used if no other route is known (no next-hop host is available from the routing table or other routing mechanisms). All packets with unknown destination addresses are sent to the default route.

Switching Methods

For systems to communicate on a network, the data needs a communication path or multiple paths on which to travel. To allow entities to communicate, these paths move the information from one location to another and back. This is the function of *switching*, which provides communication pathways between two endpoints and manages how data flows between them. Following are two of the more common switching methods used today:

▶ Packet switching

▶ Circuit switching

> **Note**
>
> When you are a network administrator, knowing the following switching methods is important, but they are not objectives on the Network+ exam, and thus you can jump to "Dynamic Routing" if pressed for time as you study.

Packet Switching

In packet switching, messages are broken into smaller pieces called *packets*. Each packet is assigned source, destination, and intermediate node addresses. Packets are required to have this information because they do not always use the same path or route to get to their intended destination. Referred to as *independent routing*, this is one of the advantages of packet switching. Independent routing enables better use of available bandwidth by letting packets travel different routes to avoid high-traffic areas. Independent routing also enables packets to take an alternative route if a particular route is unavailable for some reason.

> **Note**
>
> Packet switching is the most popular switching method for networks and is used on most WANs.

In a packet-switching system, when packets are sent onto the network, the sending device is responsible for choosing the best path for the packet. This path might change in transit, and the receiving device can receive the packets in a random or nonsequential order. When this happens, the receiving device waits until all the data packets are received, and then it reconstructs them according to their built-in sequence numbers.

Two types of packet-switching methods are used on networks:

▶ **Virtual-circuit packet switching:** A logical connection is established between the source and the destination device. This logical connection is established when the sending device initiates a conversation with the receiving device. The logical communication path between the two devices can remain active for as long as the two devices are available or can be used to send packets once. After the sending process has completed, the line can be closed.

▶ **Datagram packet switching:** Unlike virtual-circuit packet switching, datagram packet switching does not establish a logical connection between the sending and transmitting devices. The packets in datagram packet switching are independently sent, meaning that they can take different paths through the network to reach their intended destination. To do this, each packet must be individually addressed to determine its source and destination. This method ensures that packets take the easiest possible routes to their destination and avoid high-traffic areas. Datagram packet switching is mainly used on the Internet.

Circuit Switching

In contrast to the packet-switching method, circuit switching requires a dedicated physical connection between the sending and receiving devices. The most commonly used analogy to represent circuit switching is a telephone conversation in which the parties involved have a dedicated link between them for the duration of the conversation. When either party disconnects, the circuit is broken, and the data path is lost. This is an accurate representation of how circuit switching works with network and data transmissions. The sending system establishes a physical connection, and the data is transmitted between the two. When the transmission is complete, the channel is closed.

Some clear advantages to the circuit-switching technology make it well suited for certain applications, such as *public switched telephone network (PSTN)* and *Integrated Services Digital Network (ISDN)*. The primary advantage is that after a connection is established, a consistent and reliable connection exists between the sending and receiving devices. This allows for transmissions at a guaranteed rate of transfer.

Like all technologies, circuit switching has its downsides. As you might imagine, a dedicated communication line can be inefficient. After the physical connection is established, it is unavailable to any other sessions until the transmission completes. Again, using the phone call analogy, this would be like a caller trying to reach another caller and getting a busy signal. Circuit switching therefore can be fraught with long connection delays.

Comparing Switching Methods

Table 3.6 provides an overview of the various switching technologies.

TABLE 3.6 **Comparison of Switching Methods**

Switching Method	Pros	Cons	Key Features
Packet switching	Packets can be routed around network congestion. Packet switching makes efficient use of network bandwidth.	Packets can become lost while taking alternative routes to the destination. Messages are divided into packets that contain source and destination information.	The two types of packet switching are datagram and virtual circuit. Datagram packets are independently sent and can take different paths throughout the network. Virtual circuit uses a logical connection between the source and destination devices.
Circuit switching	Circuit switching offers a dedicated transmission channel that is reserved until it is disconnected.	Dedicated channels can cause delays because a channel is unavailable until one side disconnects. Circuit switching uses a dedicated physical link between the sending and receiving devices.	Circuit switching offers the capability of storing messages temporarily to reduce network congestion.

Dynamic Routing

In a *dynamic routing* environment, routers use special routing protocols to communicate. The purpose of these protocols is simple: they enable routers to pass on information about themselves to other routers so that other routers can build routing tables.

> **Note**
>
> The use of any routing protocol to advertise routes that have been learned (through another protocol, through static configuration, and so on) is known as *route redistribution*.

With distance-vector router communications, each router on the network communicates all the routes it knows about to the routers to which it is directly attached. In this way, routers communicate only with their router neighbors and are unaware of other routers that may be on the network. Distance-vector routing protocols operate by having each router send updates about all the other routers it knows about to the routers directly connected to it. The routers use these updates to compile their routing tables. The updates are sent automatically every 30 or 60 seconds. The interval depends on the routing protocol used. Apart from the periodic updates, routers can also be configured to send a triggered update if a change in the network topology is detected. The process by which routers learn of a change in the network topology is called convergence.

A router that uses a link-state protocol differs from a router that uses a distance-vector protocol because it builds a map of the entire network and then holds that map in memory. On a network that uses a link-state protocol, routers send link-state advertisements (LSAs) that contain information about the networks to which they connect. The LSAs are sent to every router on the network, thus enabling the routers to build their network maps.

Both distance-vector and link-stated based routers can be considered dynamic. The most popular routing protocols in use today are

- ▶ **BGP:** When you want the best of both worlds, distance vector and link state, you can turn to a hybrid, the Border Gateway Protocol (BGP). BGP can be used between gateway hosts on the Internet. BGP examines the routing table, which contains a list of known routers, the addresses they can reach, and a cost metric associated with the path to each router so that the best available route is chosen. BGP communicates between the routers using TCP. BGP supports the use of autonomous system numbers (ASNs), which are globally unique numbers used by connected groups of IP networks that share the same routing policy.

- ▶ **EIGRP:** This protocol enables routers to exchange information more efficiently than earlier network protocols. EIGRP uses its neighbors to help determine routing information. Routers configured to use EIGRP keep copies of their neighbors' routing information and query these tables to help find the best possible route for transmissions to follow. EIGRP uses *Diffusing Update Algorithm (DUAL)* to determine the best route to a destination.

- ▶ **Open Shortest Path First (OSPF):** This link-state routing protocol is based on the *shortest path first (SPF)* algorithm to find the least-cost path to any destination in the network. In operation, each router using OSPF sends a list of its neighbors to other routers on the network. From this information, routers can determine the network design and the shortest path for data to travel.

> **ExamAlert**
>
> Be sure that you can identify the dynamic routing protocols discussed here.

> **Note**
>
> A time-to-live (TTL) value can be set with routing equal to the number of hops that a packet can reach at a maximum before being discarded by a router.

The communication between distance-vector routers is known as *hops*. On the network, each router represents one hop, so a network using six routers has five hops between the first and last router.

The **tracert** command is used in a Windows environment to see how many hops a packet takes to reach a destination (the same functionality exists in macOS and Linux with the **traceroute** command). To try this at the command prompt, enter **tracert comptia.org**. Figure 3.4 shows an example of the output on a Windows workstation.

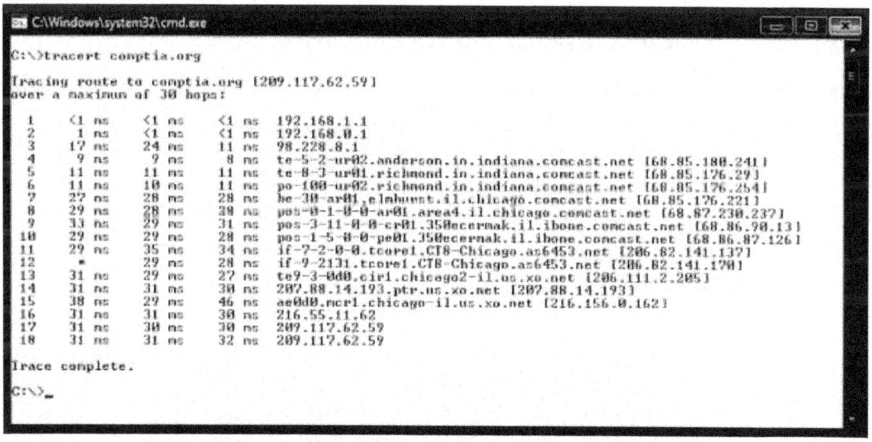

```
C:\Windows\system32\cmd.exe

C:\>tracert comptia.org

Tracing route to comptia.org [209.117.62.59]
over a maximum of 30 hops:

  1    <1 ms    <1 ms    <1 ms  192.168.1.1
  2     1 ms    <1 ms    <1 ms  192.168.0.1
  3    1? ms    24 ms    11 ms  98.228.8.1
  4     9 ms     9 ms     8 ms  te-5-2-ur02.anderson.in.indiana.comcast.net [68.85.180.241]
  5    11 ms    11 ms    11 ms  te-8-3-ur01.richmond.in.indiana.comcast.net [68.85.176.29]
  6    11 ms    10 ms    11 ms  po-100-ur02.richmond.in.indiana.comcast.net [68.85.176.254]
  7    27 ms    28 ms    28 ms  be-30-ar01.elmhurst.il.chicago.comcast.net [68.85.176.221]
  8    29 ms    28 ms    39 ms  pos-0-1-0-0-ar01.area4.il.chicago.comcast.net [68.87.230.237]
  9    33 ms    29 ms    31 ms  pos-3-11-0-0-cr01.350ecermak.il.ibone.comcast.net [68.86.90.13]
 10    29 ms    29 ms    28 ms  pos-1-5-0-0-pe01.350ecermak.il.ibone.comcast.net [68.86.87.126]
 11    29 ms    35 ms    34 ms  if-7-2-0-0.tcore1.CT8-Chicago.as6453.net [206.82.141.137]
 12     *        29 ms    28 ms  if-9-2131.tcore1.CT8-Chicago.as6453.net [206.82.141.170]
 13    31 ms    29 ms    27 ms  te9-3-0d0.cir1.chicago2-il.us.xo.net [206.111.2.205]
 14    31 ms    31 ms    30 ms  207.88.14.193.ptr.us.xo.net [207.88.14.193]
 15    38 ms    29 ms    46 ms  ae0d0.ncr1.chicago-il.us.xo.net [216.156.0.162]
 16    31 ms    31 ms    30 ms  216.55.11.62
 17    31 ms    30 ms    30 ms  209.117.62.59
 18    31 ms    31 ms    32 ms  209.117.62.59

Trace complete.

C:\>_
```

FIGURE 3.4   **The Results of Running tracert on a Windows System**

In addition to the **tracert** command in IPv4, you can get similar functionality in IPv6 with **tracert -6**, **traceroute6**, and **traceroute -6**.

Routing loops can occur on networks with slow convergence. Routing loops occur when the routing tables on the routers are slow to update and a redundant communication cycle is created between routers. Two strategies can combat potential routing loops:

▶ **Split horizon:** Works by preventing the router from advertising a route back to the other router from which it was learned. This prevents two nodes from bouncing packets back and forth between them, creating a loop.

▶ **Poison reverse (also called split horizon with poison reverse):** Dictates that the route is advertised back on the interface from which it was learned, but it has a hop count of infinity, which tells the node that the route is unreachable.

ExamAlert

If a change in the routing is made, it takes some time for the routers to detect and accommodate this change. This is known as *convergence*.

Although distance-vector protocols can maintain routing tables, they have three problems:

▶ The periodic update system can make the update process slow.

▶ The periodic updates can create large amounts of network traffic—much of the time unnecessarily, because the network's topology should rarely change.

▶ Perhaps the most significant problem is that because the routers know about only the next hop in the journey, incorrect information can be propagated between routers, creating routing loops.

ExamAlert

Know that "next hop" in routing is the next closest router that a packet can go through.

When the network maps on each router are complete, the routers update each other at a given time, just like with a distance-vector protocol; however, the updates occur much less frequently with link-state protocols than with distance-vector protocols. The only other circumstance under which updates are sent is if a change in the topology is detected, at which point the routers use LSAs to detect the change and update their routing tables. This mechanism, combined with the fact that routers hold maps of the entire network, makes convergence on a link-state-based network quickly occur.

Although it might seem as though link-state protocols are an obvious choice over distance-vector protocols, routers on a link-state-based network require more powerful hardware and more RAM than those on a distance-vector-based network. Not only do the routing tables need to be calculated, but they must also be stored. A router that uses distance-vector protocols need only maintain a small database of the routes accessible by the routers to which it is directly connected. A router that uses link-state protocols must maintain a database of all the routers in the entire network.

Route Selection

Route selection variables play crucial roles in determining the best route for forwarding packets to their destinations. Following are several metrics related to routing that you should know for the exam:

▶ *Administrative distance* is a numerical value assigned to a route based on its perceived quality (trustworthiness). The number may be manu-ally assigned or assigned based on an algorithm employed by a routing protocol. The lower the number, the better the route is believed to be: 0 is the best and 255 is the worst, so the router selects the route with the lowest administrative distance as the best route to the destination. When a router receives multiple routing updates for the same destination from different sources (such as different routing protocols or static routes), it uses the administrative distance to prioritize the routes.

▶ *Prefix length* (also known as subnet mask or prefix) defines the length of the network prefix in the destination IP address. With IPv4 routing, the prefix length indicates the number of leading bits in the destination IP address that represent the network portion of the address. For example, a prefix length of 24 indicates a 24-bit network prefix (or subnet mask of 255.255.255.0). When multiple routes exist for the same destination net-work, the router selects the route with the longest prefix length (i.e., the most specific match) as the best route: this is known as the longest prefix match algorithm and ensures that the router forwards packets to the most specific destination network entry in its routing table.

▶ *Metric* is a quantitative value assigned to each route to measure the desir-ability or cost of the path to the destination network. The metric value represents various factors such as hop count, bandwidth, delay, reliability, or administrative preference. Each routing protocol uses its own metric calculation method to evaluate the cost of reaching a destination network based on these factors. When multiple routes to the same destination exist with different metrics, the router selects the route with the lowest metric value as the best route.

▶ *Hop counts* are the number of hops necessary to reach a node. A hop count of infinity means the route is unreachable.

▶ The *maximum transmission unit (MTU)* defines the largest data unit that can be passed without fragmentation.

▶ *Bandwidth* specifies the maximum packet size permitted for Internet transmission.

▶ *Latency* is the amount of time it takes for a packet to travel from one location to another.

Address Translation

The purpose of address translation is to enable communication between devices on different networks with incompatible addressing schemes, primarily between private (or internal) networks and the public Internet. This can be achieved by translating IP addresses and/or port numbers in the IP header of packets as they traverse a device enabled to perform this function—typically a router or firewall.

NAT

The basic principle of *Network Address Translation (NAT)* is that many computers can "hide" behind a single IP address. The main reason you need to do this is that there aren't enough IPv4 addresses to go around: NAT allows multiple devices within a private network to share a single public IP address, thus conserving public IPv4 address space and extending the usability of the limited IPv4 address pool.

Using NAT means that only one registered IP address is needed on the system's external interface, acting as the gateway between the internal and external networks. Figure 3.5 shows an example of enabling NAT on a SOHO router.

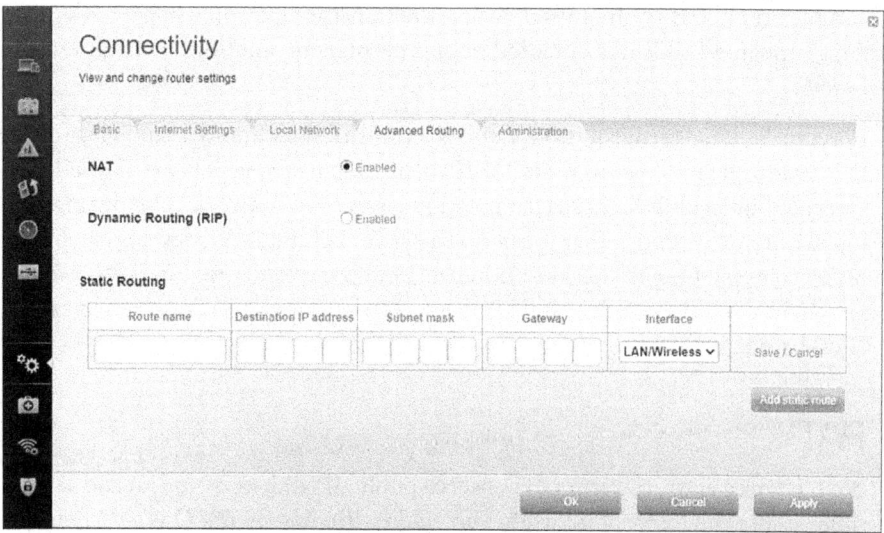

FIGURE 3.5   **NAT Configuration on a SOHO Router**

> **Note**
>
> Don't confuse NAT with proxy servers. The proxy service is different from NAT, but many proxy server applications do include NAT functionality.

NAT enables you to use whatever addressing scheme you like on your internal networks; however, it is common practice to use the private address ranges, which were discussed earlier.

When a system is performing NAT, it funnels the requests given to it to the Internet. To the remote host, the request looks like it is originating from a single address. The system performing the NAT function keeps track of who asked for what and makes sure that when the data is returned, it is directed to the correct system. Servers that provide NAT functionality do so in different ways. For example, you can statically map a specific internal IP address to a specific external one (known as the *one-to-one NAT method*) so that outgoing requests are always tagged with the same IP address. Alternatively, if you have a group of public IP addresses, you can have the NAT system assign addresses to devices on a first-come, first-served basis. Either way, the basic function of NAT is the same.

Tunneling can be used for transmitting packets of one type (such as IPv6) over another network (such as IPv4). *6to4* is one such tunneling technology, allowing IPv6 packets to be transmitted over an IPv4 network without having to create a complex tunnel. It is often used during the transition period when a network is being updated and is not intended to be a permanent solution. Its counterpart is *4to6*.

For a more long-term solution, there is a transition technology known as *Teredo* that gives full IPv6 connectivity for IPv6-capable hosts, which are on the IPv4 Internet but lack direct native connection to an IPv6 network. The distinguishing feature of Teredo is that it can do this from behind NAT devices (such as home routers). One of the most popular Teredo implementations is *Miredo*; it is a client designed to allow full IPv6 connectivity to systems that are strictly IPv4-based.

PAT

NAT enables administrators to conserve public IP addresses and, at the same time, secure the internal network. *Port Address Translation (PAT)* is a variation on NAT. With PAT, all systems on the LAN are translated to the same IP address, but with a different port number assignment. PAT is used when multiple clients want to access the Internet. However, with not enough available public IP addresses, you need to map the inside clients to a single public IP address. When packets come back into the private network, they are routed to their destination with a table within PAT that tracks the public and private port numbers.

When PAT is used, there is typically only a single IP address exposed to the public network, and multiple network devices access the Internet through this exposed IP address. The sending devices, IP address, and port number are not exposed. For example, an internal computer with the IP address of 192.168.2.2 wants to access a remote web server at address 204.23.85.49. The request goes to the PAT router, where the sender's private IP and port number are modified, and a mapping is added to the PAT table. The remote web server sees the request coming from the IP address of the PAT router and not the computer actually making the request. The web server sends the reply to the address and port number of the router. When received, the router checks its table to see the packet's actual destination and forwards it.

ExamAlert

PAT enables nodes on a LAN to communicate with the Internet without revealing their IP address. All outbound IP communications are translated to the router's external IP address. Replies come back to the router, which then translates them back into the private IP address of the original host for final delivery.

First Hop Redundancy Protocol (FHRP)

First Hop Redundancy Protocol (FHRP) is designed to ensure high availability and fault tolerance by providing redundancy for the default gateway/router. FHRP allows multiple routers to share a virtual IP address and MAC address, ensuring uninterrupted connectivity for hosts in the event of a router failure. The primary purpose of FHRP is to prevent a single point of failure at the default gateway, which could result in network downtime or connectivity issues for hosts on the LAN.

Some important features of FHRP include

▶ **Virtual IP Address:** FHRP assigns a virtual IP address to a group of routers participating in the redundancy protocol. This virtual IP address serves as the default gateway for hosts on the LAN. If the primary router fails, another router in the FHRP group takes over the virtual IP address, ensuring uninterrupted connectivity for hosts without requiring manual reconfiguration.

▶ **Virtual MAC Address:** Along with the virtual IP address, FHRP also assigns a virtual MAC address to the group of routers. The virtual MAC address is associated with the virtual IP address and is used by hosts on the LAN to resolve the Layer 2 (MAC) address of the default gateway. Like the virtual IP address, the virtual MAC address facilitates seamless failover between routers in the FHRP group.

▶ **Redundancy Protocols:** Common FHRP protocols include Hot Standby Router Protocol (HSRP), Virtual Router Redundancy Protocol (VRRP), and Gateway Load Balancing Protocol (GLBP). These protocols define how routers in the FHRP group communicate, elect a primary router, and handle failover events. Each protocol has its own features, advantages, and configuration options.

▶ **Router Election:** FHRP protocols typically involve the election of a primary router within the group, which actively forwards traffic and maintains the virtual IP and MAC addresses. Backup routers in the group monitor the primary router's status and take over its responsibilities if it becomes unavailable. This ensures continuous operation and fault tolerance for the default gateway.

Leveraging FHRP, it is possible to minimize network downtime, improve resilience, and enhance the overall reliability of the network.

> **ExamAlert**
>
> Know that the purpose of FHRP is to improve network reliability and resilience, ensuring continuous access to network resources and services.

Virtual IP

A *virtual IP address (VIP)* is an IP address assigned to multiple applications and is often used in high availability implementations. Data packets coming in are sent to the address and that routes them to the correct network interfaces. This allows hosting of different applications and virtual appliances on servers with only one (logical) IP address.

Subinterfaces

In the wonderful world of routing, *subinterfaces* are logical interfaces that are configured on a physical interface of a network device, such as a router or a switch. Subinterfaces are commonly used in scenarios where a single physical interface needs to be divided into multiple virtual interfaces to support multiple virtual local-area networks (VLANs) or multiple IP subnets. Each virtual interface has its own configuration settings, such as IP address, VLAN membership, and encapsulation type.

Each subinterface is associated with a specific VLAN and can be configured to add VLAN tags (802.1Q tags) to incoming and outgoing packets. Subinterfaces are commonly used in a router-on-a-stick configuration, where a single router interface is used to route traffic between multiple VLANs. Each subinterface on the router represents a different VLAN, and the router performs inter-VLAN routing between the subinterfaces. Subinterfaces enable routers to perform inter-VLAN routing by allowing them to separate traffic from different VLANs into distinct logical interfaces. This enables communication between hosts in different VLANs while maintaining network segmentation and security.

Subinterfaces are typically configured with a combination of physical interface configuration settings and additional parameters specific to each subinterface (such as the VLAN assignment, IP addressing, and encapsulation type) using router configuration commands or through an interface provided by the network device's management software.

Virtual Local-Area Networks

The word *virtual* is used a lot in the computing world—perhaps too often. For virtual local-area networks (VLANs), the word *virtual* does little to help explain the technology. Perhaps a more descriptive name for the VLAN concept might have been *segmented*. For now at least, use *virtual*.

> **Tip**
>
> 802.1Q is the Institute of Electrical and Electronics Engineers (IEEE) specification developed to ensure interoperability of VLAN technologies from the various vendors.

VLANs are used for network segmentation, a strategy that significantly increases the network's performance capability, removes potential performance bottlenecks, and can even increase network security. A VLAN is a group of connected computers that act as if they are on their own network segment, even though they might not be. For instance, suppose that you work in a three-story building in which the advertising employees are spread over all three floors. A VLAN can enable all the advertising personnel to be combined and access network resources as if they were connected on the same physical segment. This virtual segment can be isolated from other network segments. In effect, it would appear to the advertising group that they were on a network by themselves.

> **ExamAlert**
>
> VLANs enable you to create multiple broadcast domains on a single switch. In essence, this is the same as creating separate networks for each VLAN.

VLANs offer some clear advantages. Logically segmenting a network gives administrators flexibility beyond the restrictions of the physical network design and cable infrastructure. VLANs enable easier administration because the network can be divided into well-organized sections. Furthermore, you can increase security by isolating certain network segments from others. For example, you can segment the marketing personnel from finance or the administrators from the students. VLANs can ease the burden on overworked routers and reduce broadcast storms. Table 3.7 summarizes the benefits of VLANs.

TABLE 3.7 **Benefits of VLANs**

Advantage	Description
Increased security	With the creation of logical (virtual) boundaries, network segments can be isolated.
Increased performance	By reducing broadcast traffic throughout the network, VLANs free up bandwidth.
Organization	Network users and resources that are linked and that communicate frequently can be grouped in a VLAN.
Simplified administration	With a VLAN, the network administrator's job is easier when moving users between LAN segments, recabling, addressing new stations, and reconfiguring switches and routers.

VLAN Trunking Protocol (VTP), a Cisco proprietary protocol, is used to reduce administration in the switched network. You can, for example, put all switches in the same VTP domain and reduce the need to configure the same VLAN everywhere.

> **Note**
>
> A *VLAN database* is a feature found in some network devices, particularly older Cisco switches, that stores VLAN configuration information separately from the device's running configuration. It was commonly used in older Cisco switches that ran the Catalyst operating system (CatOS), which is now largely replaced by Cisco IOS on modern switches.

Trunking falls under 802.1Q, and a trunk port is one that is assigned to carry traffic for a specific switch (as opposed to an access port). The trunk port is usually fiber optic and used to interconnect switches to make a network, to interconnect LANs to make a WAN, and so on.

> **ExamAlert**
>
> IEEE 802.1Q also focuses on tagging and untagging in VLANs. *Tagging* means that the port will send out a packet with a header that has a tag number that matches its VLAN tag number. On any given port you can have just one *untagged* VLAN, and that will be the default port that traffic will go to unless it is tagged to go elsewhere.

> **ExamAlert**
>
> VLAN tags are used to identify packets as belonging to a particular VLAN. VLANs are used to segment a network into logical sub-networks, and each VLAN is assigned a unique VLAN tag. If the VLAN tag is not configured correctly, the computer may not be able to access network resources.

Port binding determines whether and how a port is bound. This can be done in one of three ways: static, dynamic, or ephemeral. Conversely, *port aggregation* is the combining of multiple ports on a switch, and it can be done in one of three ways: auto, desirable, or on.

The *Link Aggregation Control Protocol (LACP)* is a common aggregation protocol that enables multiple physical ports to be bound together. Most devices allow you to bind up to four, but some go up to eight.

VLAN Membership

You can use several methods to determine VLAN membership or how devices are assigned to a specific VLAN. The following sections describe the common methods to determine how VLAN membership is assigned:

▶ **Protocol-based VLANs:** With protocol-based VLAN membership, computers are assigned to VLANs using the protocol in use and the Layer 3 address. For example, this method enables a particular IP subnet to have its own VLAN.

The term *Layer 3 address* refers to one of the most important networking concepts, the Open Systems Interconnection (OSI) reference model. This conceptual model, created by the *International Organization for Standardization (ISO)* in 1978 and revised in 1984, describes a network architecture that enables data to be passed between computer systems. There are seven layers in total, which are discussed in detail in Chapter 2, "Network Topologies, Architectures, and Types." In brief, Layer 3, known as the *network layer*, identifies the mechanisms by which data can be moved between two networks or systems, such as transport protocols, which in the case of TCP/IP is IP.

Although VLAN membership may be based on Layer 3 information, this has nothing to do with routing or routing functions. The IP numbers are used only to determine the membership in a particular VLAN, not to determine routing.

▶ **Port-based VLANs:** Port-based VLANs require that specific ports on a network switch be assigned to a VLAN. For example, ports 1 through 4 may be assigned to marketing, ports 5 through 7 may be assigned to sales, and so on. Using this method, a switch determines VLAN membership by taking note of the port used by a particular packet. Figure 3.6 shows how the ports on a server could be used for port-based VLAN membership.

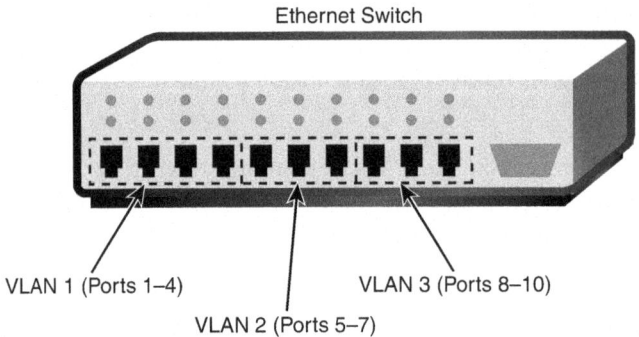

FIGURE 3.6 **Port-Based VLAN Membership**

▶ **MAC address-based VLANs:** The *Media Access Control (MAC)* address is a unique 12-digit hexadecimal number that is stamped into every network interface card. Every device used on a network has this unique address built in to it. It cannot be modified in any way. As you may have guessed, the MAC address type of a VLAN assigns membership according to the workstation's MAC address. To do this, the switch must keep track of the MAC addresses that belong to each VLAN. The advantage of this method is that a workstation computer can be moved anywhere in an office without needing to be reconfigured. Because the MAC address does not change, the workstation remains a member of a particular VLAN. Table 3.8 provides examples of the membership of MAC address-based VLANs.

TABLE 3.8 **MAC Address-Based VLANs**

MAC Address	VLAN	Description
44-45-53-54-00-00	1	Sales
44-45-53-54-13-12	2	Marketing
44-45-53-54-D3-01	3	Administration
44-45-53-54-F5-17	1	Sales

VLAN Segmentation

The capability to logically segment a LAN provides a level of administrative flexibility, organization, and security. Whether the LAN is segmented using the protocol, MAC address, or port, the result is the same: the network is segmented. The segmentation is used for several reasons, including security, organization, and performance. To give you a better idea of how this works, Figure 3.7 shows a network that doesn't use a VLAN.

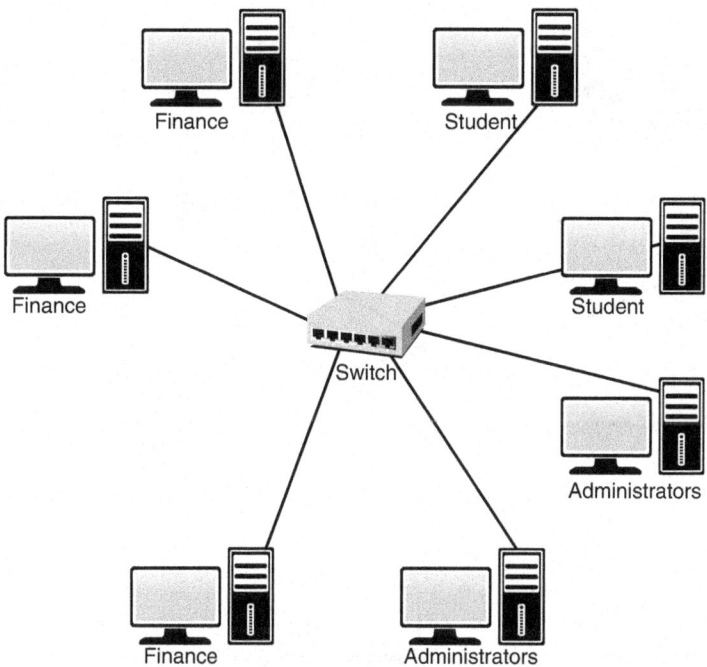

FIGURE 3.7   **Network Configuration Without Using a VLAN**

In Figure 3.7, all systems on the network can see each other. That is, the students can see the finance and administrator computers. Figure 3.8 shows how this network may look using a VLAN.

ExamAlert

Remember that one of the primary purposes of segmentation is to protect sensitive information from other hosts or the rest of the network in general.

Switch Virtual Interface (SVI)

A *Switch Virtual Interface (SVI)*, also known as a VLAN interface or a Layer 3 interface, is a virtual interface configured on a Layer 2 switch to enable routing between VLANs. SVIs are used to route traffic between VLANs at Layer 3 of the OSI model, providing inter-VLAN routing functionality within the switch itself. Each SVI (configured as a logical interface associated with a specific VLAN on a Layer 2 switch) represents a VLAN interface and is assigned an IP address and subnet mask.

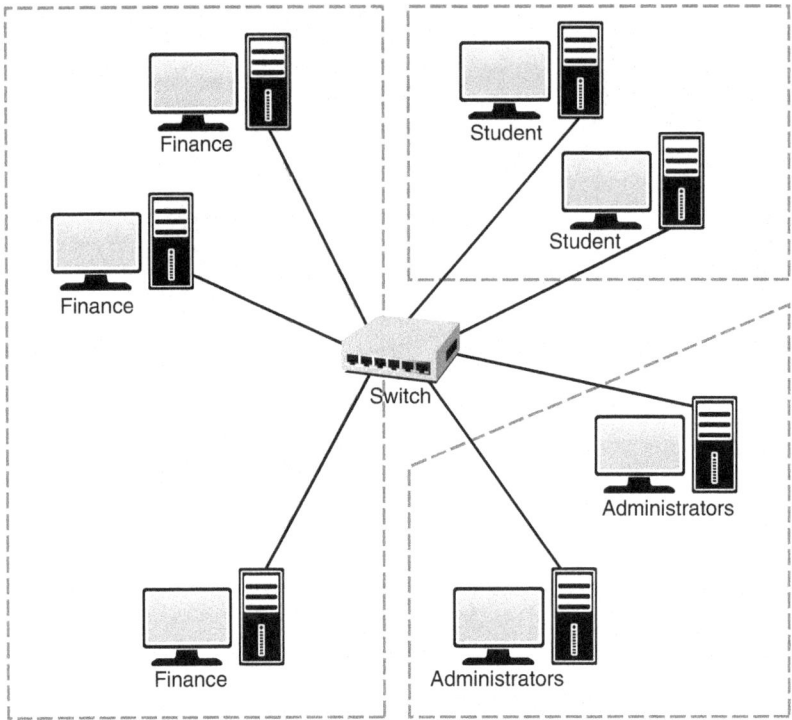

FIGURE 3.8 **Network Configuration Using a VLAN**

When traffic needs to be routed between hosts in different VLANs, it is forwarded to the SVI associated with the ingress VLAN. The SVI then routes the traffic to the appropriate destination VLAN based on the routing table. SVIs are often used as default gateways for hosts within their respective VLANs. Hosts in a VLAN are configured with the SVI's IP address as their default gateway, allowing them to communicate with devices outside their own VLAN. Each SVI is associated with a specific VLAN and serves as the Layer 3 representation of that VLAN. Multiple SVIs can be configured on a switch, with each SVI representing a different VLAN.

SVIs are typically configured using the switch's command-line interface (CLI) or a management interface. Administrators can assign IP addresses, configure routing protocols, apply access control lists (ACLs), and perform other configuration tasks on SVIs. In switches that support Layer 3 functionality, SVIs allow the switch to perform routing functions in addition to its traditional switching capabilities. This enables the switch to act as a router for inter-VLAN communication without the need for an external router.

Interface Configuration and Switch Management

Aside from VLAN trunking (802.1Q), binding, and a number of other possibilities previously discussed in this chapter, when you configure a switch interface, there are often other options that you can choose or tweak. They include the following:

▶ **Tag versus untag VLANs:** *802.1Q tagging* should be used if you are trunking. Because trunking combines VLANs, you need a way to identify which packet belongs to which VLAN; this is easily accomplished by placing a VLAN header (a *tag*) in the data packet. The only VLAN that is not tagged in a trunk is the *native VLAN*, and frames are transmitted to it unchanged.

> **ExamAlert**
>
> In the context of IEEE 802.1Q trunking, the native VLAN refers to the VLAN that is not tagged with a VLAN identifier when transmitted over the trunk link.

▶ **Default VLAN:** The *default VLAN* is mandatory (cannot be deleted) and is used for communication between switches (such as configuring STP). In the Cisco world, the default VLAN is VLAN 1.

▶ **Flow control:** Ethernet provides a means of temporarily stopping the transmission of data to ensure zero packet loss in the presence of network congestion. This is accomplished using *flow control* and the pause frame. First appearing as a part of the IEEE 802.3x standard, it was further expanded upon in the IEEE 802.1Qbb standard.

▶ **Port mirroring:** Also known as port spanning, this is used to monitor network traffic and monitor how well a switch works. Port mirroring copies the traffic from all ports to a single port and disallows bidirectional traffic on that port. There are a number of reasons why port mirroring can be used (duplicating the data for one port and sending it to another). One of the most common is to monitor the traffic. This can be done locally or remotely—the latter using a remote protocol such as Remote Switched Port Analyzer (RSPAN) instead of Switched Port Analyzer (SPAN). To use port mirroring, administrators configure a copy of all inbound and outbound traffic to go to a certain port. A protocol analyzer examines the data sent to the port and therefore does not interrupt the flow of regular traffic.

▶ **Port security:** Port security works at Layer 2 of the OSI model and allows an administrator to configure switch ports so that only certain MAC addresses can use the port. This essentially differentiates so-called dumb switches from managed (or intelligent) switches. Three main areas of port security are (1) MAC limiting and filtering (limit access to the network to MAC addresses that are known, and filter out those that are not); (2) 802.1X (adding port authentication to MAC filtering takes security for the network down to the switch port level and increases your security exponentially); and (3) blocking unused ports (all ports not in use should be disabled).

▶ **Authentication, accounting, and authorization (AAA):** AAA overrides can also be configured for network security parameters as needed. AAA is the primary method for access control and often uses RADIUS, TACACS+, or Kerberos to accomplish integrated security.

▶ **Usernames/passwords:** It is possible to configure, without AAA, local username authentication using a configured username and password. This does not provide the same level of access control as AAA does and is not recommended.

▶ **Virtual consoles and terminals:** The console port (often called the *virtual console* or *VC*) is often a serial or parallel port, and it is possible for virtual ports to connect to physical ports. The *virtual terminal* (vt or vty) is a remote port connected to through Telnet or a similar utility and, as an administrator, you will want to configure an access list to limit who can use it.

ExamAlert

Know that the simplest way to protect a virtual terminal interface is to configure a username and password for it and prevent unauthorized logins.

▶ **Jumbo Frames:** One of the biggest issues with networking is that data of various sizes is crammed into packets and sent across the medium. Each time this is done, headers are created (more data to process), along with any filler needed, creating additional overhead. To get around this, the concept of *jumbo frames* is used to allow for very large Ethernet frames; when a lot of data is sent at once, the number of packets is reduced, and the data sent is less processor intensive.

▶ **Other:** Other common configuration parameters include the speed, whether duplexing will be used or not, IP addressing, and the default gateway. Duplexing determines the direction in which data can flow through the network media and is discussed in Chapter 5, "Cabling Solutions and Issues."

Voice VLAN

A *Voice VLAN* is a type of VLAN configuration commonly used in VoIP (voice over Internet Protocol) networks to ensure the efficient and secure transmission of voice traffic. It is designed to separate voice traffic from data traffic on the network, prioritizing voice traffic to ensure high-quality voice communication and minimizing potential interference from other network traffic.

Trunking

In computer networking, the term *trunking* refers to the use of multiple network cables or ports in parallel to increase the link speed beyond the limits of any one cable or port. Sound confusing? If you have network experience, you might have heard the term *link aggregation*, which is essentially the same thing. It is using multiple cables to increase the throughput. The higher-capacity trunking link is used to connect switches to form larger networks.

> **Note**
>
> *Aggregation* is a popular term any time multiples are combined. The term *route aggregation* applies when specific routes are combined into one route, and this is accomplished in BGP with the aggregate-address command.

VLAN trunking is the application of trunking to the virtual LAN—now common with routers, firewalls, VMware hosts, and wireless access points. VLAN trunking provides a simple and cheap way to offer a nearly unlimited number of virtual network connections. The requirements are only that the switch, the network adapter, and the OS drivers all support VLANs. The *VLAN Trunking Protocol (VTP)* is a proprietary protocol from Cisco for just such a purpose.

The Spanning Tree Protocol (STP)

An Ethernet network can have only a single active path between devices on a network. When multiple active paths are available, switching loops can occur. Switching loops are the result of having more than one path between two switches in a network. *Spanning Tree Protocol (STP)* is designed to prevent these loops from occurring.

STP is used with network bridges and switches. With the help of *Spanning Tree Algorithm (STA)*, STP avoids or eliminates loops on a Layer 2 bridge.

> **Note**
>
> As a heads up, talking about STP refers to Layer 2 of the OSI model. Both bridges and most switches work at Layer 2; routers work at Layer 3, as do Layer 3 switches.

STA enables a bridge or switch to dynamically work around loops in a network's topology. Both STA and STP were developed to prevent loops in the network and provide a way to route around any failed network bridge or ports. If the network topology changes, or if a switch port or bridge fails, STA creates a new spanning tree, notifies the other bridges of the problem, and routes around it. STP is the protocol, and STA is the algorithm STP uses to correct loops.

If a particular port has a problem, STP can perform a number of actions, including blocking the port, disabling the port, or forwarding data destined for that port to another port. It does this to ensure that no redundant links or paths are found in the spanning tree and that only a single active path exists between any two network nodes.

STP uses *bridge protocol data units (BPDUs)* to identify the status of ports and bridges across the network. BPDUs are simple data messages exchanged between switches. BPDUs contain information on ports and provide the status of those ports to other switches. If a BPDU message finds a loop in the network, it is managed by shutting down a particular port or bridge interface.

Redundant paths and potential loops can be avoided within ports in several ways:

▶ **Blocking:** A blocked port accepts BPDU messages but does not forward them.

▶ **Disabled:** The port is offline and does not accept BPDU messages.

▶ **Forwarding:** The port is part of the active spanning tree topology and forwards BPDU messages to other switches.

▶ **Learning:** In a learning state, the port is not part of the active spanning tree topology but can take over if another port fails. Learning ports receive BPDUs and identify changes to the topology when made.

▶ **Listening:** A listening port receives BPDU messages and monitors for changes to the network topology.

Most of the time, ports are in either a forwarding or blocked state. When a disruption to the topology occurs or a bridge or switch fails for some reason, listening and learning states are used.

> **ExamAlert**
>
> STP actively monitors the network, searching for redundant links. When it finds some, it shuts them down to prevent switching loops. STP uses STA to create a topology database to find and then remove the redundant links. With STP operating from the switch, data is forwarded on approved paths, which limits the potential for loops.

Maximum Transmission Unit (MTU)

The maximum transmission unit (MTU) is the largest size of a data packet that can be transmitted over a network protocol without fragmentation. The MTU size affects the efficiency of data transmission, as larger MTU sizes generally result in fewer overheads and better throughput. If the size of the data packet exceeds the MTU of the network, it must be fragmented into smaller packets that fit within the MTU. Fragmentation adds overhead and can impact network performance.

A jumbo frame is an Ethernet frame with a payload greater than the standard MTU of 1500 bytes. Jumbo frames can be useful in certain scenarios to reduce overhead and increase network efficiency, particularly in high-performance computing (HPC) environments, storage-area networks (SANs), and certain types of network traffic such as multimedia streaming or large file transfers. It is important to know that not all network devices and protocols support jumbo frames. For jumbo frames to work effectively, all devices along the network path, including switches, routers, and network interface cards (NICs), must support them.

The opposite of jumbo frames are runts: a frame that is smaller than the minimum frame size for IEEE-802.3 standard frames. In Ethernet, that's 64 bytes.

> **ExamAlert**
>
> For the Network+ exam, know that jumbo frames are larger-than-standard Ethernet frames that can potentially improve network efficiency in certain scenarios but require careful consideration and configuration.

Cram Quiz

1. Which of the following best describes the function of the default gateway?

   ○ **A.** It provides the route for destinations outside the local network.

   ○ **B.** It enables a single Internet connection to be used by several users.

   ○ **C.** It identifies the local subnet and formulates a routing table.

   ○ **D.** It is used to communicate in a multiple-platform environment.

2. What is the term used for the trustworthiness of routing information received from different sources?

   ○ **A.** Jump list

   ○ **B.** Link stops

   ○ **C.** Connections

   ○ **D.** Administrative distance

3. Which of the following enables administrators to monitor the traffic outbound and inbound to the switch?

   ○ **A.** Spanning Tree Algorithm

   ○ **B.** Trunking

   ○ **C.** HSRP

   ○ **D.** Port mirroring

4. Which of the following is the IEEE specification developed to ensure interoperability of VLAN technologies from the various vendors?

   ○ **A.** 802.1z

   ○ **B.** 802.1s

   ○ **C.** 802.1Q

   ○ **D.** 802.1X

5. Which of the following is a proprietary protocol from Cisco used to reduce administration in the switched network?

   ○ **A.** VTP

   ○ **B.** VNMP

   ○ **C.** VCPN

   ○ **D.** VNMC

Cram Quiz Answers

1. **A.** The default gateway enables systems on one local subnet to access those on another. Answer B does not accurately describe the role of the default gateway. Answers C and D do not describe the main function of a default gateway, which is to provide the route for destinations outside the local network.

2. **D.** Administrative distance is a metric used in routing protocols to determine the trustworthiness of routing information received from different sources. It is a numerical value assigned to each routing protocol or administrative source, indicating the perceived reliability or trustworthiness of routes learned through that source.

3. **D.** Port mirroring enables administrators to monitor the traffic outbound and inbound to the switch.

4. **C.** 802.1Q is the IEEE specification developed to ensure interoperability of VLAN technologies from the various vendors.

5. **A.** *VLAN Trunking Protocol (VTP)* is used to reduce administration in the switched network.

Network Services

▶ **3.4 Given a scenario, implement IPv4 and IPv6 network services.**

CramSaver

If you can correctly answer these questions before going through this section, save time by skimming the ExamAlerts in this section and then completing the Cram Quiz at the end of the section.

1. What is the name used for ranges of IP addresses available within DHCP?

2. What is the name of the packet on a system configured to use DHCP broadcasts when it comes onto the network?

3. What is dynamic DNS?

4. Within DNS, what is the domain name, along with any subdomains, called?

Answers

1. Within DHCP, ranges of IP addresses are known as *scopes*.

2. When a system configured to use DHCP comes onto the network, it broadcasts a special packet that looks for a DHCP server. This packet is known as the DHCPDISCOVER packet.

3. Dynamic DNS is a newer system that enables hosts to be dynamically registered with the DNS server.

4. The domain name, along with any subdomains, is called the *fully qualified domain name (FQDN)* because it includes all the components from the top of the DNS namespace to the host.

Network services provide functionality enabling the network to operate. A plethora of services are available, but three you need to know for the exam are DHCP, DNS, and time-related protocols.

Dynamic Host Configuration Protocol

One method to assign IP addresses to hosts is to use static addressing. This process involves manually assigning an address from those available to you and allowing the host to always use that address. The problems with this method include the difficulty in managing addresses for a multitude of machines and efficiently and effectively issuing them.

> **ExamAlert**
>
> Be sure to know the difference between static and dynamic IP addressing as you study for the Network+ exam.

DHCP, which is defined in RFC 2131, enables ranges of IP addresses, known as *scopes* or predefined groups of addresses within *address pools* to be defined on a system running a DHCP server application. When another system configured as a DHCP client is initialized, it asks the server for an address. If all things are as they should be, the server assigns an address from the scope to the client for a predetermined amount of time, known as the *lease* or *lease time*.

At various points during the TTL of the lease time (normally the 50 percent and 85 percent points), the client attempts to renew the lease from the server. If the server cannot perform a renewal, the lease expires at 100 percent, and the client stops using the address.

In addition to an IP address and the subnet mask, the DHCP server can supply many other pieces of information; however, exactly what can be provided depends on the DHCP server implementation. In addition to the address information, the default gateway is often supplied, along with DNS information.

In addition to having DHCP supply a random address from the scope, you can configure *scope options*, such as having it supply a specific address to a client. Such an arrangement is known as a *reservation* (see Figure 3.9). Reservations are a means by which you can still use DHCP for a system but at the same time guarantee that it always has the same IP address. When based on the MAC address, this is known as *MAC reservations*. DHCP can also be configured for exclusions, also called *IP exclusions*. In this scenario, certain IP addresses are not given out to client systems.

The advantages of using DHCP are numerous. First, administrators do not need to manually configure each system. Second, human error, such as the assignment of duplicate IP addresses, is eliminated. Third, DHCP removes the need to reconfigure systems if they move from one subnet to another, or if you decide to make a wholesale change in the IP addressing structure. The downsides are that DHCP traffic is broadcast based and thus generates network traffic—albeit a small amount. Finally, the DHCP server software must be installed and configured on a server, which can place additional processor load (again, minimal) on that system. From an administrative perspective, after the initial configuration, DHCP is about as maintenance-free as a service can get, with only occasional monitoring normally required.

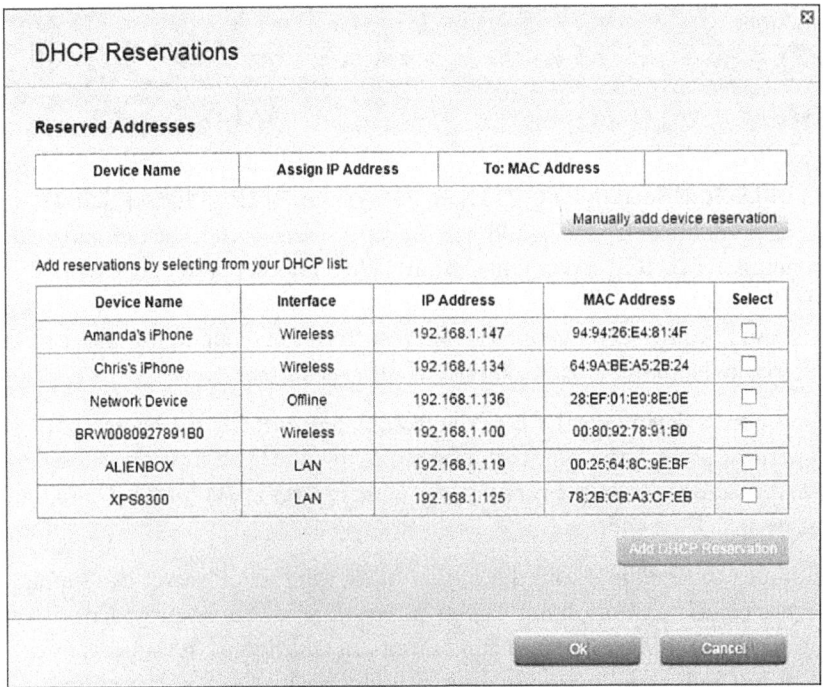

FIGURE 3.9 **DHCP Reservations**

ExamAlert

DHCP is a protocol-dependent service and is not platform dependent. This means that you can use, for instance, a Linux DHCP server for a network with Windows clients or with Linux clients. Although the DHCP server offerings in the various network operating systems might slightly differ, the basic functionality is the same across the board. Likewise, the client configuration for DHCP servers running on a different operating system platform is the same as for DHCP servers running on the same base operating system platform.

IPv6, via DHCPv6, uses *Stateless Address Auto Configuration (SLAAC)*. With SLAAC, devices send the router a request for the network prefix, and the device then uses the prefix along with its own MAC address to create an IP address.

The DHCP Process

To better understand how DHCP works, spend a few minutes looking at the processes that occur when a DHCP-enabled client connects to the network.

When a system configured to use DHCP comes onto the network, it broadcasts a special packet that looks for a DHCP server. This packet is known as the DHCPDISCOVER packet. The DHCP server, which is always on the lookout for DHCPDISCOVER broadcasts, picks up the packet and compares the request with the scopes it has defined. If it finds that it has a scope for the network from which the packet originated, it chooses an address from the scope, reserves it, and sends the address, along with any other information, such as the lease duration, to the client. This is known as the DHCPOFFER packet. Because the client still does not have an IP address, this communication is also achieved via broadcast. By default, DHCP operates on ports 67 and 68.

> **ExamAlert**
>
> Remember that DHCP operates on ports 67 and 68.

When the client receives the offer, it looks at the offer to determine if it is suitable. If more than one offer is received, which can happen if more than one DHCP server is configured, the offers are compared to see which is best. *Best* in this context can involve a variety of criteria but normally is the length of the lease. When the selection process completes, the client notifies the server that the offer has been accepted, through a packet called a DHCPREQUEST packet. At this point the server finalizes the offer and sends the client an acknowledgment. This last message, which is sent as a broadcast, is known as a DHCPACK packet. After the client system receives the DHCPACK, it initializes the TCP/IP suite and can communicate on the network.

DHCP and DNS Suffixes

In DNS, *suffixes* define the DNS servers to be used and the order in which to use them. DHCP settings can push a domain suffix search list to DNS clients. When such a list is specifically given to a client, the client uses only that list for name resolution. With Linux clients, this can occur by specifying entries in the resolve.conf file.

> **ExamAlert**
>
> Know that DHCP can provide DNS suffixes to clients.

DHCP Relays and IP Helpers

On a large network, the DHCP server can easily get bogged down trying to respond to all the requests. To make the job easier, *DHCP relays* help make the job easier. A DHCP relay is nothing more than an agent on the router that acts as a go-between for clients and the server. This feature is useful when working with clients on different subnets, because a client cannot communicate directly with the server until it has the IP configuration information assigned to it.

One level above DHCP relay is *IP helper*. These two terms are often used as synonyms, but they are not; a better way to think of it is with IP helper being a superset of DHCP relay. IP helper will, by default, forward broadcasts for DHCP/BOOTP, TFTP, DNS, TACACS/TACACS+, the time service, and the NetBIOS name/datagram service (ports 137–139). You can disable the additional traffic (or add more), but by default IP helper will do more than a DHCP relay.

> **ExamAlert**
>
> Know that an IP helper can do more than a DHCP relay agent.

Domain Name Service (DNS)

DNS performs an important function on TCP/IP-based networks. It resolves hostnames, such as www.informit.com, to IP addresses, such as 209.202.161.67. Such a resolution system makes it possible for people to remember the names of and refer to frequently used hosts using easy-to-remember hostnames rather than hard-to-remember IP addresses. By default, DNS operates on port 53.

> **Note**
>
> Like other TCP/IP-based services, DNS is a platform-independent protocol. Therefore, it can be used on Linux, UNIX, Windows, and almost every other platform.

In the days before the Internet, the network that was to become the Internet used a text file called HOSTS to perform name resolution. The *HOSTS* file was regularly updated with changes and distributed to other servers. Following is a sample of some entries from a HOSTS file:

```
192.168.3.45 server1 s1 #The main file and print server

192.168.3.223 Mail mailserver #The email server

127.0.0.1 localhost
```

Note

A comment in the HOSTS file is preceded by a hash symbol (#).

As you can see, the host's IP address is listed, along with the corresponding hostname. You can add to a HOSTS file aliases of the server names, which in this example are **s1** and **mailserver**. All the entries must be added manually, and each system to perform resolutions must have a copy of the file.

Even when the Internet was growing at a relatively slow pace, such a mechanism was both cumbersome and prone to error. It was obvious that as the network grew, a more automated and dynamic method of performing name resolution was needed. DNS became that method.

Tip

HOSTS file resolution is still supported by most platforms. If you need to resolve just a few hosts that will not change often or at all, you can still use the HOSTS file for this.

DNS solves the problem of name resolution by offering resolution through servers configured to act as name servers. The name servers run DNS server software, which enables them to receive, process, and reply to requests from systems that want to resolve hostnames to IP addresses. Systems that ask DNS servers for a hostname-to-IP address mapping are called *resolvers* or *DNS clients*. Figure 3.10 shows the DNS resolution process. In this example, the client asks to reach the first server at mycoltd.com; the router turns to the DNS server for an IP address associated with that server; and after the address is returned, the client can establish a connection.

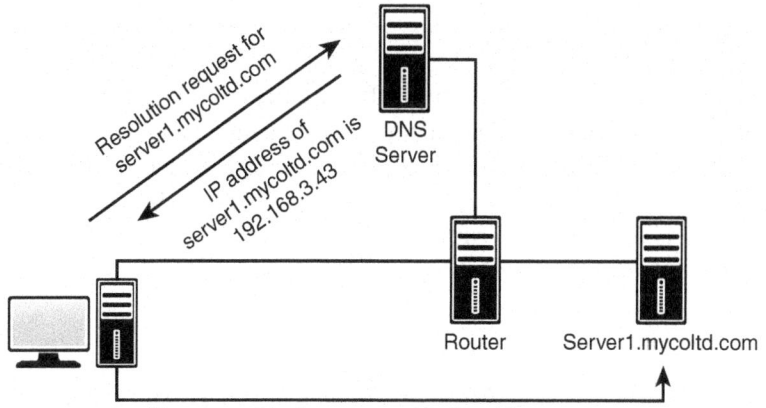

FIGURE 3.10 **The DNS Resolution Process**

Because the DNS namespace (which is discussed in the following section) is large, a single server cannot hold all the records for the entire namespace. As a result, there is a good chance that a given DNS server might not resolve the request for a certain entry. In this case, the DNS server asks another DNS server if it has an entry for the host.

> **Note**
>
> One of the problems with DNS is that, despite all its automatic resolution capabilities, entries and changes to those entries must still be manually performed. A strategy to solve this problem is to use *Dynamic DNS (DDNS)*, a newer system that enables hosts to be dynamically registered with the DNS server. When changes are made in real time to hostnames, addresses, and related information, there is less likelihood of not finding a server or site that has been recently added or changed.

> **ExamAlert**
>
> You might be asked to identify the difference between DNS and DDNS.

To speed up resolution, the client will often store the results of resolution locally (in the browser quite often) so that it does not have to query again if the same resolution needs to be done. This is known as *DNS caching*, and this is also done by caching nameservers (also known as recursive nameservers). Since it is possible that values change (a different IP address issued to a host than it previously had), caches typically come with *time-to-live (TTL)* values and time out after a while.

The DNS Namespace

DNS operates in the *DNS namespace*. This space has logical divisions *hierarchically* organized. At the top level are domains such as .com (commercial) and .edu (education), as well as domains for countries, such as .uk (United Kingdom) and .de (Germany). Below the top level are subdomains or second-level domains associated with organizations or commercial companies, such as Red Hat and Microsoft. Within these domains, hosts or other subdomains can be assigned. For example, the server ftp.redhat.com would be in the redhat.com domain. Figure 3.11 shows a DNS hierarchical namespace.

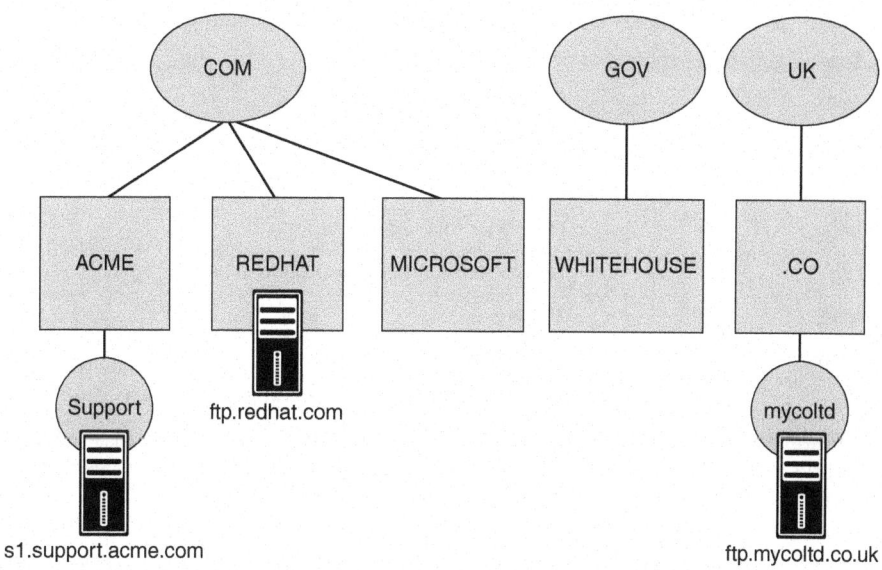

FIGURE 3.11 **A DNS Hierarchical Namespace**

ExamAlert

The domain name, along with any subdomains, is called the *fully qualified domain name (FQDN)* because it includes all the components from the top of the DNS namespace to the host.

Note

Many people refer to DNS as resolving FQDNs to IP addresses. An example of an FQDN is www.comptia.org, where www is the host, comptia is the second-level domain, and org is the top-level domain.

The lower domains are largely open to use in whatever way the domain name holder sees fit. However, the top-level domains are relatively closely controlled. Table 3.9 lists a selection of the most widely used top-level DNS domain names. Recently, a number of top-level domains were added, mainly to accommodate the increasing need for hostnames. While *root DNS servers* directly answer requests for records in the root zone, and answer other requests, they also return lists of the authoritative name servers for the top-level domain (TLD) being sought.

TABLE 3.9 **Selected Top-Level Domains in the DNS Namespace**

Top-Level Domain Name	Intended Purpose
com	Commercial organizations
edu	Educational organizations/establishments
gov	U.S. government organizations/establishments
net	Network providers/centers
org	Not-for-profit and other organizations
mil	Military
arpa	Reverse DNS lookup
de	A country-specific domain—in this case, Germany*

*In addition to country-specific domains, many countries have created subdomains that follow roughly the same principles as the original top-level domains (such as co.uk and gov.nz).

Although the assignment of domain names is supposed to conform to the structure shown in Table 3.9, the assignment of names is not as closely controlled as you might think. It's not uncommon for some domain names to be used for other purposes, such as .org or .net being used for business.

> **Note**
>
> Although the primary function of DNS is to resolve hostnames to IP addresses, you can also have DNS perform IP address-to-hostname resolution. This process is called *reverse lookup*, which is accomplished by using *pointer (PTR)* records.

> **ExamAlert**
>
> For the exam, know that PTR records are used for reverse lookup functions.

Two other words often used with DNS queries are *iterative* and *recursive*. An iterative lookup is one in which the client just keeps querying the server. A recursive lookup is one in which the server does not have the answer the client is looking for and forwards the request on to another DNS server in search of the answer. To use a silly analogy, an iterative lookup would be similar to asking your mother every five minutes if you can go outside and getting the same "no" answer over and over, while a recursive lookup would be her telling you to go ask your father.

Types of DNS Entries

Although the most common entry in a DNS database is an *A (address)* record, which maps a hostname to an IP address, DNS can hold numerous other types of entries as well. Some are the MX record, which can map entries that correspond to mail exchanger systems, and *CNAME (canonical record name)*, which can create alias records for a system. A system can have an A record and then multiple CNAME entries for its aliases. A DNS table with all these types of entries might look like this:

```
fileserve.mycoltd.com IN A 192.168.33.2

email.mycoltd.com IN A 192.168.33.7

fileprint.mycoltd.com IN CNAME fileserver.mycoltd.com

mailer.mycoltd.com IN MX 10 email.mycoltd.com
```

As you can see, rather than map to an actual IP address, the CNAME and MX record entries map to another host, which DNS in turn can resolve to an IP address.

DNS Records

Each DNS name server maintains information about its *zone*, or domain, in a series of records, known as DNS resource records. There are several DNS resource records; each contains information about the DNS domain and the systems within it. These records are text entries stored on the DNS server. Some of the DNS resource records include the following:

▶ **Address (A):** The A record maps a domain name to an IPv4 address. It is used to resolve domain names to their corresponding IPv4 (32-bit) addresses.

▶ **IPv6 Address (AAAA):** Similar to the A record, the AAAA record maps a domain name to an IPv6 address. It is used for resolving domain names to their corresponding IPv6 (128-bit) addresses.

▶ **Canonical Name (CNAME):** This record stores additional hostnames, or aliases, for hosts in the domain. A CNAME specifies an alias or nickname for a canonical hostname record in a Domain Name Service (DNS) database. CNAME records give a single computer multiple names (aliases).

▶ **Mail Exchange (MX):** The MX record specifies the mail servers responsible for receiving email messages for a domain. It points to the domain names of the mail servers that handle incoming email traffic for the domain.

▶ **Text (TXT):** The TXT record allows arbitrary text data to be associated with a domain name. It is commonly used for adding human-readable information, such as SPF (Sender Policy Framework) records, DKIM (DomainKeys Identified Mail) keys, and domain ownership verification records.

▶ **Pointer (PTR):** This record is a pointer to the canonical name, which is used to perform a reverse DNS lookup, in which case the name is returned when the query originates with an IP address.

▶ **Start of Authority (SOA):** This is a record of information containing data on DNS zones and other DNS records. A DNS zone is the part of a domain for which an individual DNS server is responsible. Each zone contains a single SOA record.

▶ **Name Server (NS):** This record stores information that identifies the name servers in the domain that store information for that domain.

▶ **Service Locator (SRV):** This is a generalized service location record, used for newer protocols instead of creating protocol-specific records such as MX.

ExamAlert

The most common type of DNS zone is the *forward lookup zone*, which allows DNS clients to obtain information such as IP addresses that correspond to DNS domain names. Remember that a *reverse lookup zone* maps from IP addresses back to DNS domain names.

ExamAlert

The exam objectives specifically list DNS records. You should expect to see a question about DNS records A, AAAA, CNAME, TXT, NS, or PTR.

Authoritative DNS servers are responsible for storing and providing authoritative DNS information for a specific domain. These servers hold the authoritative zone files that contain DNS records (such as A, AAAA, MX, NS) for the domain they are authoritative for. *Nonauthoritative DNS servers* (also known as caching resolvers or *recursive DNS servers*) are responsible for resolving DNS queries on behalf of clients, but they do not hold authoritative DNS information for any domain (instead, they query authoritative DNS servers to obtain DNS records on behalf of clients). Nonauthoritative DNS servers speed up the resolution process by passing on cached values for subsequent queries of the same record as long as it occurs within a certain period of time (determined by the TTL value). When a nonauthoritative DNS server receives a query for a domain that it does not have cached records for, it recursively queries authoritative DNS servers to obtain the required DNS information and responds to the client with the result.

DNS in a Practical Implementation

In a real-world scenario, whether you use DNS is almost a nonissue. If you have Internet access, you will most certainly use DNS, but you are likely to use the DNS facilities of your *Internet service provider (ISP)* rather than have your own internal DNS server—this is known as *external DNS*. However, if you operate a large, complex, multiplatform network, you might find that internal DNS servers are necessary. The major network operating system vendors know that you might need DNS facilities in your organization, so they include DNS server applications with their offerings, making third-party/cloud-hosted DNS a possibility. Google, for example, offers Cloud DNS, which is "low latency, high availability and is a cost-effective way to make your applications and services available to your users" (for more information, see https://cloud.google.com/dns/).

It is common practice for workstations to be configured with the IP addresses of two DNS servers for fault tolerance (configured via the Alternate Configuration tab in Windows, for example). The importance of DNS, particularly in environments in which the Internet is heavily used, cannot be overstated. If DNS facilities are not accessible, the Internet effectively becomes unusable, unless you can remember the IP addresses of all your favorite sites.

Domain Name System Security Extensions (DNSSEC) is a suite of IETF specifications for securing certain kinds of information provided by DNS. As it was originally designed, DNS did not include any security features. DNSSEC not only adds security features to DNS but is also designed to be backward compatible.

DNS over HTTPS (DoH) is a protocol that allows DNS resolution to be conducted over HTTPS to enhance privacy and security by encrypting DNS queries and responses with the intent of preventing eavesdropping and manipulation of DNS traffic by malicious actors. DoH uses port 443 for DNS resolution, allowing DNS traffic to bypass restrictive network policies or firewalls that may block traditional DNS traffic on port 53. It helps mitigate DNS spoofing, on-path attacks, and DNS hijacking by encrypting DNS traffic and authenticating DNS responses using HTTPS mechanisms, reducing the risk of DNS-related attacks.

DoH is outlined in RFC 8484 and is typically implemented in DNS client software, such as web browsers or DNS resolver applications, rather than at the DNS server level. Clients that support DoH can send DNS queries over HTTPS to DoH-compatible DNS servers, which handle DNS resolution over encrypted connections.

Instead of using HTTPS for the tunneling, as DoH does, it is possible to secure it using TLS and that is what *DNS over TLS (DoT)* does. It is a protocol that provides secure DNS resolution by encrypting DNS queries and responses using the Transport Layer Security (TLS) protocol. Like DoH, DoT is typically implemented in DNS client software, such as operating systems, DNS resolver applications, or DNS-over-TLS-capable DNS servers. Clients that support DoT can establish encrypted connections to DoT-compatible DNS servers, which handle DNS resolution securely over TLS. Outlined in RFC 7858, it has gained popularity among DNS resolver providers, operating systems, and network equipment manufacturers, leading to increasing adoption and integration into various software applications and platforms.

Time Protocols

Network Time Protocol (NTP) is one of the oldest Internet protocols in current use. It is the part of the TCP/IP protocol suite that facilitates the communication of time between systems. NTP operates over UDP port 123. The idea is that one system configured as a time provider transmits time information to other systems that can be both time receivers and time providers for other systems.

Time synchronization is important in today's IT environment because of the distributed nature of applications. Two good examples of situations in which time synchronization is important are email and directory services systems. In each of these cases, having time synchronized between devices is important because without it there would be no way to keep track of changes to data and applications.

NTP uses a hierarchical, semi-layered system of time sources wherein each level of the hierarchy is termed a *stratum*. Each stratum/level is assigned a number starting with zero for the reference clock at the top and incrementing from there with the number representing the distance from the reference clock: this means that a server synchronized to a stratum n server runs at stratum $n + 1$. This numbering is used to prevent cyclical dependencies in the hierarchy, but stratum is not always an indication of quality or reliability. It is possible to find a stratum server with a higher number (for example, 3) that is of higher quality than a stratum 2 time source.

In many environments, external time sources such as radio clocks, *Global Positioning System (GPS)* devices, and Internet-based time servers are used as sources of NTP time. In others, the system's BIOS clock is used. Regardless of what source is used, the time information is communicated between devices by using NTP.

Note

Specific guidelines dictate how NTP should be used. You can find these "rules of engagement" at http://support.ntp.org/bin/view/Servers/RulesOfEngagement. Note that the site uses HTTP, as opposed to HTTPS, and should not be considered secure.

ExamAlert

Remember that NTP is used for time synchronization and is implemented over UDP port 123.

NTP server and client software is available for a variety of platforms and devices. If you want a way to ensure time synchronization between devices, look to NTP as a solution.

A bit newer, *Precision Time Protocol (PTP)* is a standard defined by IEEE 1588 and is commonly used in applications that require precise time synchronization, such as industrial automation, telecommunications, and financial trading systems. While being a bit late might not be a problem with many applications, those that can't afford to be late rely on PTP, which can achieve sub-microsecond accuracy, making it suitable for applications that demand extremely precise timing. It operates by exchanging timing messages between one clock (known as a grandmaster clock)) that has been selected to coordinate the timekeeping across all other clocks within the network. The main clock sends synchronization messages containing precise timing information, and the other clocks adjust their time based on these messages to maintain synchronization with the main clock.

Network Time Security (NTS) can be used to protect time synchronization protocols like NTP and PTP from various security threats, such as spoofing, tampering, and man-in-the-middle attacks. Defined by RFC 8915, it employs cryptographic techniques—such as digital signatures and encryption—to secure time synchronization communications between time clients and servers in critical systems and applications.

Cram Quiz

1. One of the programmers has asked that DHCP always issue their workstation the same IP address. What feature of DHCP enables you to accomplish this?

   ○ **A.** Stipulation

   ○ **B.** Rider

   ○ **C.** Reservation

   ○ **D.** Provision

2. Which of the following is *not* a common packet sent during the normal DHCP process?

   ○ **A.** DHCPACK

   ○ **B.** DHCPPROVE

   ○ **C.** DHCPDISCOVER

   ○ **D.** DHCPOFFER

3. During a discussion, your ISP's technical support representative mentions that you might have been using the wrong FQDN. Which TCP/IP-based network service is the representative referring to?

   ○ **A.** DHCP

   ○ **B.** SNMP

   ○ **C.** SMNP

   ○ **D.** DNS

4. Which DNS record stores additional hostnames, or aliases, for hosts in the domain?

   ○ **A.** ALSO

   ○ **B.** ALIAS

   ○ **C.** CNAME

   ○ **D.** PTR

5. Which DNS record is most commonly used to map hostnames to an IP address for a host with IPv6?

   ○ **A.** A

   ○ **B.** AAAA

   ○ **C.** MX

   ○ **D.** PTR

6. Which of the following are benefits of VLANs? (Choose all that apply.)

    ○ **A.** Increased performance

    ○ **B.** Decreased security

    ○ **C.** Organization

    ○ **D.** Simplified administration

7. A DHCP server is unreachable and a user's system has been assigned an IP address on the 169.254.0.0 network and cannot access the Internet. What technology has been implemented?

    ○ **A.** APIPA

    ○ **B.** BOOTP

    ○ **C.** OSPF

    ○ **D.** Virtual-circuit packet switching

8. If multiple routes exist for data to travel, which routing table entry will be used?

    ○ **A.** Metric:

    ○ **B.** Interface:

    ○ **C.** Netmask:

    ○ **D.** Gateway:

9. Which is a protocol that provides secure DNS resolution by encrypting DNS queries and responses using the Transport Layer Security (TLS) protocol?

    ○ **A.** DNSSEC

    ○ **B.** DoH

    ○ **C.** DoT

    ○ **D.** RFC 8484

10. Which commands can you use to view MAC addresses of NICs? (Choose all that apply.)

    ○ **A. ipconfig /all**

    ○ **B. ifconfig -a**

    ○ **C. sh int interface name**

    ○ **D. route print**

Cram Quiz Answers

1. **C.** Reservations are specific addresses reserved for clients.

2. **B.** DHCPPROVE is not a common packet. The other choices presented (DHCPACK, DHCPDISCOVER, and DHCPOFFER) are part of the normal process.

3. **D.** DNS is a system that resolves hostnames to IP addresses. The term *FQDN* is used to describe the entire hostname. None of the other services use FQDNs.

4. **C.** The CNAME record stores additional hostnames, or aliases, for hosts in the domain. There is not an ALSO record or ALIAS, and PTR is used for reverse lookups.

5. **B.** The AAAA record is most commonly used to map hostnames to an IP address for a host with IPv6. The A record is not used for this purpose. MX identifies the mail exchanger, and PTR is used for reverse lookup.

6. **A, C, and D.** The benefits of VLANs include increased performance by reducing broadcast traffic throughout the network because VLANs free up bandwidth. Organization is correct because network users and resources that are linked and that communicate frequently can be grouped in a VLAN. Simplified administration is correct because with a VLAN the network administrator's job is easier when moving users between LAN segments, recabling, addressing new stations, and reconfiguring switches and routers. Answer B is incorrect because VLANs actually increase security, not decrease security; with the creation of logical (virtual) boundaries, network segments can be isolated.

7. **A.** The function of APIPA is that a system can give itself an IP address if it is incapable of receiving an address dynamically from a DHCP server. Then APIPA assigns the system an address from the 169.254.0.0 address range (between 169.254.0.1 and 169.254.255.254) and configures an appropriate subnet mask (255.255.0.0). However, it doesn't configure the system with a default gateway address. As a result, communication is limited to the local network. So, if you can connect to other devices on a local network but can't reach the Internet, for example, it is likely that your DHCP server is down and you are currently using an APIPA address. BOOTP is incorrect. It was originally created so that diskless workstations could obtain information needed to connect to the network, such as the TCP/IP address, subnet mask, and default gateway. Such a system was necessary because diskless workstations had no way to store the information. OSPF is incorrect because it is a link-state routing protocol based on the shortest path first (SPF) algorithm to find the least-cost path to any destination in the network. In operation, each router using OSPF sends a list of its neighbors to other routers on the network. From this information, routers can determine the network design and the shortest path for data to travel. Virtual-circuit packet switching is incorrect because it is a packet-switching method. With virtual-circuit packet switching, a logical connection is established between the source and the destination device. This logical connection is established when the sending device initiates a conversation with the receiving device. The logical communication path between the two devices can remain active for as long as the two devices are available or can be used to send packets once. After the sending process has completed, the line can be closed.

8. **A.** In a routing table, the metric is a measurement of the directness of a route. If multiple routes exist for data to travel, the route with the lowest metric is chosen. The lower the metric, the faster the route. The interface is the address of the interface that's used to send the packet to the destination. The netmask is the subnet mask value for the destination parameter. The gateway is where the IP address is sent. This may be a gateway server, a router, or another system acting as a gateway.

9. **C.** DNS over TLS (DoT) is a protocol that provides secure DNS resolution by encrypting DNS queries and responses using the Transport Layer Security (TLS) protocol. Like DoH, DoT is typically implemented in DNS client software, such as operating systems, DNS resolver applications, or DNS-over-TLS-capable DNS servers. Clients that support DoT can establish encrypted connections to DoT-compatible DNS servers, which handle DNS resolution securely over TLS. Domain Name System Security Extensions (DNSSEC) is a suite of IETF specifications for securing certain kinds of information provided by DNS. As it was originally designed, DNS did not include any security features. DNSSEC not only adds security features to DNS but is also designed to be backward compatible. DNS over HTTPS (DoH) is a protocol that allows DNS resolution to be conducted over HTTPS to enhance privacy and security by encrypting DNS queries and responses with the intent of preventing eavesdropping and manipulation of DNS traffic by malicious actors. DoH uses port 443 for DNS resolution, allowing DNS traffic to bypass restrictive network policies or firewalls that may block traditional DNS traffic on port 53. Answer D is not correct because DoH is outlined in RFC 8484; DoT is outlined in RFC 7858.

10. **A, B, and C.** The **ipconfig /all** command in Windows will display the MAC address and detailed information about all adapters, including IP address, subnet mask, default gateway, and DHCP and DNS servers. The **ifconfig -a** command can be used in Linux and UNIX systems for similar purposes and on Cisco devices. Entering the **sh int interface name** command displays brief information about all the available interfaces on a router or switch. Answer D is incorrect because the **route print** command is used to view the routing table on a client system.

What's Next?

Chapter 4, "Network Implementations," introduces you to commonly used networking architecture and devices. All but the most basic of networks require devices to provide connectivity and functionality. Understanding how these networking devices operate and identifying the functions they perform are essential skills for any network administrator and are requirements for a Network+ candidate.

CHAPTER 4

Network Implementations

This chapter covers the following official Network+ objectives:

▶ 1.2 Compare and contrast networking appliances, applications, and functions.

▶ 1.8 Summarize evolving use cases for modern network environments.

This chapter covers CompTIA Network+ objectives 1.2 and 1.8. For more information on the official Network+ exam topics, see the "About the Network+ Exam" section in the Introduction.

All but the most basic of networks require devices to provide connectivity and functionality. Understanding how these networking devices operate and identifying the functions they perform are essential skills for any network administrator and are requirements for a Network+ candidate.

This chapter introduces commonly used networking devices, and that is followed by a discussion of basic corporate and datacenter network architecture later in the chapter. You are not likely to encounter all the devices mentioned in this chapter on the exam, but you can expect to work with at least some of them.

Common Networking Devices

▶ **1.2 Compare and contrast networking appliances, applications, and functions.**

CramSaver

If you can correctly answer these questions before going through this section, save time by skimming the ExamAlerts in this section and then completing the Cram Quiz at the end of the section.

1. What is the difference between a router and a switch?

2. What are the types of ports found on switches?

3. What can distribute incoming data to specific application servers and help distribute the load?

4. True or false: A multilayer switch operates as both a router and a switch.

5. Your company is looking to add a hardware device to the network that can increase redundancy and data availability as it increases performance by distributing the workload. What use case might this sample technology apply to?

6. True or false: An IPS is a passive detection system that can only detect the presence of an attack and then log that information.

7. On a VLAN, what creates multiple paths to the storage resources and can be used to increase availability *and* add fault tolerance?

8. True or false: When an AP boots, it authenticates with a controller before it can start working as an AP. This is often used with VLAN pooling.

Answers

1. A router operates at the network layer (Layer 3) of the OSI model, and its primary function is to route data between different networks, making decisions based on IP addresses to determine the best path for data to reach its destination. A switch operates at the data link layer (Layer 2) of the OSI model and is designed to forward data within a local network, typically within a single local-area network (LAN) or virtual LAN (VLAN) using Media Access Control (MAC) addresses to make forwarding decisions.

2. Switches have two types of ports: *medium-dependent interface (MDI)* and *medium-dependent interface crossed (MDI-X)*.

3. A content switch can distribute incoming data to specific application servers and help distribute the load.

4. True. A multilayer switch, also known as a Layer 3 switch, operates as both a router and a switch.

5. A load balancer can be either a software or hardware component, and it increases redundancy and data availability as it increases performance by distributing the workload.

6. False. An intrusion prevention system (IPS) is an active detection system. With IPS, the device continually scans the network, looking for inappropriate activity. It can shut down any potential threats. An *intrusion detection system (IDS)* is a passive detection system. The IDS can detect the presence of an attack and then log that information.

7. On a VLAN, multipathing creates multiple paths to the storage resources and can be used to increase availability *and* add fault tolerance.

8. True. When an AP boots, it authenticates with a controller before it can start working as an AP. This is often used with *VLAN pooling*, in which multiple interfaces are treated as a single entity (usually for load balancing).

The best way to think about the first part of this chapter is as a catalog of networking devices. The first half looks at devices that you can commonly find in a network of any substantial size. The devices are discussed in objective order to simplify study.

ExamAlert

Remember this objective begins with "Compare and contrast." This means that you need to be able to distinguish one networking or networked device from another and know its appropriate placement on the network. What does it do? Where does it belong?

Router

In a common configuration, routers create larger networks by joining two network segments. A *small office/home office (SOHO)* router connects a user to the Internet. A SOHO router typically serves 1 to 10 users on the system. A router can be a dedicated hardware device or a computer system with more than one network interface and the appropriate routing software. All modern network operating systems include the functionality to act as a router.

Note

Routers normally create, add, or divide networks or network segments at the network layer of the OSI reference model because they normally are IP-based devices. Chapter 1, "Networking Models, Ports, Protocols, and Services," covers the OSI reference model in greater detail.

A router derives its name from the fact that it can route data it receives from one network to another. When a router receives a packet of data, it reads the packet's header to determine the destination address. After the router has determined the address, it looks in its routing table to determine whether it knows how to reach the destination; if it does, it forwards the packet to the next hop on the route. The next hop might be the final destination, or it might be another router. Figure 4.1 shows, in basic terms, how a router works.

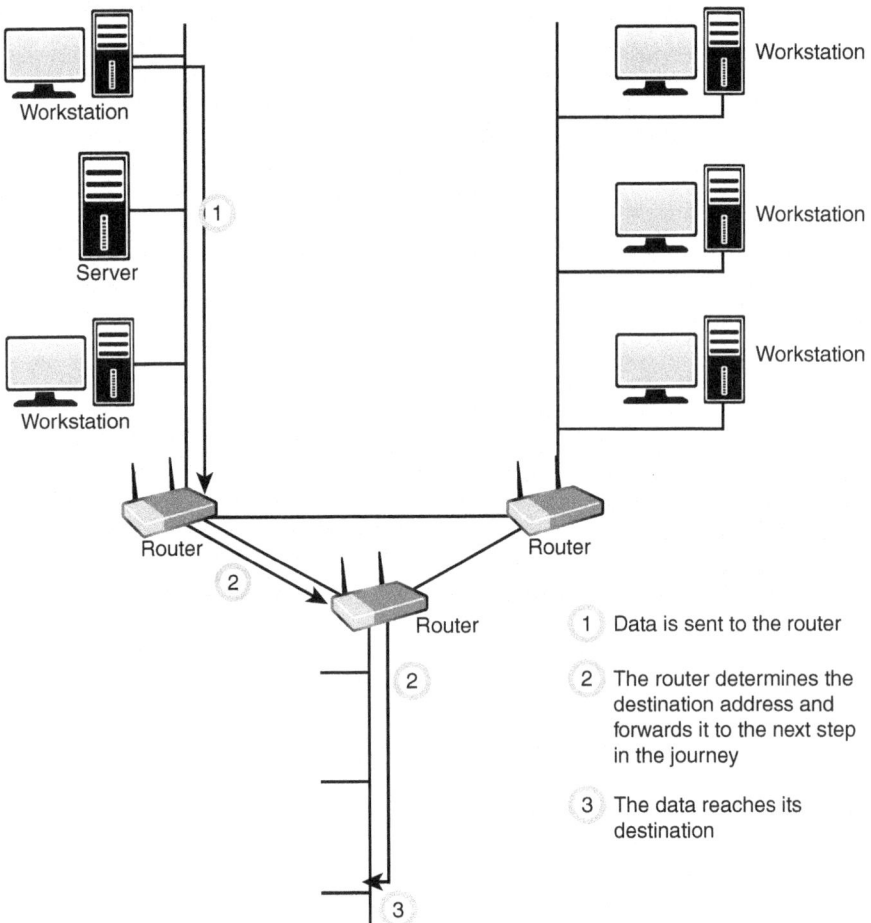

FIGURE 4.1 **How a Router Works**

> **Note**
>
> You can find more information on network routing in Chapter 3, "Network Addressing, Routing, and Switching."

A router works at Layer 3 (the network layer) of the OSI model.

ExamAlert

A router uses the destination IP address to forward packets.

Switch

Switches provide the connectivity points of an Ethernet network. Devices connect to switches via twisted-pair cabling, one cable for each device. The difference between hubs and switches is in how the devices deal with the data they receive. Whereas a hub forwards the data it receives to all the ports on the device, a switch forwards it to only the port that connects to the destination device. It does this by the MAC address of the devices attached to it and then by matching the destination MAC address in the data it receives. Figure 4.2 shows how a switch works. In this case, it has learned the MAC addresses of the devices attached to it; when the workstation sends a message intended for another workstation, it forwards the message on and ignores all the other workstations.

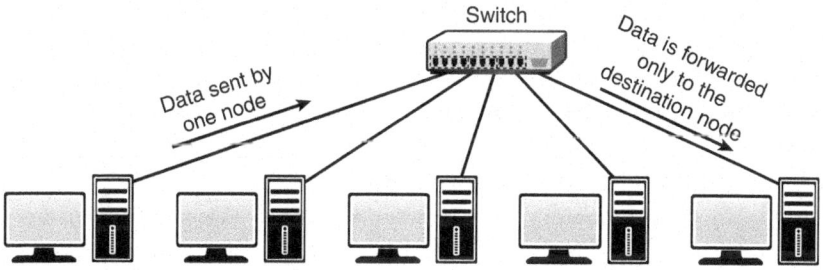

FIGURE 4.2   **How a Switch Works**

By forwarding data to only the connection that should receive it, the switch can greatly improve network performance. By creating a direct path between two devices and controlling their communication, the switch can greatly reduce the traffic on the network and therefore the number of collisions. As you might recall, collisions occur on Ethernet networks when two devices attempt to transmit at the same time. In addition, the lack of collisions enables switches to communicate with devices in full-duplex mode. In a full-duplex configuration, devices can send data to and receive data from the switch at the same time. Contrast this with half-duplex communication, in which communication can occur in only one direction at a time. Full-duplex transmission speeds are double that of a standard half-duplex connection. So, a 1000 Mbps connection becomes 2000 Mbps, and so on.

The net result of these measures is that switches can offer significant performance improvements over hub-based networks, particularly when network use is high.

Irrespective of whether a connection is at full or half duplex, the method of switching dictates how the switch deals with the data it receives. The following is a brief explanation of each method:

▶ **Cut-through:** In a cut-through switching environment, the packet begins to be forwarded as soon as it is received. This method is fast, but it creates the possibility of errors being propagated through the network because no error checking occurs.

▶ **Store-and-forward:** Unlike cut-through, in a store-and-forward switching environment, the entire packet is received and error-checked before being forwarded. The upside of this method is that errors are not propagated through the network. The downside is that the error-checking process takes a relatively long time, and store-and-forward switching is considerably slower as a result.

▶ **Fragment-free:** To take advantage of the error checking of store-and-forward switching but still offer performance levels nearing those of cut-through switching, fragment-free switching can be used. In a fragment-free switching environment, enough of the packet is read so that the switch can determine whether the packet has been involved in a collision. As soon as the collision status has been determined, the packet is forwarded.

Table 4.1 compares the various switching methods.

TABLE 4.1 **Comparing Switching Methods**

	Cut-through	**Store-and-forward**	**Fragment-free**
Description	Switch forwards a frame as soon as the destination address is read without waiting for the entire frame to be received	Switch receives the entire frame before forwarding it, verifies its integrity, and then transmits it	A variant of cut-through switching where the switch waits for the first 64 bytes of the frame (known as the frame's fragment) to arrive before forwarding
Characteristics	Low latency, ideal for high-speed networks	Provides error checking, ensures data integrity	Balances the trade-off between latency and error checking
Advantages	Minimal delay, suitable for real-time applications	More reliable data transmission, reduced chance of forwarding corrupted frames	Reduced latency compared to store-and-forward, provides basic error detection
Disadvantages	Higher risk of forwarding corrupted frames	Higher latency compared to cut-through switching	Not as thorough error checking as store-and-forward

Switch Cabling

In addition to acting as a connection point for network devices, switches can be connected to create larger networks. This connection can be achieved through standard ports with a special cable or by using special ports with a standard cable.

The ports on a switch or router to which computer systems are attached are called *medium-dependent interface crossed (MDI-X)*. The crossed designation is derived from the fact that two of the wires within the connection are crossed so that the send signal wire on one device becomes the receive signal of the other. Because the ports are crossed internally, a standard or straight-through cable can be used to connect devices.

Another type of port, called a *medium-dependent interface (MDI)* port, is often included on a hub or switch to facilitate the connection of two switches or hubs. Because the hubs or switches are designed to see each other as an extension of the network, there is no need for the signal to be crossed. If a hub or switch does not have an MDI port, hubs or switches can be connected by using a cable between two MDI-X ports. The crossover cable uncrosses the internal crossing. Auto MDI-X ports on more modern network device interfaces can detect whether the connection would require a crossover, and automatically choose the MDI or MDI-X configuration to properly match the other end of the link.

> **ExamAlert**
>
> In a crossover cable, wires 1 and 3 and wires 2 and 6 are crossed.

A switch can work at either Layer 2 (the data link layer) or Layer 3 (the network layer) of the OSI model. When it filters traffic based on the MAC address, it is called a Layer 2 switch because MAC addresses exist at Layer 2 of the OSI model (if it operated only with IP traffic, it would be a Layer 3 switch).

A multilayer switch is one that can operate at both Layer 2 and Layer 3 of the OSI model, which means that the multilayer device can operate as both a switch and a router (by operating at more than one layer, it is living up to the name of being "multilayer"). Also called a Layer 3 switch, the multilayer switch is a high-performance device that supports the same routing protocols that routers do. It is a regular switch directing traffic within the LAN; in addition, it can forward packets between subnets.

A content switch is another specialized device but is not as common on today's networks, mostly due to cost. It examines the network data it receives, decides where the content is intended to go, and forwards it. This switch can identify the application that data is targeted for by associating it with a port. For example, if data uses the Simple Mail Transfer Protocol (SMTP) port, it could be forwarded to an SMTP server.

Content servers can help with load balancing because they can distribute requests across servers and target data to only the servers that need it, or distribute data between application servers. For example, if multiple mail servers are used, the content switch can distribute requests between the servers, thereby sharing the load evenly. This is why the content switch is sometimes called a load-balancing switch.

ExamAlert

A content switch can distribute incoming data to specific application servers and help distribute the load.

As part of the troubleshooting process, a technician will often perform a **traceroute** from the client to the server, and also from the server to the client. If, while comparing the outputs, the administrator notices different hops between the hosts, a switch loop should be suspected. Simple network management protocols can be used to monitor and detect errors on switches and routers (above) and firewalls (below), and they are discussed elsewhere as they apply to the exam objectives being discussed.

Note

Chapter 10, "Network Troubleshooting," contains basic networking commands, but one you will want to know well is **show interface**. It allows you to view the status and statistics of the various interfaces on the switch, including the physical link status and the number of transmitted and received packets.

Firewall

A *firewall* is a networking device, either hardware or software based, that controls access to your organization's network. This controlled access is designed to protect data and resources from an outside threat. To provide this protection, firewalls typically are placed at a network's entry/exit points—for example, between an internal network and the Internet. After it is in place, a firewall can control access into and out of that point.

Although firewalls typically protect internal networks from public networks, they are also used to control access between specific network segments within a network. An example is placing a firewall between the Accounts and Sales departments.

As mentioned, firewalls can be implemented through software or through a dedicated hardware device. Organizations implement software firewalls through *network operating systems (NOSs)* such as Linux/UNIX, Windows servers, and macOS servers. The firewall is configured on the server to allow or block certain types of network traffic. In small offices and for regular home use, a firewall is commonly installed on the local system and is configured to control traffic. Many third-party firewalls are available.

Hardware firewalls are used in networks of all sizes today. Hardware firewalls are often dedicated network devices that can be implemented with little configuration. They protect all systems behind the firewall from outside sources. Hardware firewalls are readily available and often are combined with other devices. For example, many broadband routers and wireless access points have firewall functionality built in. In such a case, the router or AP might have a number of ports available to plug systems into. Figure 4.3 shows Windows Defender Firewall and the configured inbound and outbound rules.

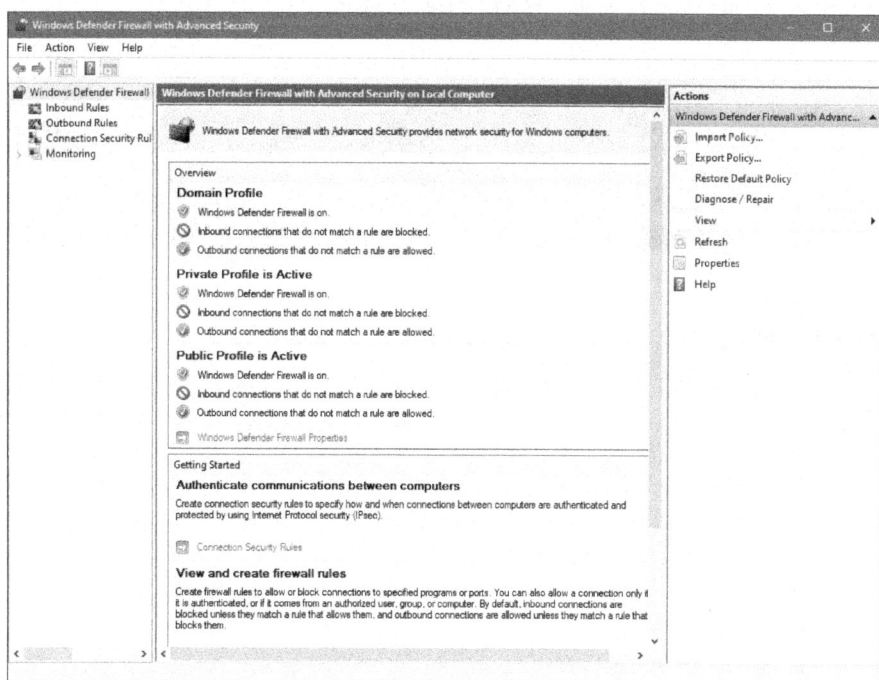

FIGURE 4.3 **Configuration of Windows Defender Firewall**

Some basic principles to keep in mind about firewalls are that they (along with encryption, and access control lists, or ACLs) are commonly used to protect confidentiality, and the last rule in the firewall access should always be implicit deny. Also, syslog server messages provide detailed information and are commonly used to determine if and how a firewall (or other network device) has been compromised. While syslog is an older standard for message logging, it is still commonly used and available on most network devices. Cloud providers often offer "security groups," which are essentially virtual firewalls that authorize traffic to and from the computing resources.

> **Note**
>
> Web application firewalls (WAFs) are sometimes referred to as deep packet inspection (DPI) firewalls because they can look at every request and response within web service layers.

> **Note**
>
> If a user cannot remotely access an internal company resource, you may have to open a port on the firewall. For example, if a user is using a remote desktop to access an internal system, you may need to permit TCP port 3389 through the firewall so that the RemoteApp server can be reached.

> **ExamAlert**
>
> Remember that a firewall uses inbound and outbound rules and can protect internal networks from public networks and control access between specific network segments.

IDS/IPS

An *intrusion detection system (IDS)* is a passive detection system. The IDS can detect the presence of an attack and then log that information. It also can alert an administrator to the potential threat. The administrator then analyzes the situation and takes corrective measures if needed.

A variation on the IDS is the *intrusion prevention system (IPS)*, which is an active detection system. With IPS, the device continually scans the network, looking for inappropriate activity. It can shut down any potential threats. The IPS looks for any known signatures of common attacks and automatically tries to prevent those attacks. An IPS is considered an active/reactive security measure because it actively monitors and can take steps to correct a potential security threat.

Following are several variations on IDSs/IPSs:

▶ **Behavior-based:** A *behavior-based system* looks for variations in behavior such as unusually high traffic, policy violations, and so on. By looking for deviations in behavior, it can recognize potential threats and quickly respond.

▶ **Signature-based:** A signature-based system, also commonly known as *misuse-detection system (MD-IDS/MD-IPS)*, is primarily focused on evaluating attacks based on attack signatures and audit trails. Attack signatures describe a generally established method of attacking a system. For example, a TCP flood attack begins with a large number of incomplete TCP sessions. If the MD-IDS knows what a TCP flood attack looks like, it can make an appropriate report or response to thwart the attack. This IDS uses an extensive database to determine the signature of the traffic.

▶ **Network-based intrusion detection/prevention system (NIDS or NIPS):** The system examines all network traffic to and from network systems. If it is software, it is installed on servers or other systems that can monitor inbound traffic. If it is hardware, it may be connected to a hub or switch to monitor traffic.

▶ **Host-based intrusion detection/prevention system (HIDS or HIPS):** These applications are spyware or virus applications that are installed on individual network systems. The system monitors and creates logs on the local system.

ExamAlert

An intrusion detection system (IDS) can detect malicious activity and send alerting messages, but it does not prevent attacks. An intrusion prevention system (IPS) protects hosts and prevents against malicious attacks from the network layer up through the application layer.

Table 4.2 compares the various intrusion detection/prevention methods.

TABLE 4.2 **Comparing Intrusion Detection/Prevention Methods**

	Behavior-based	Signature-based	Network-based	Host-based
Description	Analyzes the behavior of network traffic or system activity to detect anomalies or deviations from normal behavior	Matches patterns or signatures of known threats against network traffic or system logs	Monitors network traffic at the perimeter or within the internal network to identify malicious activity	Monitors activities and events on individual host systems to identify suspicious behavior
Characteristics	Monitors patterns and deviations from baseline behavior, often using machine learning algorithms	Relies on a database of predefined signatures, typically derived from known malware or attack patterns	Analyzes packet headers, payloads, or traffic flows to detect suspicious behavior	Examines system logs, file system changes, process activities, and registry modifications
Advantages	Effective at detecting previously unseen threats, adaptable to evolving attack methods	Highly accurate at identifying known threats, relatively low false positive rate	Provides visibility into networkwide threats and attacks, can detect unauthorized access or data exfiltration	Provides detailed insights into host-level activity, effective at detecting local threats and malware infections
Disadvantages	May generate false positives, requires ongoing tuning and refinement of detection algorithms	Limited effectiveness against zero-day attacks or polymorphic malware, requires frequent updates to signature databases	Limited visibility into encrypted traffic, may miss attacks that do not involve network traffic (e.g., insider threats)	Resource-intensive, may impact system performance, limited to protecting individual hosts

Load Balancer

Network servers are the workhorses of the network. They are relied on to hold and distribute data, maintain backups, secure network communications, and more. The load of servers is often a lot for a single server to maintain. This is where load balancing comes into play. *Load balancing* is a technique in which the workload is distributed among several servers. This feature can take networks to the next level; it increases network performance, reliability, and availability.

> **ExamAlert**
>
> Remember that load balancing increases redundancy and therefore data availability. Also, load balancing increases performance by distributing the workload.

A load balancer can be either a hardware device or software specially config-ured to balance the load. A virtual IP address (VIP) is the address clients use to access a cluster of servers, and that VIP refers to both the physical IP address and also to the logical load-balancer configuration. Quite often, reverse proxy servers are configured in a cluster to provide scalability and high availability for the network. Two popular load-balancing strategies are round-robin (evenly distributing requests among servers) and weighted round-robin (which allows administrators to adjust the allocation of requests based on server capabilities or priorities).

> **Note**
>
> Multilayer switches and DNS servers can serve as load balancers.

Proxy Server

Proxy servers typically are part of a firewall system. They have become so integrated with firewalls that the distinction between the two can sometimes be lost.

However, proxy servers perform a unique role in the network environment—a role that is separate from that of a firewall. For the purposes of this book, a proxy server is defined as a server that sits between a client computer and the Internet and looks at the web page requests the client sends. For example, if a client computer wants to access a web page, the request is sent to the proxy server rather than directly to the Internet. The proxy server first determines whether the request is intended for the Internet or for a web server locally. If the request is intended for the Internet, the proxy server sends the request *as if it originated the request*. When the Internet web server returns the informa-tion, the proxy server returns the information to the client. Although a delay might be induced by the extra step of going through the proxy server, the pro-cess is largely transparent to the client that originated the request. Because each request a client sends to the Internet is channeled through the proxy server, the proxy server can provide certain functionality over and above just forwarding requests.

One of the most notable extra features is that proxy servers can greatly improve network performance through a process called *caching*. When a caching proxy server answers a request for a web page, the server makes a copy of all or part of that page in its cache. Then, when the page is requested again, the proxy server answers the request from the cache rather than going back to the Internet. For example, if a client on a network requests the web page www.comptia.org, the proxy server can cache the contents of that web page. When a second client computer on the network attempts to access the same site, that client can grab it from the proxy server cache, and accessing the Internet is unnecessary. This feature greatly increases the response time to the client and can significantly reduce the bandwidth needed to fulfill client requests.

Nowadays, speed is everything, and the capability to quickly access information from the Internet is a crucial concern for some organizations. Proxy servers and their capability to cache web content accommodate this need for speed.

An example of this speed might be found in a classroom. If a teacher asks 30 students to access a specific *Uniform Resource Locator (URL)* without a proxy server, all 30 requests would be sent into cyberspace and subjected to delays or other issues that could arise. The classroom scene with a proxy server is quite different. Only one request of the 30 finds its way to the Internet; the other 29 are filled by the proxy server's cache. Web page retrieval can be almost instantaneous.

However, this caching has a potential drawback. When you log on to the Internet, you get the latest information, but this is not always so when information is retrieved from a cache. For some web pages, it is necessary to go directly to the Internet to ensure that the information is up to date. Some proxy servers can update and renew web pages, but they are always one step behind.

The second key feature of proxy servers is allowing network administrators to filter client requests. If a server administrator wants to block access to certain websites, a proxy server enables this control, making it easy to completely disallow access to some websites. This is okay, but what if it were necessary to block numerous websites? In this case, maintaining proxy servers gets a bit more complicated.

Determining which websites users can or cannot access is usually done through something called an *access control list (ACL)*. An ACL can be used to provide rules for which port numbers or IP addresses are allowed access. An ACL can also be a list of allowed or nonallowed websites; as you might imagine, compiling such a list can be a monumental task. Given that millions of websites exist, and new ones are created daily, how can you target and disallow access to the "questionable" ones? One approach is to reverse the situation and deny access to all pages except those that appear in an "allowed" list. This approach has high administrative overhead and can greatly limit the productive benefits available from Internet access.

Understandably, it is impossible to maintain a list that contains the locations of all sites with questionable content. In fairness, that is not what proxy servers were designed to do. However, by maintaining a list, proxy servers can better provide a greater level of control than an open system. Along the way, proxy servers can make the retrieval of web pages far more efficient.

A *reverse proxy server* is one that resides near the web servers and responds to requests. These are often used for load-balancing purposes because each proxy can cache information from a number of servers.

ExamAlert

Remember that a proxy server can be used if users within a corporate network need to connect to the Internet but corporate network policy does not allow direct connections.

Network-Attached Storage (NAS)

Storage is always a big issue, and the best answer is always a storage-area network (discussed next). Unfortunately, a SAN can be costly and difficult to implement and maintain. That is where *network-attached storage (NAS)* comes in. NAS is easier than SAN and uses TCP/IP. It offers file-level access, and a client sees the shared storage as a file server.

Note

On a VLAN, multipathing creates multiple paths to the storage resources and can be used to increase availability *and* add fault tolerance.

NAS systems typically consist of one or more hard drives organized into logical storage volumes, accessible over the network using standard file-sharing protocols such as Network File System (NFS) or Server Message Block (SMB). NAS devices come in various form factors, from standalone appliances to rack-mounted servers, with different capacities and features such as Redundant Array of Independent Disks (RAID) for data redundancy and scalability. Once the NAS is connected to the network, storage configuration involves creating logical volumes or disk arrays, configuring RAID levels for data protection and performance, and allocating storage space for different file shares or directories.

From an administration standpoint, it is important to monitor the NAS device for performance, capacity utilization, and potential issues using management tools provided by the NAS vendor. It is also important to perform routine maintenance tasks such as firmware updates, disk health checks, and capacity planning to ensure optimal NAS performance and reliability.

Storage-Area Networks

When it comes to data storage in the cloud, encryption is one of the best ways to protect it (keeping it from being of value to unauthorized parties), and virtual private network (VPN) routing and forwarding can help. Backups should be performed regularly (and encrypted and stored in safe locations), and access control should be a priority.

The consumer retains the ultimate responsibility for compliance. Per NIST SP 800-144:

> The main issue centers on the risks associated with moving important applications or data from within the confines of the organization's computing center to that of another organization (i.e., a public cloud), which is readily available for use by the general public. The responsibilities of both the organization and the cloud provider vary depending on the service model. Reducing cost and increasing efficiency are primary motivations for moving towards a public cloud, but relinquishing responsibility for security should not be. Ultimately, the organization is accountable for the choice of a public cloud and the security and privacy of the outsourced service.

For more information, see http://nvlpubs.nist.gov/nistpubs/Legacy/SP/nistspecialpublication800-144.pdf.

Storage can be shared on storage-area networks (SANs), network-attached storage (NAS), and so on; the virtual machine sees only a "physical disk." With clustered storage, you can use multiple devices to increase performance. A handful of technologies exist in this realm, and the following are those that you need to know for the Network+ exam.

> **Tip**
>
> Look to CompTIA's Cloud+ certification for more specialization in cloud and virtualization technologies.

iSCSI

The *Small Computer Systems Interface (SCSI)* standard has long been the language of storage. The *Internet Small Computer Systems Interface (iSCSI)* expands this through Ethernet, allowing IP to be used to send SCSI commands.

Logical unit numbers (LUNs) came from the SCSI world and carry over, acting as unique identifiers for devices. Both NAS and SAN use "targets" that hold up to eight devices.

Using iSCSI for a virtual environment gives users the benefits of a file system without the difficulty of setting up Fibre Channel. Because iSCSI works both at the hypervisor level and in the guest operating system, the rules that govern the size of the partition in the OS are used rather than those of the virtual OS (which are usually more restrictive).

The disadvantage of iSCSI is that users can run into IP-related problems if configuration is not carefully monitored.

Fibre Channel and FCoE

Instead of using an older technology and trying to adhere to legacy standards, *Fibre Channel (FC)* is an option providing a higher level of performance than anything else. It utilizes FCP, the Fibre Channel Protocol, to do what needs to be done, and *Fibre Channel over Ethernet (FCoE)* can be used in high-speed (100 GB and higher) implementations, though most implementations today are not that fast.

The big advantage of Fibre Channel is its scalability. FCoE encapsulates FC over the Ethernet portions of connectivity, making it easy to add into an existing network. As such, FCoE is an extension to FC intended to extend the scalability and efficiency associated with Fibre Channel.

> **ExamAlert**
>
> Know that FCoE allows Fibre Channel to use 100 Gigabit Ethernet (or even higher) networks. This solves the problem of enterprises having to run parallel infrastructures for both LANs and SANs.

> **ExamAlert**
>
> For the exam, you should know the difference between NAS and SAN technologies and how to apply them.

Wireless Access Point

The term *access point (AP)* technically can be used for either a wired or wireless connection, but in reality it is almost always associated only with a wireless-enabling device. A *wireless access point (WAP)* is a transmitter and receiver (transceiver) device used to create a *wireless LAN (WLAN)*. WAPs typically are separate network devices with a built-in antenna, transmitter, and adapter. WAPs use the wireless infrastructure network mode to provide a connection point between WLANs and a wired Ethernet LAN. WAPs also usually have several ports, giving you a way to expand the network to support additional clients.

Depending on the size of the network, one or more WAPs might be required. Additional WAPs are used to allow access to more wireless clients and to expand the range of the wireless network. Each WAP is limited by a transmission range—the distance a client can be from a WAP and still obtain a usable signal. The actual distance depends on the wireless standard used and the obstructions and environmental conditions between the client and the WAP.

> **ExamAlert**
>
> An AP or WAP can operate as a bridge connecting a standard wired network to wireless devices or as a router passing data transmissions from one access point to another.

Saying that a WAP is used to extend a wired LAN to wireless clients does not give you the complete picture. A wireless AP today can provide different services in addition to just an access point. Today, the APs might provide many ports that can be used to easily increase the network's size. Systems can be added to and removed from the network with no effect on other systems on the network. Also, many APs provide firewall capabilities and *Dynamic Host Configuration Protocol (DHCP)* service. When they are hooked up, they give client systems a private IP address and then prevent Internet traffic from accessing those systems. So, in effect, the AP is a switch, DHCP server, router, and firewall.

APs come in all shapes and sizes. Many are cheaper and are designed strictly for home or small office use. Such APs have low-powered antennas and limited expansion ports. Higher-end APs used for commercial purposes have high-powered antennas, enabling them to extend how far the wireless signal can travel.

> **Note**
>
> Antenna placement and types are covered in Chapter 9, "Network Security," and Chapter 10, "Network Troubleshooting."

> **Note**
>
> APs are used to create a wireless LAN and to extend a wired network. APs are an integral component of the infrastructure wireless topology.

An AP works at Layer 2 (the data link layer) of the OSI model.

Wireless LAN Controller

Wireless LAN controllers, or just *controllers*, are often used with branch/remote office deployments for wireless authentication. When an AP boots, it authenticates with a controller before it can start working as an AP. This is often used with *VLAN pooling*, in which multiple interfaces are treated as a single entity (usually for load balancing).

Applications/Content Delivery Network

Just as you can spread storage around the network, so too is it possible to do with applications and content. A *content delivery network (CDN)* is a geographically distributed network of servers and datacenters that work together to deliver web content, such as images, videos, CSS files, JavaScript, and other static or dynamic content, to users more efficiently and reliably. CDNs are designed to reduce latency, improve website performance, and enhance the user experience by caching content closer to the end users and serving it from the nearest edge server.

A CDN typically involves

▶ **Content replication:** When a website or web application is integrated with a CDN, copies of its static content are replicated and stored on multiple servers distributed across various geographic locations (cached).

▶ **Request routing:** When a user requests content from the website, that request is directed to the nearest edge server in the CDN network based on factors such as the user's location, network latency, and server availability. This routing ensures that content is delivered from the closest server to the user, minimizing latency and reducing the time it takes for the content to load.

▶ **Content delivery:** The edge server that receives the user's request serves the requested content directly to the user's device. Since the content is cached locally on the edge server, it can be delivered quickly without having to retrieve it from the origin server where the website is hosted.

▶ **Dynamic content acceleration:** In addition to caching static content, some CDNs also offer features for accelerating the delivery of dynamic content, such as personalized web pages, e-commerce transactions, or real-time data. These features may involve techniques like intelligent caching, load balancing, and route optimization to ensure fast and reliable delivery of dynamic content.

▶ **Load balancing and failover:** CDNs often include load-balancing and failover mechanisms to distribute incoming traffic across multiple servers and datacenters and ensure high availability and reliability. If one server or datacenter becomes unavailable, requests are automatically rerouted to alternate locations to minimize downtime and disruption.

▶ **Security:** Many CDNs also offer security features such as distributed denial-of-service (DDoS) protection, SSL/TLS encryption, and web application firewalls to protect websites and web applications from malicious attacks and unauthorized access.

Overall, a CDN enhances website performance, scalability, and reliability by caching content closer to end users, reducing latency, and improving the overall user experience. Organizations of all sizes use it widely to accelerate content delivery, optimize bandwidth usage, and improve website availability and security.

> **ExamAlert**
>
> Know that a content delivery network (CDN) can improve the performance, reliability, and scalability of delivering web content to users by caching and distributing content across multiple servers located in different geographic locations. This helps reduce latency, improves load times, and enhances the user experience.

VPNs

A *virtual private network (VPN)* encapsulates encrypted data inside another datagram that contains routing information. The connection between two computers establishes a switched connection dedicated to the two computers. The encrypted data is encapsulated inside *Point-to-Point Protocol (PPP)*, and that connection is used to deliver the data.

A VPN enables users with an Internet connection to use the infrastructure of the public network to connect to the main network and access resources as if they were logged on to the network locally. It also enables two networks to be connected to each other securely.

To put it more simply, a VPN extends a LAN by establishing a remote connection using a public network such as the Internet. A VPN provides a point-to-point dedicated link between two points over a public IP network. For many companies, the VPN link provides the perfect method to expand their networking capabilities and reduce their costs. By using the public network (Internet), a company does not need to rely on expensive private leased lines to provide corporate network access to its remote users. Using the Internet to facilitate the remote connection, the VPN enables network connectivity over a possibly long physical distance. In this respect, a VPN is a form of wide-area network (WAN).

> **Note**
>
> Many companies use a VPN to provide a cost-effective method to establish a connection between remote clients and a private network. There are other times a VPN link is handy. You can also use a VPN to connect one private LAN to another, known as LAN-to-LAN internetworking. For security reasons, you can use a VPN to provide controlled access within an intranet. As an exercise, try drawing what the VPN would look like in these two scenarios.

Components of the VPN Connection

A VPN enables anyone with an Internet connection to use the infrastructure of the public network to dial in to the main network and access resources as if the user were locally logged on to the network. It also enables two networks to securely connect to each other.

Many elements are involved in establishing a VPN connection, including the following:

▶ **VPN client:** This computer initiates the connection to the VPN server.

▶ **VPN server:** This server authenticates connections from VPN clients.

▶ **VPN headend:** The headend is the termination point for the VPN tunnels.

▶ **VPN concentrator:** The concentrator is a dedicated network device that provides secure connections between remote users and a company network. VPN concentrators tend to be enterprise-grade devices capable of handling a large number of simultaneous Internet connections.

▶ **Access method:** As mentioned, a VPN is most often established over a public network such as the Internet; however, some VPN implementations use a private intranet. The network used must be IP-based.

▶ **VPN protocols:** These protocols are required to establish, manage, and secure data over the VPN connection. Point-to-Point Tunneling Protocol (PPTP) and Layer 2 Tunneling Protocol (L2TP) are commonly associated with VPN connections. These protocols enable authentication and encryption in VPNs. Authentication enables VPN clients and servers to correctly establish the identity of people on the network. Encryption enables potentially sensitive data to be guarded from the general public.

VPNs have become popular because they enable the public Internet to be safely used as a WAN connectivity solution.

> **ExamAlert**
>
> VPNs support analog modems, Integrated Services Digital Network (ISDN) wireless connections, and dedicated broadband connections, such as cable and digital subscriber lines (DSL). Remember these details for the exam.

VPN Connection Types

Two types of client-to-site VPN connections are possible: full tunnel and split tunnel. When full tunnel is used, all requests through the VPN are encrypted regardless of where the service is hosted, and it is not possible to access local network resources. With split tunnel, all incoming requests are encrypted over the VPN, but traffic going to sites outside the client network (including Zoom, Office 365, and Google) does not go through the VPN server. Full tunnel is more secure and generally recommended, but know that the split tunnel allows for more efficient use of bandwidth and reduces the load on the VPN server, allowing the client to continue accessing local resources while connected to the remote network.

When a choice between the two exists, it is recommended that you use full tunnel if you are connecting from an untrusted network such as in a coffee shop or hotel. While split tunnel may be necessary if there is a need to access both local resources and the organization's resources, you should always recognize that it is less secure, so a better option is to disconnect from the VPN first before accessing those other sites.

VPN Pros and Cons

As with any technology, VPN has both pros and cons. Fortunately with VPN technology, these are clear cut, and even the cons typically do not prevent an organization from using VPNs in its networks. Using a VPN offers two primary benefits:

▶ **Cost:** If you use the infrastructure of the Internet, you do not need to spend money on dedicated private connections to link remote clients to the private network. Furthermore, when you use the public network, you do not need to hire support personnel to support those private links.

▶ **Easy scalability:** VPNs make it easy to expand the network. Employees who have a laptop with wireless capability can simply log on to the Internet and establish the connection to the private network.

Table 4.3 outlines some of the advantages and potential disadvantages of using a VPN.

TABLE 4.3 **Pros and Cons of Using a VPN**

Advantage	Description
Reduced cost	When you use the Internet, you do not need to rent dedicated lines between remote clients and a private network. In addition, a VPN can replace remote-access servers and long-distance dial-up network connections that were commonly used in the past by business travelers who needed access to their company intranet. This way, you eliminate long-distance phone charges.
Network scalability	The cost to an organization to build a dedicated private network may be reasonable at first, but it increases exponentially as the organization grows. The Internet enables an organization to grow its remote client base without having to increase or modify an internal network infrastructure.
Reduced support	Using the Internet, organizations do not need to employ support personnel to manage a VPN infrastructure.
Simplicity	With a VPN, a network administrator can easily add remote clients. All authentication work is managed from the VPN authentication server, and client systems can be easily configured for automatic VPN access.
Disadvantage	Description
Security	Using a VPN, data is sent over a public network, so data security is a concern. VPNs use security protocols to address this shortcoming, but VPN administrators must understand data security over public networks to ensure that data is not tampered with or stolen.
Reliability	The reliability of the VPN communication depends on the public network and is not under an organization's direct control. Instead, the solution relies on an Internet service provider (ISP) and its Quality of Service (QoS).

A site-to-site VPN can be used to connect two datacenters securely in different regions, whereas a client-to-site VPN is best for a user in a remote location who needs to connect to the corporate LAN.

IPsec

The *IP Security (IPsec)* protocol is designed to provide secure communications between systems. This includes system-to-system communication in the same network, as well as communication to systems on external networks. IPsec is an IP layer security protocol that can both encrypt and authenticate network transmissions. In a nutshell, IPsec is composed of two separate protocols: *Authentication Header (AH)* and *Encapsulating Security Payload (ESP)*. AH provides the authentication and integrity checking for data packets, and ESP provides encryption services.

> **Note**
>
> Next-generation firewalls (NGFWs) can provide content filtering and threat protection and also manage multiple IPsec site-to-site connections.

> **ExamAlert**
>
> IPsec relies on two underlying protocols: AH and ESP. AH provides authentication services, and ESP provides encryption services.

By using both AH and ESP, data traveling between systems can be secured, ensuring that transmissions cannot be viewed, accessed, or modified by those who should not have access to them. It might seem that protection on an internal network is less necessary than on an external network; however, much of the data you send across networks has little or no protection, allowing unwanted eyes to see it.

> **Note**
>
> The Internet Engineering Task Force (IETF) created IPsec, which you can use on both IPv4 and IPv6 networks.

IPsec provides three key security services:

▶ **Data verification:** Verifies that the data received is from the intended source

▶ **Protection from data tampering:** Ensures that the data has not been tampered with or changed between the sending and receiving devices

▶ **Private transactions:** Ensures that the data sent between the sending and receiving devices is unreadable by any other devices

IPsec operates at the network layer of the Open Systems Interconnection (OSI) reference model and provides security for protocols that operate at the higher layers. Thus, by using IPsec, you can secure practically all TCP/IP-related communications.

SSL/TLS/DTLS

Security is often provided by working with the Transport Layer Security (TLS) protocol (which has replaced Secure Sockets Layer in most implementations). SSL VPN, also marketed as WebVPN and OpenVPN, can be used to connect locations that would run into trouble with firewalls and NAT when used with IPsec. It is known as an SSL VPN whether the encryption is done with TLS or truly with the older SSL.

> **Note**
>
> SSL was first created for use with the Netscape web browser and is used with a limited number of TCP/IP protocols (such as HTTP and FTP). TLS is not only an enhancement to SSL but also a replacement for it, working with almost every TCP/IP protocol. Because of this, TLS is popular with VPNs and VoIP applications. Just as the term *Kleenex* is often used to represent any paper tissue, whether or not it is made by Kimberly-Clark, *SSL* is often the term used to signify the confidentiality function, whether it is actually SSL in use or TLS (the latest version of which is 1.3).

The Datagram Transport Layer Security (DTLS) protocol is a derivation of SSL/TLS by the OpenSSL project; it provides the same security services but strives to increase reliability.

The National Institute of Standards and Technology (NIST) publishes the *Guide to SSL VPNs*, which you can access at https://nvlpubs.nist.gov/nistpubs/ Legacy/SP/nistspecialpublication800-113.pdf.

Quality of Service

Quality of Service (QoS) describes the strategies used to manage and increase the flow of network traffic. QoS features enable administrators to predict bandwidth use, monitor that use, and control it to ensure that bandwidth is available to the applications that need it. These applications generally can be broken into two categories:

▶ **Latency sensitive:** These applications need bandwidth for quick delivery where network lag time impacts their effectiveness. This includes voice and video transfer. For example, *voice over IP (VoIP)* would be difficult to use if there were a significant lag time in the conversation.

▶ **Latency insensitive:** Controlling bandwidth also involves managing latency-insensitive applications. This includes bulk data transfers such as huge backup procedures and *File Transfer Protocol Secure (FTPS)* transfers.

With bandwidth limited, and networks becoming increasingly congested, it becomes more difficult to deliver latency-sensitive traffic. If network traffic continues to increase and you cannot always increase bandwidth, the choice is to prioritize traffic to ensure timely delivery. This is where QoS comes into play. QoS ensures the delivery of applications, such as video conferencing (and related video applications), VoIP telephony, and unified communications without adversely affecting network throughput. QoS achieves more efficient use of network resources by differentiating between latency-insensitive traffic such as email data and latency-sensitive streaming media.

> **Tip**
>
> Jitter and QoS misconfiguration often affect VoIP service. This can often manifest itself as the caller hearing echoes of the receiver's voice.

Two important components of QoS are DSCP and CoS. *Differentiated services code point (also known as Diffserv)* is an architecture that specifies a simple and coarse-grained mechanism for classifying and managing network traffic and providing QoS on modern networks. *Class of service (CoS)* is a parameter that is used in data and voice to differentiate the types of payloads being transmitted.

One important strategy for QoS is priority queuing. Essentially, traffic is placed in order based on its importance of delivery time. All data is given access, but the more important and latency-sensitive data is given higher priority.

> **ExamAlert**
>
> Be sure that you understand QoS and the methods used to ensure QoS on networks. Know that it is used with high-bandwidth applications such as VoIP, video applications, and unified communications.

Time To Live (TTL)

Time to live (TTL) is a value assigned to data packets in a network that specifies the maximum amount of time or number of hops (routers or network devices through which the packet can pass) that the packet is allowed to remain active or travel before it is discarded. In other words, it specifies the maximum lifespan before the packet is discarded.

The TTL field is typically found in the IP header of a packet and is used primarily for two purposes:

▶ **Preventing routing loops:** TTL helps prevent routing loops by ensuring that packets do not circulate indefinitely in the network. Each time a router forwards a packet, it decrements the TTL value by one. If the TTL value reaches zero before the packet reaches its destination, the router discards the packet and sends an ICMP (Internet Control Message Protocol) Time Exceeded message back to the sender.

▶ **Determining packet lifetime:** TTL can also be used to estimate the round-trip time (RTT) of packets and determine their approximate lifetime in the network. By examining the TTL value in the ICMP Time Exceeded message returned by routers, network administrators can gauge the number of hops and the approximate path that packets take through the network.

TTL is discussed in Chapter 3 in relation to routing, but here are some key points to know about it:

▶ TTL is measured in seconds or hops, depending on the implementation and network configuration.

▶ The initial TTL value is set by the sender of the packet and typically varies between different operating systems and network devices.

▶ The TTL value is decremented by one each time a packet traverses a router or network device.

▶ If the TTL value reaches zero, the packet is considered expired and is discarded by the router.

▶ The TTL field helps ensure the efficient and reliable delivery of packets in computer networks and plays a crucial role in preventing network congestion and routing anomalies.

One tricky aspect to understand about TTL is that it is not a measure of time in the conventional sense, despite the term *time* in its name. Instead, TTL is measured in terms of the number of hops that a packet can traverse before it is discarded.

Cram Quiz

1. Users are complaining that the network's performance is unsatisfactory. It takes a long time to pull files from the server, and, under heavy loads, workstations can become disconnected from the server. The network is heavily used, and a new video-conferencing application is about to be installed. The network is a 1000BASE-T system created with Ethernet hubs. Which device are you most likely to install to alleviate the performance problems?

   - ○ **A.** Switch
   - ○ **B.** Router
   - ○ **C.** NAS
   - ○ **D.** Firewall

2. Which of the following devices passes data based on the MAC address?

   - ○ **A.** Hub
   - ○ **B.** Switch
   - ○ **C.** MSAU
   - ○ **D.** Router

3. Which of the following can serve as load balancers?

   - ○ **A.** IDS and DNS servers
   - ○ **B.** Multilayer switches and IPS
   - ○ **C.** Multilayer switches and DNS servers
   - ○ **D.** VoIP PBXs and UTM appliances

4. Which of the following is the best answer for a device that continually scans the network, looking for inappropriate activity?

   - ○ **A.** IPS
   - ○ **B.** NGFW
   - ○ **C.** VCPN
   - ○ **D.** AAA

5. An administrator needs to secure two datacenters in different regions. Which would be the best option?

   ○ **A.** Site-to-site VPN

   ○ **B.** Client-to-site VPN

   ○ **C.** A WAP

   ○ **D.** A WLAN

Cram Quiz Answers

1. **A.** Replacing Ethernet hubs with switches can yield significant performance improvements. Of the devices listed, switches are also the only ones that can be substituted for hubs. A router is used to separate networks, not as a connectivity point for workstations. Network-attached storage (NAS) systems typically consist of one or more hard drives organized into logical storage volumes, accessible over the network using standard file-sharing protocols such as Network File System (NFS) or Server Message Block (SMB). NAS devices come in various form factors, from standalone appliances to rack-mounted servers, with different capacities and features such as Redundant Array of Independent Disks (RAID) for data redundancy and scalability. A firewall is not a solution to the problem presented.

2. **B.** When determining the destination for a data packet, the switch learns the MAC address of all devices attached to it and then matches the destination MAC address in the data it receives. None of the other devices listed pass data based solely on the MAC address.

3. **C.** Multilayer switches and DNS servers can serve as load balancers.

4. **A.** An intrusion prevention system (IPS) is a device that continually scans the network, looking for inappropriate activity.

5. **A.** If you need to connect two datacenters securely in different regions, a site-to-site VPN is the best option. If a user in a remote location needs to connect to the corporate LAN, a client-to-site VPN is the best way to go. A wireless access point (WAP) is a transmitter and receiver (transceiver) device used to create a wireless LAN (WLAN). WAPs typically are separate network devices with a built-in antenna, transmitter, and adapter.

Networking Use Cases

▶ **1.8 Summarize evolving use cases for modern network environments.**

CramSaver

If you can correctly answer these questions before going through this section, save time by skimming the ExamAlerts in this section and then complete the Cram Quiz at the end of the section.

1. True or false: VXLAN is a network virtualization technology designed to address the scalability limitations of traditional VLANs by allowing the creation of logical Layer 2 networks over Layer 3 infrastructure.

2. What architecture emphasizes a granular approach to access control, where access decisions are made based on contextual factors such as user identity, device posture, location, and behavior. Implementing fine-grained access controls requires a comprehensive understanding of the organization's network, applications, and user requirements?

3. What allows network environments to be defined and managed using code, typically in configuration files or scripts?

Answers

1. True. VXLAN is a network virtualization technology designed to address the scalability limitations of traditional VLANs by allowing the creation of logical Layer 2 networks over Layer 3 infrastructure.

2. Zero trust architecture emphasizes a granular approach to access control, where access decisions are made based on contextual factors such as user identity, device posture, location, and behavior. Implementing fine-grained access controls requires a comprehensive understanding of the organization's network, applications, and user requirements.

3. Infrastructure as Code (IaC) allows network environments to be defined and managed using code, typically in configuration files or scripts.

The networking devices discussed previously in this chapter are used to build networks. For this particular objective, CompTIA wants you to be aware of how to implement a modern network using some of those devices and incorporating current technologies. Whether you're putting together a datacenter or a corporate office, a great deal of planning should always be involved, and no network should be allowed to haphazardly sprout without management and oversight.

Software-Defined Networking

Software-defined networking (SDN) is a dynamic approach to computer networking intended to allow administrators to get around the static limitations of physical architecture associated with traditional networks. They can do so through the implementation of technologies such as the Cisco Systems Open Network Environment.

The goal of SDN is not only to add dynamic capabilities to the network but also to reduce IT costs through implementation of cloud architectures. SDN combines network and application services into centralized platforms that can automate provisioning and configuration of the entire infrastructure.

The SDN architecture, from the top down, consists of the application layer, control layer, and infrastructure layer. CompTIA also adds the management plane as an objective, and a discussion of each of these components follows.

Application Layer

The *application layer* is the top of the SDN stack, and this is where load balancers, firewalls, intrusion detection, and other standard network applications are located. While a standard (non-SDN) network would use a specialized appliance for each of these functions, with an SDN network, an application is used in place of a physical appliance.

Control Layer

The *control layer* is the place where the SDN controller resides; the controller is software that manages policies and the flow of traffic throughout the network. This controller can be thought of as the brains behind SDN, making it all possible. Applications communicate with the controller through a northbound interface, and the controller communicates with switching using southbound interfaces.

Infrastructure Layer

The physical switch devices themselves reside at the *infrastructure layer*. This is also known as the control plane when breaking the architecture into "planes" because this is the component that defines the traffic routing and network topology.

Management Plane

With SDN, the management plane allows administrators to see their devices and traffic flows and react as needed to manage data plane behavior. This can be done automatically through configuration apps that can, for example, add more bandwidth if it looks as if edge components are getting congested. The management plane manages and monitors processes across all layers of the network stack.

> **ExamAlert**
>
> A major benefit of SDN is that it replaces traditional, dedicated hardware services with virtual infrastructure and software-based solutions.

SDNs and the Software-Defined Wide-Area Network (SD-WAN)

A *software-defined wide-area network (SD-WAN)* is an extension of the SDN that is commonly used in telco and datacenters on a large scale. The concept behind it is to take many of the principles that make cloud computing so attractive and make them accessible at the WAN level. This is done by adopting a virtual WAN architecture leveraging a combination of transport services (MPLS, 5G, LTE, broadband, and so on) to connect users to applications.

Moving away from the router-centric WAN architecture that has always been used, SD-WANs support applications hosted pretty much anywhere: public or private clouds or on-premises datacenters. As with an SDN, an SD-WAN enables services on demand, reduces operational costs, and is intended to improve network scalability and performance.

The SD-WAN evolved from Multiprotocol Label Switching (MPLS) technology and implements a centralized controller for setting and maintaining policies—managing the implementation. The ability of the SD-WAN solution to intelligently utilize multiple types of network transport technologies, such as MPLS, broadband Internet, Long-Term Evolution (LTE), 5G, and even satellite links, to efficiently transport data traffic across the wide-area network means that the solution is *transport agnostic*—possessing the intelligence to dynamically select the best path for each application's traffic based on real-time network conditions, performance metrics, and application requirements (which can include such factors as latency, packet loss, jitter, available bandwidth, and link reliability). Hospitals, for example, can use 5G SD-WAN architecture to guarantee reliable service to life saving equipment, provide secure access to clinical apps, and deploy IoT sensors across the entire campus.

> **Tip**
>
> A software-defined wide-area network (SD-WAN) is a virtual WAN architecture that uses software to manage connectivity, devices, and services and can make changes in the network based on current operations. MPLS is a switching mechanism that imposes labels (numbers) to data and then uses those labels to forward data when it arrives at the MPLS network.

> **ExamAlert**
>
> For the exam, think of SD-WAN as a software abstraction of MPLS technology.

An SDN is a dynamic approach to computer networking intended to allow administrators to get around the static limitations of physical architecture associated with traditional networks. The goal of SDN is to not only add dynamic capabilities to the network but to also reduce IT costs through implementation of cloud architectures. SDN combines network and application services into centralized platforms that can automate provisioning and configuration of the entire infrastructure.

An SD-WAN is considered "application aware" because it has the ability to identify, classify, and prioritize network traffic based on the specific applications or services being used. This application-awareness is a key feature of SD-WAN technology and enables several benefits for performance, security, and optimization:

- ▶ **Deep packet inspection (DPI):** SD-WAN solutions use DPI to analyze the contents of network packets at the application layer (Layer 7 of the OSI model). By inspecting packet headers and payload data, SD-WAN devices can identify the applications or services generating the traffic.

- ▶ **Application recognition:** Once packets are inspected, SD-WAN devices can use built-in application recognition algorithms and signatures to identify known applications. This allows SD-WAN controllers to categorize traffic flows based on the specific applications or services being used, such as web browsing, video conferencing, voice over IP (VoIP), file transfers, or cloud applications.

- ▶ **Dynamic policies and QoS:** With knowledge of the applications traversing the network, SD-WAN controllers can dynamically apply policies and Quality of Service (QoS) settings to prioritize critical applications and ensure optimal performance. For example, real-time applications like VoIP or video conferencing can be given higher priority to minimize latency and packet loss, while less critical applications may be allocated lower bandwidth.

▶ **Path selection and routing optimization:** SD-WAN solutions leverage application awareness to make intelligent routing decisions based on the performance requirements of different applications. By considering factors such as latency, jitter, packet loss, and available bandwidth, SD-WAN controllers can dynamically select the best path for each application's traffic, whether it's utilizing MPLS, broadband Internet, LTE, or other network links.

▶ **Traffic steering and load balancing:** Application-aware SD-WAN solutions can steer traffic across multiple WAN links based on the specific requirements of each application. This includes load balancing traffic across multiple links to optimize bandwidth utilization and ensure reliable connectivity for mission-critical applications.

▶ **Security and access control:** Application-awareness enables SD-WAN solutions to enforce security policies and access controls based on the type of application traffic. For example, traffic to sanctioned applications can be allowed, while traffic to unauthorized or risky applications can be blocked or redirected for inspection.

The application-awareness of SD-WAN technology allows organizations to gain granular visibility and control over network traffic, optimize application performance, enhance user experience, and improve overall network efficiency and security by intelligently identifying and prioritizing applications.

ExamAlert

For the exam, know that the application-awareness of SD-WAN technology allows organizations to optimize application performance, gain granular visibility and control, and improve network efficiency and security.

One key feature of the SD-WAN is *zero-touch provisioning*—the ability to automate deployment and configuration of SD-WAN devices with minimal manual intervention. It enables organizations to rapidly deploy and scale their SD-WAN infrastructure across distributed locations without the need for on-site technicians or extensive manual configuration. Devices—such as edge routers or appliances—can be shipped directly from the vendor or distributor to the deployment site (branch office or remote locations), and upon booting up, the SD-WAN device automatically initiates a process to discover and connect to the SD-WAN orchestration or controller platform (this can involve query-ing for network configuration information or using preconfigured discovery mechanisms, such as DHCP options or DNS records). Once the device estab-lishes connectivity to the controller, it downloads its initial configuration and policy settings from the centralized management platform (network topology information, security policies, QoS settings, routing parameters, and so on), and the controller orchestrates the provisioning process by pushing configura-tion templates and policies to the new device based on predefined templates or profiles. Before activating the device in the production network, the control-ler can perform validation checks to ensure the device is properly configured and compliant with organizational policies. Once provisioned, the SD-WAN device becomes operational and begins forwarding traffic according to the con-figured policies. The SD-WAN controller continues to monitor the device's performance, health, and compliance status, providing centralized visibility and management.

Central policy management is a crucial aspect of SD-WANs, enabling organizations to define, enforce, and manage network policies consistently across distributed network environments from a centralized location. Here are some reasons why this is key:

▶ **Consistency and uniformity:** Centralized policy management ensures that network policies, including QoS, security, routing, and application performance, are defined consistently and uniformly across all SD-WAN devices deployed throughout the network. This helps maintain opera-tional consistency and simplifies management tasks, reducing the risk of misconfigurations and ensuring compliance with organizational policies.

▶ **Simplified configuration:** With central policy management, network administrators can define and configure network policies using a single, centralized interface or management console. This simplifies the con-figuration process and eliminates the need to manually configure policies on individual devices, reducing the potential for errors and speeding up deployment times.

▶ **Granular control and visibility:** Centralized policy management provides granular control and visibility into network traffic, allowing administrators to define policies based on specific criteria such as applica-tion type, user identity, device type, and network location. This enables fine-tuning of policies to meet the unique requirements of different applications, users, and business units while providing comprehensive visibility into network traffic and performance.

▶ **Dynamic policy enforcement:** SD-WAN solutions with central policy management capabilities can dynamically enforce policies based on real-time network conditions, application requirements, and security threats. Policies can be adjusted dynamically to optimize network performance, prioritize critical applications, and mitigate security risks, ensuring consistent and reliable network performance across the organization.

▶ **Scalability and flexibility:** Central policy management scales efficiently to support large and geographically distributed network environments. It enables organizations to define and manage policies centrally while deploying SD-WAN devices across multiple branch offices, remote locations, and cloud environments. This scalability and flexibility allow organizations to adapt quickly to changing business requirements and network conditions without sacrificing management simplicity or control.

▶ **Policy orchestration and automation:** Central policy management enables policy orchestration and automation, allowing organizations to automate policy deployment, updates, and enforcement processes across the SD-WAN infrastructure. Automation reduces manual intervention, speeds up policy changes, and ensures consistency and compliance across the network.

Virtual Extensible Local-Area Network (VXLAN)

A *virtual extensible local-area network (VXLAN)* is an extension of a VLAN that utilizes network virtualization technology in such a way as to overcome the scalability limitations of traditional VLANs by allowing the creation of logical Layer 2 networks over Layer 3. The VXLAN first creates an overlay network on top of an existing IP network infrastructure, allowing virtual networks to be established across physical network boundaries. It then uses a 24-bit virtual network identifier (VNI) to identify and segregate virtual networks (each VXLAN segment is associated with a unique VNI, enabling network segmentation and multitenancy within the same physical infrastructure).

VXLAN tunnel endpoints (VTEPs) are responsible for encapsulating and decapsulating VXLAN packets. They reside on network devices such as switches, routers, or hypervisors and handle the encapsulation and forwarding of VXLAN traffic. When a host in a segment sends a Layer 2 Ethernet frame, the frame is encapsulated within a VXLAN header (known as *Layer 2 encapsulation*). This header includes the VNI, source and destination IP addresses of the VTEPs, and other metadata necessary for routing and forwarding traffic.

Once encapsulated, VXLAN packets are transmitted over the underlying IP network infrastructure. Routers and switches within the IP network treat VXLAN packets as regular IP traffic and forward them based on IP routing principles. Upon reaching the destination VTEP, VXLAN packets are decapsulated, and the original Layer 2 Ethernet frame is extracted. The frame is then forwarded to the appropriate destination host within the VXLAN segment.

Some key features and benefits of VXLAN include

▶ **Scalability:** VXLAN supports a large number of virtual networks (up to 16 million VNIs), enabling extensive network segmentation and multitenancy in cloud and datacenter environments.

▶ **Flexibility:** VXLAN operates over any IP-based network infrastructure, making it highly flexible and interoperable with existing network technologies and equipment.

▶ **Isolation:** VXLAN segments provide logical isolation between virtual networks, allowing different tenants or applications to coexist securely within the same physical infrastructure.

▶ **Efficiency:** By encapsulating Layer 2 traffic within Layer 3 packets, VXLAN reduces the complexity of Layer 2 networks and avoids the limitations of VLANs, such as the 4,096 VLAN ID limitation.

The *Data Center Interconnect (DCI)* is what connects multiple datacenters together in a VXLAN, enabling seamless communication and data exchange between them. DCI solutions facilitate workload mobility, disaster recovery, load balancing, and data replication across geographically dispersed datacenters. VXLAN networks are typically established within individual datacenters to create virtualized Layer 2 networks.

> **ExamAlert**
>
> Know that the DCI is what extends this virtual network beyond the boundaries of a single datacenter, allowing VXLAN segments to communicate with each other seamlessly and employ Layer 2 encapsulation.

Zero Trust Architecture (ZTA)

Zero trust architecture (ZTA) is a security framework and approach to network computing that assumes no implicit trust within the network, regardless of whether the connection is internal or external. In a traditional security model, once a user or device is inside the network perimeter, it is often trusted implicitly, leading to potential security vulnerabilities and threats. ZTA challenges this assumption by enforcing strict access controls and verification mechanisms for all users, devices, and applications, regardless of their location or network segment.

Key principles of zero trust architecture include

▶ **Policy-based authentication:** Policy-based authentication in ZTA considers contextual factors such as user identity, device health, location, time of access, and behavior to make access control decisions. Access policies are defined based on these contextual attributes, allowing organizations to tailor authentication requirements to the specific context of each access request.

▶ **Authorization:** ZTA requires continuous verification and authentication of users, devices, and applications before granting access to network resources. This includes multifactor authentication (MFA), identity verification, and strong authentication mechanisms to ensure the legitimacy of access requests.

▶ **Least privilege access:** ZTA follows the principle of least privilege, granting users and devices only the minimum level of access required to perform their tasks. Access permissions are based on the principle of need-to-know and are dynamically adjusted based on changing user roles, responsibilities, and context.

▶ **Microsegmentation:** ZTA advocates for network microsegmentation, dividing the network into smaller, isolated segments or zones based on user roles, applications, and sensitivity levels. Each segment is protected by access controls and firewalls, restricting lateral movement and containing potential security breaches.

▶ **Continuous monitoring and inspection:** ZTA emphasizes continuous monitoring and inspection of network traffic, user behavior, and security events to detect and respond to threats in real time. Advanced threat detection technologies, anomaly detection, and behavioral analytics are used to identify suspicious activities and unauthorized access attempts.

▶ **Encryption and data protection:** ZTA promotes the use of encryption and data protection mechanisms to secure data in transit and at rest. All network communications, including internal traffic, are encrypted using strong cryptographic protocols to prevent eavesdropping and data interception.

▶ **Dynamic trust assessment:** ZTA employs dynamic trust assessment to evaluate the trustworthiness of users, devices, and applications based on contextual factors such as device health, user behavior, location, and network conditions. Access decisions are made dynamically in real time based on the current context, rather than relying solely on static trust assumptions.

ExamAlert

Know that zero trust architecture (ZTA) fundamentally shifts the traditional perimeter-based security model by adopting a "never trust, always verify" approach. In ZTA, trust is never assumed, regardless of whether the user is inside or outside the network perimeter. Instead, access to resources and assets is continuously verified based on factors such as user identity, device health, context, and behavior, both at the time of authentication and throughout the session.

Secure Access Secure Edge (SASE)/ Security Service Edge (SSE)

Secure Access Service Edge (SASE) is a concept/architecture that combines network security functions with wide-area networking (WAN) capabilities to provide secure and scalable access to cloud-based applications and resources. It converges network security functions (such as firewall, secure web gateway, CASB, DLP, and zero trust access) with WAN capabilities (such as SD-WAN, WAN optimization, and routing) into a unified cloud-native platform in an effort to simplify security and networking architecture, reduce complexity, and improve operational efficiency.

SASE solutions leverage cloud infrastructure and software-defined networking (SDN) to deliver security and networking capabilities as a service, accessible from anywhere and on any device. They also adopt a zero trust security model, assuming no implicit trust and verifying every user, device, and application attempting to access the network, ensuring that access is granted based on strict authentication and authorization policies. SASE prioritizes identity-centric security, focusing on user and device identities rather than network perimeters. Access policies are enforced based on user identity, device posture, and contextual factors such as location, time of access, and security posture, ensuring that only authorized users and devices can access corporate resources.

SASE extends security and networking capabilities to the *Security Service Edge (SSE)*, providing secure access and low-latency connectivity to cloud-based applications and services. Such edge computing capabilities enable organizations to enforce security policies closer to the users and devices, reducing latency and improving performance for cloud-based applications. Lastly, SASE platforms deploy Global Secure Access Points (GSAPs) at strategic locations worldwide, providing secure connectivity and access to cloud-based applications and resources. GSAPs serve as points of presence (PoPs) for SASE services, enabling organizations to connect users and devices securely to the closest and most optimal access point.

> **ExamAlert**
>
> Know that this approach converges network security and networking functionalities into a cloud-native architecture, enabling organizations to secure and optimize access to applications, data, and services regardless of user location or device type. By integrating various security services such as secure web gateways, cloud access security brokers, firewalls, and software-defined wide-area networking (SD-WAN) capabilities into a unified cloud-based platform, SASE/SSE not only simplifies security management but also enhances visibility, agility, and scalability.

Infrastructure as Code (IaC)

Infrastructure as Code (IaC) involves managing and provisioning computing infrastructure using machine-readable configuration files or scripts, rather than manual processes or physical hardware configuration. With IaC, infrastructure resources such as virtual machines, networks, storage, and containers are defined and deployed programmatically using code. By treating infrastructure as code, organizations can achieve greater agility, scalability, reliability, and efficiency in their IT operations, accelerating the delivery of applications and services while reducing operational overhead and risk.

Some key aspects of the approach include

▶ **Declarative configuration:** IaC allows infrastructure to be defined in a declarative manner, where the desired state of the infrastructure is specified in configuration files or scripts. Instead of specifying step-by-step instructions for provisioning infrastructure resources, users declare the desired configuration, and the IaC tooling takes care of implementing it.

▶ **Version control:** IaC configurations are treated as code and managed using version control systems (such as Git). This enables developers and operations teams to track changes to infrastructure configurations, collaborate on infrastructure changes, and roll back to previous versions if needed. Version control also provides auditability and traceability for infrastructure changes.

▶ **Automated provisioning:** IaC automates the process of provisioning and configuring infrastructure resources, reducing manual intervention and minimizing the risk of human error. IaC tools interpret configuration files and execute the necessary actions to create, modify, or delete infrastructure resources based on the desired configuration.

▶ **Scalability and consistency:** By using code to define infrastructure, IaC enables organizations to scale their infrastructure rapidly and consistently across different environments (e.g., development, staging, production). Infrastructure configurations can be easily replicated and reused, ensuring consistency and reducing the time and effort required for deployment.

▶ **Immutable infrastructure:** IaC promotes the concept of immutable infrastructure, where infrastructure resources are treated as disposable and are not modified once deployed. Instead of making changes to exist- ing infrastructure, new infrastructure is provisioned from scratch with the desired configuration. This approach improves reliability, simplifies troubleshooting, and reduces the risk of *configuration drift* (thus increasing the likelihood of *compliance*).

▶ **Integration with continuous integration/continuous deployment (CI/CD):** IaC integrates seamlessly with CI/CD pipelines, enabling automated testing, validation, and deployment of infrastructure changes alongside application code changes. Infrastructure changes can be tested in preproduction environments before being promoted to production, ensuring reliability and minimizing disruptions.

IaC automation typically leverages playbooks, templates, and reusable tasks to streamline the provisioning and management of infrastructure resources. *Playbooks* are high-level definitions of infrastructure configurations and work- flows written in a declarative or imperative language, such as YAML or Ansible, that define the desired state of the infrastructure, including which resources should be provisioned, how they should be configured, and any dependencies or sequencing of tasks. Playbooks provide a structured and reusable format for defining infrastructure configurations, making it easier to manage and automate complex deployment processes. Similarly, *templates* are predefined configurations or blueprints for infrastructure resources, often expressed using a templating language or format such as JSON or YAML, that define the structure and properties of infrastructure resources, such as virtual machines, networks, storage, and security settings. Templates serve as reusable building blocks for provisioning infrastructure, allowing users to define common con- figurations once and reuse them across multiple environments or deployments. Conversely, *reusable tasks* are modular units of automation logic that perform specific actions or operations on infrastructure resources and can include provisioning virtual machines, configuring network settings, installing software packages, and applying security policies. By encapsulating common automation logic into reusable tasks, users can build modular and scalable automation workflows that can be easily adapted and reused across different environments and use cases.

IaC automation plays a crucial role in managing *upgrades* and *dynamic inventories* by providing automated and repeatable processes for updating infrastructure configurations and managing inventory changes. Upgrades to infrastructure components such as operating systems, software packages, and configuration settings can be automated using scripts or playbooks to define the necessary steps to perform the upgrade, including downloading and installing updates, restarting services, and validating the successful completion of the upgrade process. By automating upgrade processes, organizations can ensure consistency, reliability, and efficiency in upgrading their infrastructure components across multiple environments. Similarly, IaC automation tools support dynamic inventory management, allowing infrastructure inventories to be dynamically updated based on changes in the environment. Dynamic inventory plugins or scripts can query infrastructure resources such as cloud providers, virtualization platforms, or configuration management databases (CMDBs) to generate up-to-date inventory lists automatically. This ensures that automation scripts and playbooks always operate on the latest inventory information, reducing the risk of errors or inconsistencies due to outdated inventory data.

IaC automation tools typically integrate with *version control* systems such as Git, allowing infrastructure configurations and automation scripts to be managed and tracked over time. This enables organizations to maintain a version history of infrastructure changes, including upgrades, and facilitates rollback to previous configurations in case of issues or failures during the upgrade process. Version control provides visibility, auditability, and traceability for upgrade activities, ensuring that changes can be managed and tracked effectively.

When multiple users or teams collaborate on infrastructure configurations, conflicts may arise when conflicting changes are made to the same files or lines of code. Version control systems provide mechanisms for *conflict identification* and resolution by highlighting conflicting changes and allowing users to review, merge, or revert changes as needed. This enables smooth collaboration and ensures that conflicts are identified and resolved effectively, minimizing the risk of errors or inconsistencies in infrastructure configurations.

Equally important, version control systems support *branching*, which allows users to create separate branches or copies of the codebase to work on independent features, fixes, or experiments without affecting the main codebase. Branching is particularly useful in IaC workflows for managing different environments (e.g., development, staging, production) or implementing new features or changes in isolation. Users can create feature branches to develop and test changes independently, and then merge them back into the main branch (e.g., master) once they are ready. This enables parallel development, experimentation, and testing while maintaining a clean and stable main codebase.

Finally, IaC encourages the use of *central repositories* to store and manage infrastructure configurations, providing a centralized location for collaboration, version control, and change management. Central repositories serve as the single source of truth for infrastructure configurations, ensuring that all changes are tracked, documented, and auditable. They enable users to access, review, and contribute to infrastructure configurations from anywhere, facilitating distributed collaboration and ensuring consistency across environments. Central repositories also provide access controls and permissions management to regulate who can access, modify, and approve changes to infrastructure configurations, ensuring security and compliance with organizational policies.

ExamAlert

Know that IAC enables the automation and management of infrastructure deployments through code rather than manual configuration. With IAC, infrastructure resources such as virtual machines, networks, and storage are defined and provisioned using code scripts, templates, or configuration files

---

IPv6 Addressing

Internet Protocol version 4 (IPv4) was addressed in Chapter 3 and has served as the Internet's protocol for decades. When IPv4 was in development all those years ago, it would have been impossible for its creators to imagine or predict the future demand for IP devices and therefore IP addresses.

Note

There was an IPv5 after IPv4 and before IPv6, but it was an experimental protocol that never went anywhere.

Mitigating Address Exhaustion

IPv4 uses a 32-bit addressing scheme. This gives IPv4 a total of 4,294,967, 296 possible unique addresses that can be assigned to IP devices. More than 4 billion addresses might sound like a lot, and it is. However, the number of IP-enabled devices increases daily at a staggering rate. Not all these addresses can be used by public networks. Many of these addresses are reserved and are unavailable for public use. Reserving these addresses reduces the number of addresses that can be allocated as public Internet addresses.

The IPv6 project started in the mid-1990s, well before the threat of IPv4 limitations. Now network hardware and software are equipped for and ready to deploy IPv6 addressing. IPv6 offers a number of improvements. The most notable is its capability to handle growth in public networks. IPv6 uses a 128-bit addressing scheme, enabling a huge number of possible addresses:

> 340,282,366,920,938,463,463,374,607,431,768,211,456

Identifying IPv6 Addresses

As previously discussed, IPv4 uses a dotted-decimal format: 8 bits converted to its decimal equivalent and separated by periods. An example of an IPv4 address is 192.168.2.1.

Because of the 128-bit structure of the IPv6 addressing scheme, it looks quite a bit different. An IPv6 address is divided along 16-bit boundaries, and each 16-bit block is converted into a four-digit hexadecimal number and separated by colons. The resulting representation is called colon hexadecimal. Now look at how it works. Figure 4.4 shows the IPv6 address 2001:0:4137:9e50:2811:34ff: 3f57:febc from a Windows system.

An IPv6 address can be simplified by removing the leading 0s within each 16-bit block. Not all the 0s can be removed, however, because each address block must have at least a single digit. Removing the 0 suppression, the address representation becomes

> 2001:0000:4137:9e50:2811:34ff:3f57:febc

Some of the IPv6 addresses you will work with have sequences of 0s. When this occurs, the number is often abbreviated to make it easier to read. In the preceding example, you saw that a single 0 represented a number set in hexadecimal form. To further simplify the representation of IPv6 addresses, a contiguous sequence of 16-bit blocks set to 0 in colon hexadecimal format can be compressed to ::, known as the double colon.

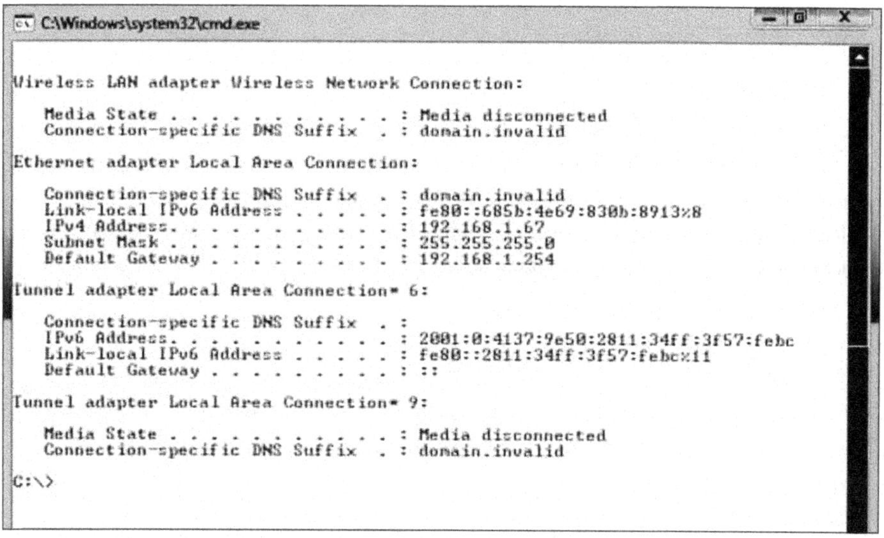

FIGURE 4.4   **An IPv6 Address in a Windows Dialog Screen**

For example, the IPv6 address

2001:0000:0000:0000:3cde:37d1:3f57:fe93

can be compressed to

2001::3cde:37d1:3f57:fe93

However, there are limits on how the IPv6 0s can be reduced. Within the IPv6 address, 0s cannot be eliminated when they are not first in the number sequence. For instance, 2001:4000:0000:0000:0000:0000:0000:0003 cannot be compressed as 2001:4::3. This would actually appear as 2001:4000::3.

When you look at an IPv6 address that uses a double colon, how do you know exactly what numbers are represented? The formula is to subtract the number of blocks from 8 and then multiply that number by 16. For example, the address 2001:4000::3 uses three blocks: 2001, 4000, and 3. So the formula is as follows:

$$(8 - 3) \times 16 = 80$$

Therefore, the total number of bits represented by the double colon in this example is 80.

> **Note**
>
> You can remove 0s only once in an IPv6 address. Using a double colon more than once would make it impossible to determine the number of 0 bits represented by each instance of ::.

IPv6 Address Types

Another difference between IPv4 and IPv6 is in the address types. IPv4 addressing was discussed in detail earlier. IPv6 addressing offers several types of addresses, as detailed in this section.

Unicast IPv6 Addresses

As you might deduce from the name, a unicast address specifies a single interface. Data packets sent to a unicast destination travel from the sending host to the destination host. It is a direct line of communication. A few types of addresses fall under the unicast banner, as discussed next.

Global Unicast Addresses

Global unicast addresses are the equivalent of IPv4 public addresses. These addresses are routable and travel throughout the network.

Link-Local Addresses

Link-local addresses are designated for use on a single local network. Link-local addresses are automatically configured on all interfaces. This automatic configuration is comparable to the 169.254.0.0/16 APIPA automatically assigned IPv4 addressing scheme (discussed shortly). The prefix used for a link-local address is fe80::/64. On a single-link IPv6 network with no router, link-local addresses are used to communicate between devices on the link.

Site-Local Addresses

Site-local addresses are equivalent to the IPv4 private address space (10.0.0.0/8, 172.16.0.0/12, and 192.168.0.0/16). As with IPv4, in which private address ranges are used in private networks, IPv6 uses site-local addresses that do not interfere with global unicast addresses. In addition, routers do not forward site-local traffic outside the site. Unlike link-local addresses, site-local addresses are not automatically configured and must be assigned through either stateless or stateful address configuration processes. The prefix used for the site-local address is feC0::/10.

Multicast Addresses

As with IPv4 addresses, multicasting sends and receives data between groups of nodes. It sends IP messages to a group rather than to every node on the LAN (broadcast) or just one other node (unicast).

Anycast Addresses

Anycast addresses represent the middle ground between unicast addresses and multicast addresses. Anycast delivers messages to any one node in the multicast group.

> **Note**
>
> You might encounter the terms *stateful* and *stateless* configuration. *Stateless* refers to IP autoconfiguration, in which administrators need not manually input configuration information. In a *stateful* configuration network, devices obtain address information from a server.

> **ExamAlert**
>
> Similar to stateful/stateless, *classful* and *classless* are address adjectives that are often used. Classful means that the address falls into one of the five IPv4 classes (A, B, C, D, or E), whereas classless uses the CIDR notation previously discussed.
>
> Remember that fe80:: is a private link-local address.

> **ExamAlert**
>
> Earlier, you read that IPv4 reserves 127.0.0.1 as the loopback address. IPv6 has the same reservation. IPv6 addresses 0:0:0:0:0:0:0:0 and 0:0:0:0:0:0:0:1 are reserved as the loopback addresses. 0:0:0:0:0:0:0:1 shortened is ::1. In CIDR format, the loopback address for IPv4 is 127.0. 0.1/8; for IPv6, it is ::1/128.

Neighbor Discovery

IPv6 supports the Neighbor Discovery Protocol (NDP). Operating at the network layer, it is responsible for autoconfiguring node addresses, discovering other nodes on the link, determining the addresses of other nodes, detecting duplicate addresses, finding available routers and DNS servers, discovering address prefixes, and maintaining reachability information of other active neighbor nodes.

Comparing IPv4 and IPv6 Addressing

Table 4.4 compares IPv4 and IPv6 addressing.

TABLE 4.4 **Comparing IPv4 and IPv6 Addressing**

Address Feature	IPv4 Address	IPv6 Address
Loopback address	127.0.0.1	0:0:0:0:0:0:0:1 (::1)
Networkwide addresses	IPv4 public address ranges	Global unicast IPv6 addresses
Private network addresses	10.0.0.0 172.16.0.0 192.168.0.0	Site-local address ranges (feC0::)
Autoconfigured addresses	IPv4 automatic private IP addressing (169.254.0.0)	Link-local addresses of the fe80:: prefix

ExamAlert

Make sure that you know the information provided in Table 4.4.

Note

IPv6 supports *dual stack*: this means that both IPv4 and IPv6 can run on the same network. This capability is extremely useful when transitioning from one to the other during the adoption and deployment phases. It also enables the network to continue to support legacy devices that may not be able to transition.

Compatibility Requirements

IPv6 *compatibility requirements* refer to the measures and technologies used to ensure that networks, devices, and applications can effectively support and interoperate with IPv6. Not only does IPv6 compatibility include dual stack capability (referenced above), but also tunneling and NAT64.

In this case, *tunneling* is used to encapsulate IPv6 packets within IPv4 packets, allowing IPv6 traffic to traverse IPv4 networks that do not natively support IPv6. Tunneling mechanisms such as 6to4, Teredo, and Generic Routing

Encapsulation (GRE) enable the creation of IPv6-over-IPv4 tunnels, allowing IPv6 traffic to be transmitted over IPv4 networks. IPv6 compatibility requirements may include support for tunneling protocols to ensure seamless communication between IPv6-enabled and IPv4-only networks.

NAT64 is a translation mechanism used to facilitate communication between IPv6-only and IPv4-only devices by mapping IPv6 addresses to IPv4 addresses and vice versa. NAT64 translates IPv6 addresses to IPv4 addresses (and IPv4 addresses to IPv6 addresses) at the network edge, allowing IPv6-only devices to access IPv4-only services and vice versa. IPv6 compatibility requirements may include support for NAT64 translation mechanisms to enable seamless communication between IPv6 and IPv4 networks, particularly during the transition period from IPv4 to IPv6.

ExamAlert

IPv6 compatibility requirements encompass the deployment of tunneling mechanisms, dual-stack implementation, and NAT64 translation to ensure that networks, devices, and applications can effectively support and interoperate with IPv6. By addressing these compatibility requirements, organizations can facilitate the transition to IPv6, enable seamless communication between IPv6 and IPv4 networks, and ensure sustainability and growth.

Cram Quiz

1. Logical unit numbers (LUNs) came from the SCSI world and use "targets" that hold up to how many devices?

   ○ **A.** 4
   ○ **B.** 6
   ○ **C.** 8
   ○ **D.** 128

2. What is the IPv6 equivalent of 127.0.0.1? (Choose two.)

   ○ **A.** 0:0:0:0:0:0:0:1
   ○ **B.** 0:0:0:0:0:0:0:24
   ○ **C.** ::1
   ○ **D.** ::24

3. On a VLAN, what creates multiple paths to the storage resources and can be used to increase availability and add fault tolerance?

   - ○ **A.** FCoE
   - ○ **B.** Adding a management plane
   - ○ **C.** Colocating
   - ○ **D.** Multipathing

4. Which of the following is a key advantage of SD-WAN technology?

   - ○ **A.** Improved physical security
   - ○ **B.** Reduced network complexity
   - ○ **C.** Higher latency
   - ○ **D.** Limited scalability

5. Which of the following components is responsible for encapsulating and decapsulating VXLAN packets?

   - ○ **A.** VTEPs
   - ○ **B.** SDN Controller
   - ○ **C.** VXLAN Gateway
   - ○ **D.** DHCP Server

6. What does zero trust architecture (ZTA) prioritize in access control decisions?

   - ○ **A.** Static network segmentation
   - ○ **B.** Implicit trust in user credentials
   - ○ **C.** Least privilege access
   - ○ **D.** Open access policies

7. Which is a translation mechanism used to facilitate communication between IPv6-only and IPv4-only devices by mapping IPv6 addresses to IPv4 addresses and vice versa?

   - ○ **A.** Tunneling
   - ○ **B.** Stateless configuration
   - ○ **C.** NDP
   - ○ **D.** NAT64

8. What does fe80::/64 represent?

   - ○ **A.** A link-local address
   - ○ **B.** A site-local address
   - ○ **C.** A multicast address
   - ○ **D.** An anycast address

9. Which allows users to create separate copies of the codebase to work on independent features, fixes, or experiments without affecting the main codebase?

○ **A.** Central repositories

○ **B.** Version control

○ **C.** Branching

○ **D.** Immutable infrastructure

10. Which is an extension of a VLAN that utilizes network virtualization technology
in such a way as to overcome the scalability limitations of traditional VLANs by
allowing the creation of logical Layer 2 networks over Layer 3?

○ **A.** VTEPS

○ **B.** VXLAN

○ **C.** Zero-touch provisioning

○ **D.** DCI

Cram Quiz Answers

1. **C.** LUNs came from the SCSI world and carry over, acting as unique identifiers for
devices. Both NAS and SAN use "targets" that hold up to eight devices.

2. **A and C.** The IPv4 address 127.0.0.1 is reserved as the loopback address, and
IPv6 has the same reservation. IPv6 addresses 0:0:0:0:0:0:0:0 and 0:0:0:0:0:0:0:1
are reserved as the loopback addresses. The address 0:0:0:0:0:0:0:1 can be
shown using the :: notation with the 0s removed, resulting in ::1.

3. **D.** On a VLAN, multipathing creates multiple paths to the storage resources and
can be used to increase availability and add fault tolerance.

4. **B.** SD-WAN technology simplifies network management and configuration by centralizing control and providing a software-defined approach to WAN connectivity.
It abstracts underlying network hardware and enables automated provisioning,
policy-based routing, and dynamic traffic management, leading to reduced
complexity and improved agility.

5. **A.** VXLAN tunnel endpoints (VTEPs) are responsible for encapsulating outgoing
Layer 2 Ethernet frames into VXLAN packets and decapsulating incoming VXLAN
packets to extract the original Layer 2 Ethernet frames. VTEPs reside on network
devices such as switches, routers, or hypervisors and handle the encapsulation
and forwarding of VXLAN traffic.

6. **C.** ZTA follows the principle of least privilege, granting users and devices only the
minimum level of access required to perform their tasks. Access decisions are
based on the principle of need-to-know and are dynamically adjusted based on
changing user roles, responsibilities, and contextual factors. By enforcing least
privilege access, ZTA minimizes the risk of unauthorized access and lateral
movement within the network.

7. **D.** NAT64 is a translation mechanism used to facilitate communication between IPv6-only and IPv4-only devices by mapping IPv6 addresses to IPv4 addresses and vice versa. Tunneling is used to encapsulate IPv6 packets within IPv4 packets, allowing IPv6 traffic to traverse IPv4 networks that do not natively support IPv6. Tunneling mechanisms such as 6to4, Teredo, and Generic Routing Encapsulation (GRE) enable the creation of IPv6-over-IPv4 tunnels, allowing IPv6 traffic to be transmitted over IPv4 networks. Stateless configuration refers to IP autoconfiguration, in which administrators need not manually input configuration information. In a stateful configuration network, devices obtain address information from a server. IPv6 supports the Neighbor Discovery Protocol (NDP). Operating at the network layer, it is responsible for autoconfiguring node addresses, discovering other nodes on the link, determining the addresses of other nodes, detecting duplicate addresses, finding available routers and DNS servers, discovering address prefixes, and maintaining reachability information of other active neighbor nodes.

8. **A.** The prefix used for a link-local address is fe80::/64. On a single-link IPv6 network with no router, link-local addresses are used to communicate between devices on the link. Unlike link-local addresses, site-local addresses are not automatically configured and must be assigned through either stateless or stateful address configuration processes. The prefix used for the site-local address is feC0::/10. As with IPv4 addresses, multicasting sends and receives data between groups of nodes. It sends IP messages to a group rather than to every node on the LAN (broadcast) or just one other node (unicast). Anycast addresses represent the middle ground between unicast addresses and multicast addresses. Anycast delivers messages to any one node in the multicast group.

9. **C.** Branching allows users to create separate branches or copies of the codebase to work on independent features, fixes, or experiments without affecting the main codebase. Branching is particularly useful in IaC workflows for managing different environments (e.g., development, staging, production) or implementing new features or changes in isolation. IaC encourages the use of central repositories to store and manage infrastructure configurations, providing a centralized location for collaboration, version control, and change management. Central repositories serve as the single source of truth for infrastructure configurations, ensuring that all changes are tracked, documented, and auditable. IaC automation tools typically integrate with version control systems such as Git, allowing infrastructure configurations and automation scripts to be managed and tracked over time. Version control provides visibility, auditability, and traceability for upgrade activities, ensuring that changes can be managed and tracked effectively. IaC promotes the concept of immutable infrastructure, where infrastructure resources are treated as disposable and are not modified once deployed. Instead of making changes to existing infrastructure, new infrastructure is provisioned from scratch with the desired configuration.

10. **B.** A virtual extensible local-area network (VXLAN) is an extension of a VLAN that utilizes network virtualization technology in such a way as to overcome the scalability limitations of traditional VLANs by allowing the creation of logical Layer 2 networks over Layer 3. The VXLAN first creates an overlay network on top of an existing IP network infrastructure, allowing virtual networks to be

established across physical network boundaries. VXLAN tunnel endpoints (VTEPs) are responsible for encapsulating and decapsulating VXLAN packets. They reside on network devices such as switches, routers, or hypervisors and handle the encapsulation and forwarding of VXLAN traffic. One key feature of the SD-WAN is zero-touch provisioning—the ability to automate deployment and configuration of SD-WAN devices with minimal manual intervention. The Data Center Interconnect (DCI) is what connects multiple datacenters together in a VXLAN, enabling seamless communication and data exchange between them.

What's Next?

For the Network+ exam, and for routinely working with an existing network or implementing a new one, you need to identify the characteristics of network media and their associated cabling. Chapter 5, "Cabling Solutions and Issues," focuses on the media and connectors used in today's networks and what you are likely to find in wiring closets.

CHAPTER 5

Cabling Solutions and Issues

This chapter covers the following official Network+ objectives:

▶ 1.5 Compare and contrast transmission media and transceivers.

▶ 5.2 Given a scenario, troubleshoot common cabling and physical interface issues.

This chapter covers CompTIA Network+ objectives 1.5 and 5.2. For more information on the official Network+ exam topics, see the "About the Network+ Exam" section in the Introduction.

When working with an existing network or implementing a new one, you need to identify the characteristics of network media and their associated cabling. This chapter focuses on the media and connectors used in today's networks and how they fit into wiring closets and beyond.

General Media Considerations

▶ **1.5 Compare and contrast transmission media and transceivers.**

CramSaver

If you can correctly answer these questions before going through this section, save time by skimming the ExamAlerts in this section and then completing the Cram Quiz at the end of the section.

1. What are the two main types of twisted-pair wiring used today?

2. What is the name of the wiring standard that offers a minimum of 500 MHz of bandwidth and specifies transmission distances up to 100 meters with 10 Gbps?

3. What is the difference between RJ-11 and RJ-45 connectors?

4. What are the two most common connectors used with fiber-optic cabling?

5. What are F-type connectors used for?

6. True or false: Category 5e UTP was created for use where distances are short (such as between switches and servers in a datacenter) because it can obtain speeds up to 40 Gbps at 2000 MHz and only for distances up to 30 meters (approximately 98 feet).

Answers

1. The two main types of twisted-pair cabling in use today are unshielded twisted-pair (UTP) and shielded twisted-pair (STP).

2. Category 6a (Cat 6a) offers improvements over Category 6 (Cat 6) by offering a minimum of 500 MHz of bandwidth. It specifies transmission distances up to 100 meters with 10 Gbps networking speeds.

3. RJ-11 connectors are used with standard phone lines and are similar in appearance to RJ-45 connectors used in networking. However, RJ-11 connectors are smaller. RJ-45 connectors are used with UTP cabling.

4. Fiber-optic cabling uses a variety of connectors, but SC and ST are more commonly used than others. ST connectors offer a twist-type attachment, whereas SCs have a push-on connector. LC and MTRJ are other types of fiber-optic connectors. In environments where vibration can be a problem, FC connectors can be used and feature a threaded body.

5. F-type connectors are used to connect coaxial cable to devices such as Internet modems.

6. False. Category 8 UTP was created for use where distances are short (such as between switches and servers in a datacenter). While it was not specifically intended for general office use (primarily due to cost), it will work great if used in a SOHO network because it can obtain speeds up to 40 Gbps at 2000 MHz and only for distances up to 30 meters (approximately 98 feet).

In addition to identifying the characteristics of network media and their associated cabling, the Network+ exam requires knowledge of some general terms and concepts that are associated with network media. Before you look at the individual media types, it is a good idea to first have an understanding of some general media considerations.

> **ExamAlert**
>
> Remember that this objective begins with "Compare and contrast" and focuses on transmission media and transceivers This means that you will need to be able to explain which is the appropriate cable or connector type for a solution.

Broadband Versus Baseband Transmissions

Networks employ two types of signaling methods/modulation techniques:

▶ **Baseband transmissions:** Baseband transmissions use digital signaling over a single wire. Communication on baseband transmissions is bidirectional, enabling signals to be sent and received, but not at the same time. To send multiple signals on a single cable, baseband uses something called *time-division multiplexing (TDM)*. TDM divides a single channel into time slots. The key thing about TDM is that it does not change how baseband transmission works—only how data is placed on the cable.

> **ExamAlert**
>
> Most networks use baseband transmissions. (Notice the word *base*.) Examples are 1000BASE-T and 10GBASE-T.

▶ **Broadband transmissions:** In terms of LAN network standards, broadband transmissions use analog transmissions. For broadband transmissions to be sent and received, the medium must be split into two channels. (Alternatively, two cables can be used: one to send and one to receive transmissions.) Multiple channels are created using *frequency-division multiplexing (FDM)*. FDM enables broadband media to accommodate traffic going in different directions on a single medium at the same time.

Simplex, Half-Duplex, and Full-Duplex Modes

Simplex, half-duplex, and full-duplex modes are referred to as *dialog modes*, and they determine the direction in which data can flow through the network media:

▶ Simplex mode enables one-way communication of data through the network, with the full bandwidth of the cable used for the transmitting signal. One-way communication is of little use on LANs, making it unusual at best for network implementations.

▶ Far more common is half-duplex mode, which accommodates transmitting and receiving on the network, but not at the same time. Many networks are configured for half-duplex communication.

▶ The preferred dialog mode for network communication is full-duplex mode. To use full-duplex, both the network card and the hub or switch must support full duplexing. Devices configured for full duplexing can simultaneously transmit and receive. This means that 100 Mbps network cards theoretically can transmit at 200 Mbps using full-duplex mode.

Data Transmission Rates

One of the more important media considerations is the supported data transmission rate or speed. Different media types are rated to certain maximum speeds, but whether they are used to this maximum depends on the networking standard used and the network devices connected to the network.

> ### Note
>
> The transmission rate of media is sometimes incorrectly called the *bandwidth*. But the term *bandwidth* refers to the width of the range of electrical frequencies or the number of channels that the medium can support.

Transmission rates normally are measured by the number of data bits that can traverse the medium in a single second. In the early days of data communications, this measurement was expressed in bits per second (bps), but today's networks are measured in *megabits per second (Mbps)* and *gigabits per second (Gbps)*.

The different network media vary greatly in the transmission speeds they support. Many of today's application-intensive networks require more than the 10 Mbps or 100 Mbps offered by the older networking standards. In some cases, even 1 Gbps, which is found in many modern LANs, is not enough to meet current network needs. For this reason, many organizations now deploy 10 Gbps or higher implementations.

Wired Versus Wireless

As you might imagine, the primary difference between a wireless computer network and a wired computer network lies in the method used for data transmission. A wireless computer network uses radio frequency (RF) signals to transmit data between devices without the need for physical cables, whereas a wired computer network relies on physical cables, such as Ethernet cables, to transmit data between devices. Devices in a wireless network communicate over the airwaves through wireless access points (WAPs, or simply, APs) or routers while devices in a wired network are connected to each other and to network infrastructure devices, such as switches and routers, through Ethernet cables. Wireless networks provide flexibility and mobility, allowing devices to connect to the network from anywhere within the coverage area of the wireless signal, whereas wired networks typically offer higher reliability, stability, and data transfer speeds compared to wireless networks. Common wireless networking standards include Wi-Fi (802.11), which operates in various frequency bands such as 2.4 GHz and 5 GHz, and are discussed in detail in Chapter 6, "Wireless Solutions." Common wired networking standards include Ethernet (IEEE 802.3), which uses twisted-pair copper cables or fiber-optic cables for data transmission, and are focused on in this chapter.

> **Tip**
>
> It is very important to know wireless standards for network administration today. Be sure to pay attention to this topic in Chapter 6.

Cellular Technology Access

One reason why cellular access is an important topic from the perspective of this exam is that when devices (smartphones, tablets, and so on) are accessing the network outside of a Wi-Fi connection, they are often doing so through a cellular network and that cellular network becomes the WAN. As a network administrator, you are dependent on the cellular network your users are using (and the security, or lack thereof, inherent in it) to protect your data and resources.

The *Global System for Mobile Communications (GSM)* initially used *time-division multiple access (TDMA)* to provide multiuser access by chopping up the channel into sequential time slices. Each user of the channel takes turns to transmit and receive signals and, ideally, this happens so quickly that the user is unaware of it. TDMA was replaced in later implementations by *code-division multiple access (CDMA)*, which (instead of splitting the channel into time slices) uses different frequencies for each user to provide various means of cell phone coverage.

The individual methods that can be used for cellular access include 5G, LTE/4G, or 3G, and they represent enhancements to the technology over time. Each generation represents new frequency bands and higher data rates. The original GSM access (with both TDMA and CDMA) was labeled 2G. As standards that became available focused on increasing speeds and enabling the sending of images, this morphed into 3G (which, initially, was more marketing hype than anything else). 4G added the capability to implement mobile broadband Internet access (not just for smartphones but also laptops with wireless modems and other similar devices). Long-Term Evolution (LTE) was based on Enhanced Data rates for GSM Evolution (EDGE) and high-speed packet access (HSPA) technologies, which increased the capacity and speed by using a different radio interface together with core network improvements. The newest iteration, 5G, not only provides faster speeds but is also needed to meet the needs of Internet of Things (IoT) sensors and other communication-intensive devices.

For purposes of comparison, a typical download speed of basic 3G would be 0.0375 Mbps; 4G would be 150 Mbps; LTE would be approximately 600 Mbps; and 5G is estimated to be between 1–10 Gbps.

Satellite Access

Satellite technology is used for wireless computer network access through satellite communication systems, which provide Internet connectivity to remote locations or areas where traditional wired infrastructure is unavailable or impractical. Satellite ISPs operate ground stations equipped with satellite dishes that communicate with geostationary satellites orbiting the Earth, and users subscribe to satellite Internet services provided by these ISPs, which include a satellite dish installed at the user's location to establish communication with the satellite.

Satellite Internet operates using a two-way communication process: upstream and downstream. Upstream communication involves transmitting data from the user's location to the satellite in space, typically done through a small satellite dish installed at the user's premises. Downstream communication involves receiving data from the satellite to the user's location, allowing users to access Internet services, websites, and other online content.

Satellite Internet connections often have higher latency compared to terrestrial connections due to the distance that signals must travel between the Earth and the satellite in space. Signal propagation delay, also known as ping or round-trip time, can affect real-time applications such as online gaming and video conferencing.

Satellite Internet services can provide coverage to remote and rural areas where terrestrial infrastructure, such as fiber-optic cables or DSL lines, is not available. They offer an alternative Internet connectivity option for users in regions with limited access to traditional wired broadband services.

> **ExamAlert**
>
> For the exam, be ready to compare and contrast the various wireless 802.11 standards, cellular, and satellite. Focus on their use cases and types of transmission media.

Types of Wired Network Media

Whatever type of network is used, some type of network medium is needed to carry signals between computers. Two types of media are used in networks: cable-based media, such as twisted-pair, and the media types associated with wireless networking, such as radio waves.

In networks using cable-based media, there are two basic choices:

► Copper

► Fiber-optic

Copper wire is used with both twisted-pair and coaxial cables to conduct the signals electronically; fiber-optic cable uses a glass or plastic conductor and transmits the signals as light.

For many years, coaxial was the cable of choice for most LANs. Today, twisted-pair has proven to be the cable medium of choice, thus retiring coaxial to the confines of storage closets. Fiber-optic cable has seen a rise in popularity, but cost slowed its adoption to the home (although it is common today). It is widely used as a network backbone where segment length and higher speeds are needed and is common in server room environments as a server-to-switch connection method and in building-to-building connections in *metropolitan-area networks (MANs)*.

The following sections summarize the characteristics of each of these cable types.

Twisted-Pair Cabling (Copper)

Twisted-pair cabling has been around for a long time. It was originally created for voice transmissions and has been widely used for telephone communication. Today, in addition to telephone communication, twisted-pair is the most widely used medium for networking.

The popularity of twisted-pair can be attributed to the fact that it is lighter, more flexible, and easier to install than coaxial or fiber-optic cable. It is also cheaper than other media alternatives and can achieve greater speeds than its coaxial competition. These factors make twisted-pair the ideal solution for most network environments.

Two main types of twisted-pair cabling are in use today: *unshielded twisted-pair (UTP)* and *shielded twisted-pair (STP)*. UTP is significantly more common than STP and is used for most networks. Shielded twisted-pair is used in environments in which greater resistance to EMI and attenuation is required. The greater resistance comes at a price, however. The additional shielding, plus the need to ground that shield (which requires special connectors), can significantly add to the cost of a cable installation of STP.

STP provides the extra shielding by using an insulating material that is wrapped around the wires within the cable. This extra protection increases the distances that data signals can travel over STP but also increases the cost of the cabling. Figure 5.1 shows UTP and STP cabling.

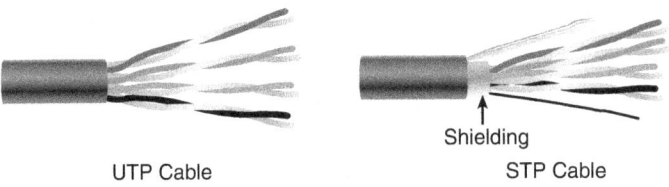

Shielding

UTP Cable   STP Cable

FIGURE 5.1 **UTP and STP Cabling**

There are several categories of twisted-pair cabling. The early categories are most commonly associated with voice transmissions. The categories are specified by the *Electronic Industries Alliance/Telecommunications Industry Association (EIA/TIA)*. EIA/TIA is an organization that focuses on developing standards for electronic components, electronic information, telecommunications, and Internet security. These standards are important to ensure uniformity of components and devices.

> **Note**
>
> When learning about cabling, you need to understand the distinction between hertz and bits per second in relation to bandwidth. When you talk about bandwidth and a bits-per-second rating, you refer to a rate of data transfer.

EIA/TIA has specified a number of categories of twisted-pair cable, some of which are now obsolete. Those still in use today include the following:

▶ **Category 5:** This data-grade cable typically was used with Fast Ethernet operating at 100 Mbps with a transmission range of 100 meters. Although Category 5 was a popular media type, this cable is an outdated standard. Newer implementations use the 5e or greater standards, and the IEEE 802.11ae standard specifies 1000 Mbps over Category 5 cable.

▶ **Category 5e:** This data-grade cable is used on networks that run at 10/100 Mbps and even up to 1000 Mbps. Category 5e cabling can be used up to 100 meters, depending on the implementation and standard used. Category 5e cable provides a minimum of 100 MHz of bandwidth.

▶ **Category 6:** This high-performance UTP cable can transmit data up to 10 Gbps. Category 6 has a minimum of 250 MHz of bandwidth and specifies cable lengths up to 100 meters with 10/100/1000 Mbps transfer, along with 10 Gbps over shorter distances. Category 6 cable typically is made up of four twisted pairs of copper wire, but its capabilities far exceed those of other cable types. Category 6 twisted-pair uses a longitudinal separator, which separates each of the four pairs of wires from each other. This extra construction significantly reduces the amount of crosstalk in the cable and makes the faster transfer rates possible.

▶ **Category 6a:** Also called augmented 6, this cable offers improvements over Category 6 by offering a minimum of 500 MHz of bandwidth. It specifies transmission distances up to 100 meters with 10 Gbps networking speeds.

▶ **Category 7:** The big advantage to this cable is that shielding has been added to individual pairs and to the cable as a whole to greatly reduce crosstalk. It is rated for transmission of 600 MHz and is backward compatible with Category 5 and Category 6. Category 7 differs from the other cables in this group in that it is not recognized by the EIA/TIA and that it is shielded twisted-pair, whereas all others listed beneath the exam objectives are unshielded.

▶ **Category 8:** This standard was created for use where distances are short (such as between switches and servers in a datacenter). While it was not specifically intended for general office use (primarily due to cost), it will work great if used in a SOHO network because it can obtain speeds up to 40 Gbps at 2000 MHz and only for distances up to 30 meters (approximately 98 feet).

ExamAlert

On the exam, you might see these categories as Cat 5, Cat 5e, Cat 6, Cat 6a, Cat 7, and Cat 8. Remember their characteristics, such as cable length, speed, and bandwidth.

Tip

If you work on a network that is a few years old, you might need to determine which category of cable it uses. The easiest way to do this is to read the cable. The category number should be clearly printed on it.

Table 5.1 summarizes the categories and the speeds they support in common network implementations.

TABLE 5.1 **Twisted-Pair Cable Categories**

Category	Common Application
5	100 Mbps
5e	1000 Mbps (1 Gbps)
6	10/100/1000 Mbps plus 10 Gbps
6a	10 Gbps and beyond networking
7	10 Gbps and beyond networking
8	Up to 40 Gbps

Note

The numbers shown in Table 5.1 refer to speeds these cables are commonly used to support. Ratified standards for these cabling categories might actually specify lower speeds than those listed, but cable and network component manufacturers are always pushing the performance envelope in the quest for greater speeds. The ratified standards define minimum specifications. For more information on cabling standards, visit the TIA website at www.tiaonline.org/.

Coaxial Cables

Coaxial cable, or *coax* as it is commonly called, has been around for a long time. Coax found success in both TV signal transmission and network implementations. As shown in Figure 5.2, coax is constructed with a copper core at the center (the main wire) that carries the signal, insulation (made of plastic), ground (braided metal shielding), and insulation on the outside (an outer plastic covering).

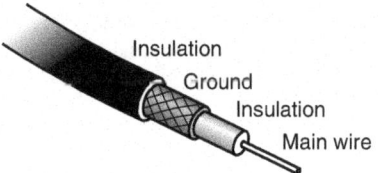

Insulation
Ground
Insulation
Main wire

FIGURE 5.2   **Coaxial Cabling**

Coaxial cable is constructed in this way to add resistance to *attenuation* (the loss of signal strength as the signal travels over distance), *crosstalk* (the degradation of a signal caused by signals from other cables running close to it), and EMI. Two types of coax are used in networking: thin coax, also known as thinnet or 10BASE2, and thick coax, also known as *thicknet*. Neither is particularly popular anymore, but you are most likely to encounter thin coax. Thick coax was used primarily for backbone cable. It could be run through plenum spaces because it offered significant resistance to EMI and crosstalk and could run in lengths up to 500 meters. Thick coax offers speeds up to 10 Mbps, far too slow for today's network environments.

> **Note**
>
> The plenum is the space between the structural ceiling and a drop-down ceiling. It is commonly used for heating, ventilation, and air-conditioning systems and to run network cables.

Thin coax is much more likely to be seen than thick coax in today's networks, but it isn't common. Thin coax is only 0.25 inch in diameter, making it fairly easy to install. Unfortunately, one of the disadvantages of all thin coax types is that they are prone to cable breaks, which increase the difficulty when installing and troubleshooting coaxial-based networks.

Several types of thin coax cable exist, each of which has a specific use. Table 5.2 summarizes these categories.

> **ExamAlert**
>
> For the exam, you should focus on RG-6 and know the difference between it and RG-59.

TABLE 5.2 **Thin Coax Categories**

Cable Type	Description
RG-59	Used to generate low-power video connections. The RG-59 cable cannot be used over long distances because of its high-frequency power losses. In such cases, RG-6 cables are used instead.
RG-6	Often used for cable TV and cable modems.

Direct Attach Copper

Direct attach copper (DAC) cabling is a type of high-speed, short-distance cabling commonly used to connect networking equipment such as switches, routers, and servers. The cables consist of *twinaxial* copper cable assemblies with connectors, usually SFP+ (small form-factor pluggable) or QSFP (quad small form-factor pluggable), attached at both ends.

> **ExamAlert**
>
> Remember that DAC cables typically consist of copper conductors enclosed in a protective jacket and are terminated with standard connectors, such as SFP (small form-factor pluggable) or QSFP (quad small form-factor pluggable) connectors.

Twinaxial cable, or *twinax*, has two inner conductors instead of one. As shown in Figure 5.3, twinax is constructed with two wires at the center, insulation, ground (braided metal shielding), and insulation on the outside (an outer plastic covering). These cables are commonly used for short distances (10 meters or less). The cables support high-speed data transmission rates, making them suitable for applications requiring high bandwidth, such as storage-area networks (SANs), server clustering, and high-performance computing (HPC) environments, and they are commonly used with Ethernet protocols such as 10 Gigabit Ethernet (10 GbE), 25 Gigabit Ethernet (25 GbE), and 40 Gigabit Ethernet (40 GbE), as well as Fibre Channel protocols.

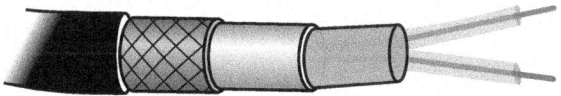

FIGURE 5.3   **Twinaxial Cabling**

Fiber-Optic Cables

In many ways, fiber-optic media address the shortcomings of copper-based media. Because fiber-based media use light transmissions instead of electronic pulses, threats such as EMI, crosstalk, and attenuation become nonissues. Fiber is well suited for the transfer of data, video, and voice transmissions. In addition, fiber-optic is the most secure of all cable media. Anyone trying to access data signals on a fiber-optic cable must physically tap into the medium. Given the composition of the cable, this is a particularly difficult task.

Unfortunately, despite the advantages of fiber-based media over copper, it still does not enjoy the popularity of twisted-pair cabling. The moderately difficult installation and maintenance procedures of fiber often require skilled technicians with specialized tools. Furthermore, the cost of a fiber-based solution limits the number of organizations that can afford to implement it. Another sometimes hidden drawback of implementing a fiber solution is the cost of retrofitting existing network equipment. Fiber is incompatible with most electronic network equipment. This means you have to purchase fiber-compatible network hardware.

> **ExamAlert**
>
> Fiber-optic cable, although still more expensive than other types of cable, is well suited for high-speed data communications. It eliminates the problems associated with copper-based media, such as near-end crosstalk, EMI, and signal tampering.

As shown in Figure 5.4, fiber-optic cable is composed of a core (glass fiber) that is surrounded by *cladding* (silica). A silicone coating is next, followed by a buffer jacket. There are strength members next, and then a protective sheath (polyurethane outer jacket) surrounds everything.

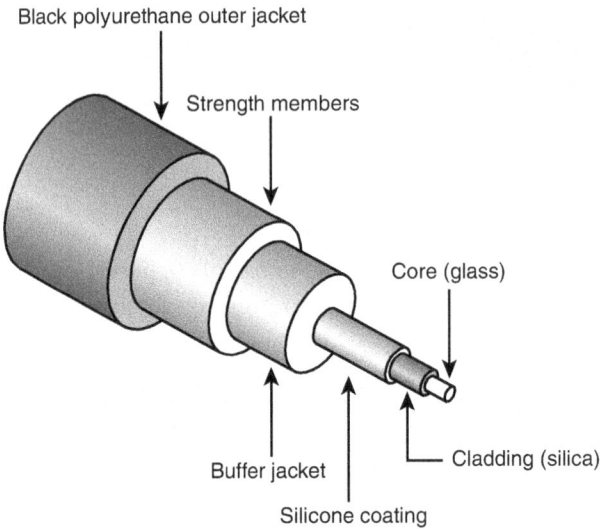

FIGURE 5.4 **Fiber-Optic Cabling**

Two types of fiber-optic cable are available:

▶ **Multimode fiber:** Many beams of light travel through the cable, bouncing off the cable walls. This strategy actually weakens the signal, reducing the length and speed at which the data signal can travel.

▶ **Single-mode fiber:** This type uses a single direct beam of light, thus allowing for greater distances and increased transfer speeds.

Some common types of fiber-optic cable include the following:

▶ 62.5-micron core/125-micron cladding multimode

▶ 50-micron core/125-micron cladding multimode

▶ 8.3-micron core/125-micron cladding single mode

In the ever-increasing search for bandwidth that can keep pace with the demands of modern applications, fiber-optic cables are sure to continue to play a key role.

> **ExamAlert**
>
> Understanding the types of fiber optics available focusing on single-mode and multimode, as well as their advantages and limitations, is important for real-world applications as well as the Network+ exam.

Plenum Versus PVC Cables

A plenum is the space that resides between the false, or drop, ceiling and the true ceiling. This space typically is used for air conditioning and heating ducts. It might also hold a myriad of cables, including telephone, electrical, and networking. The cables that occupy this space must be plenum-rated rather than the standard PVC cables. Plenum cables are coated with a nonflammable material, often Teflon or Kynar, and they do not give off toxic fumes if they catch fire. As you might imagine, plenum-rated cables cost more than regular (PVC-based) cables, but they are mandatory when cables are not run through a conduit. As a bonus, plenum-rated cables suffer from less attenuation than nonplenum cables.

> **ExamAlert**
>
> Cables run through the plenum areas must have two important characteristics: They must be fire resistant, and they must not produce toxic fumes if exposed to intense heat.

Types of Media Connectors

Various connectors are used with the associated network media. Media connectors attach to the transmission media and allow the physical connection into the computing device. For the Network+ exam, you need to identify the connectors associated with a specific medium. The following sections describe the connectors and associated media.

BNC Connectors

Bayonet Neill–Concelman (BNC) connectors are associated with coaxial media and 10BASE2 networks. BNC connectors are not as common as they previously were, but they still are used on some networks, older network cards, and older hubs. Common BNC connectors include a barrel connector, T-connector, and terminators. Figure 5.5 shows two terminators (top and bottom) and two T-connectors (left and right).

Terminators

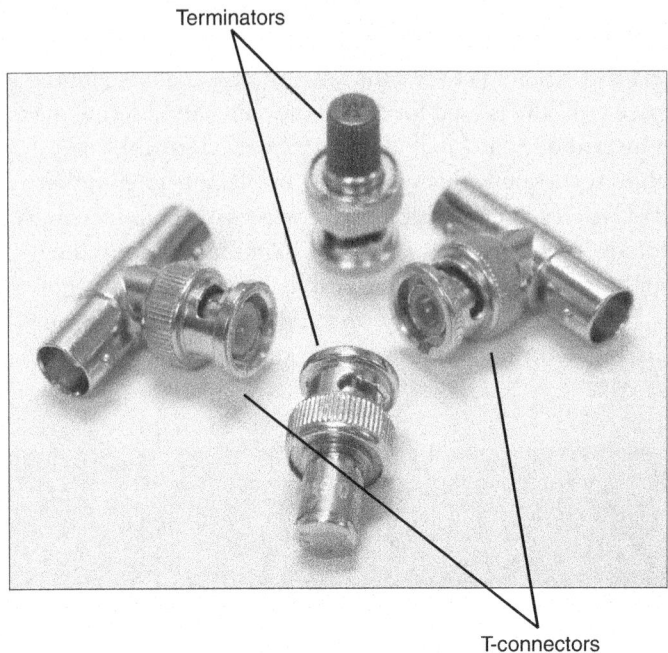

T-connectors

FIGURE 5.5 **BNC Connectors**

> **ExamAlert**
>
> *Connectors* are sometimes referred to as *couplers*. For exam purposes, consider the two words to be synonyms.

RJ-11 Connectors

RJ-11 (Registered Jack) connectors are small plastic connectors used on telephone cables. They have capacity for six small pins. However, in many cases, not all the pins are used. For example, a standard telephone connection uses only two pins, and a cable used for a *digital subscriber line (DSL)* modem connection uses four.

RJ-11 connectors are somewhat similar to RJ-45 connectors, which are discussed next, although they are a little smaller. Both RJ-11 and RJ-45 connectors have a small plastic flange on top of the connector to ensure a secure connection. Figure 5.6 shows two views of an RJ-11 connector.

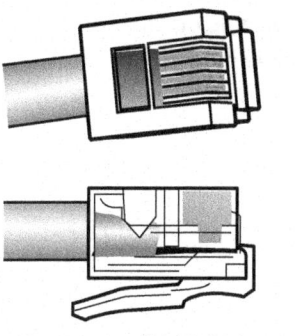

FIGURE 5.6 **RJ-11 Connectors**

RJ-45 Connectors

RJ-45 connectors, as shown in Figure 5.7, are the ones you are most likely to encounter in your network travels. RJ-45 connectors are used with twisted-pair cabling, the most prevalent network cable in use today. RJ-45 connectors resemble the aforementioned RJ-11 phone jacks, but they support up to eight wires instead of the six supported by RJ-11 connectors. RJ-45 connectors are also larger than RJ-11 connectors.

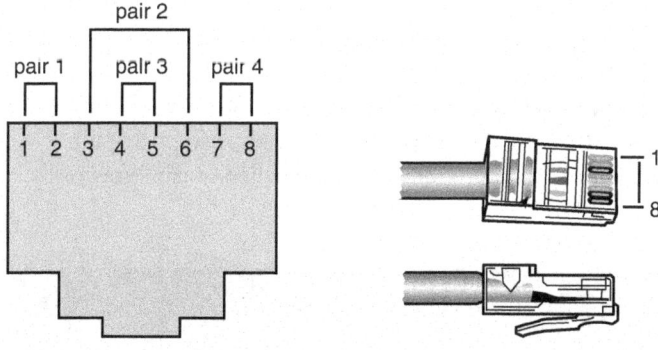

FIGURE 5.7 **RJ-45 Connectors**

F-Type Connectors

F-type connectors, as shown in Figure 5.8, are screw-on connections used to attach coaxial cable to devices. This includes RG-59 and RG-6 cables. In the world of modern networking, F-type connectors are most commonly associated with connecting Internet modems to equipment from a cable or satellite *Internet service provider (ISP)*. However, F-type connectors are also used to connect to some proprietary peripherals.

FIGURE 5.8   **F-Type Connector**

F-type connectors have a "nut" on the connection that provides something to grip as the connection is tightened by hand. If necessary, this nut can also be lightly gripped with pliers to aid disconnection.

> **ExamAlert**
>
> For the Network+ exam, you will be expected to identify and compare and contrast the connectors discussed in this chapter by their appearance.

Fiber Connectors

A variety of connectors are associated with fiber cabling, and there are several ways of connecting them. They include bayonet, snap-lock, and push-pull connectors. Figure 5.9 shows the fiber connectors identified in the Network+ objectives.

> **ExamAlert**
>
> As with the other connectors discussed in this section, be prepared to identify fiber connectors by their appearance and by how they are physically connected. Focus on SC, LC, ST, and MPO fiber connector types.

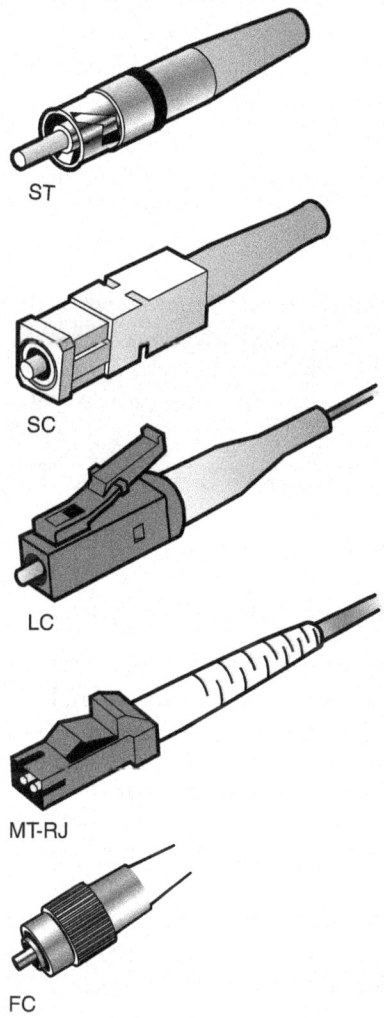

The ST connector uses a half-twist bayonet type of lock.

ST

The SC uses a push-pull connector similar to common audio and video plugs and sockets.

SC

LC connectors have a flange on top, similar to an RJ-45 connector, that aids secure connection.

LC

MT-RJ is a popular connector for two fibers in a very small form factor.

MT-RJ

FC connectors have a threaded body and are used in environments where vibration is a problem.

FC

FIGURE 5.9 **Fiber Connectors**

Within the various types of connectors (ST, SC, LC, MPO, and so on), you can choose to purchase ones that are either *angled physical contact (APC)* or *ultra-physical contact (UPC)*. The biggest difference between these two is the "angle" present in APC. UPC connectors have an endface polished at a zero-degree angle (flat), whereas APC is eight degrees. As a general rule, the more polished (UPC) gives less insertion loss.

Multifiber push-on (MPO) connectors are optical fiber connectors widely used in high-density fiber-optic cabling systems, particularly in datacenters and telecommunications networks. As the name implies, MPO connectors utilize a push-on coupling mechanism, which allows for quick and easy connections without the need for screws or latches. This simplifies installation and maintenance tasks, especially in dense cabling environments. MPO connectors are compatible with single-mode (SM) and multimode (MM) and are available in various configurations, including 12-fiber, 24-fiber, and 48-fiber variants (the most common configuration is the 12-fiber MPO connector, which consists of 12 fibers aligned in a single row).

> **ExamAlert**
>
> Remember that MPO connectors utilize a push-on coupling mechanism, which allows for quick and easy connections without the need for screws or latches. This simplifies installation and maintenance tasks, especially in dense cabling environments.

Transceivers

On routers, *small form-factor pluggable (SFP)* modules and *gigabit interface converter (GBIC)* modules are often used to link a gigabit Ethernet port with a fiber network (often 1000BASE-X). Both SFPs and GBICs exist for technologies other than fiber (Ethernet and SONET/SDH are usual), but connecting to fiber has become the most common use.

> **Note**
>
> *SFP+* is an enhanced small form-factor pluggable module; it is a newer version of SFP that supports data rates up to 16 Gbps. Quad small form-factor pluggable (QSFP) is a different transceiver that is both compact and hot-pluggable; it has been jointly developed by many networking vendors. Similarly, *Enhanced QSFP+* is an evolution of QSFP that supports four channels.

Fiber transceivers are *bidirectional* and capable of operating in *duplex* mode. With either an SFP or GBIC, there is a *receiver port (RX)* and *transmitter port (TX)*. These devices are static-sensitive as well as dust-sensitive, and dirty connectors can cause intermittent problems. Care should be taken to not remove them more often than absolutely necessary to keep from shortening their life. After a module goes bad, they can be swapped for a new one to resolve the problem.

Note

Cisco has a great post on the care and maintenance of SFPs at www.cisco.com/en/US/products/hw/modules/ps4999/products_tech_note09186a00807a30d6.shtml.

ExamAlert

When transmitter (TX) and receiver (RX) ports are transposed, communication errors and performance degradation typically occur. One device may be configured for full duplex while the other is set to half duplex. In this case, one side of the communication expects to send and receive data simultaneously (full duplex), while the other side can only send or receive data one at a time (half duplex).

Signal loss can occur not only from unclean connectors, but also from *connector mismatch*. Improper alignment and differences in core diameters contribute to signal loss.

When troubleshooting an SFP or GBIC, you should make sure that you do not have a *cable mismatch* or a *bad cable/transceiver*. As simple as it may sound, it is important to verify that you are using a single-mode fiber with a single-mode interface and a multimode fiber cable for a multimode interface. Such a *fiber type mismatch* can cause the physical link to go completely down but does not always do so, thus making troubleshooting this problem difficult.

Transceivers are designed to support specific protocols, data rates, and transmission distances. The choice of transceivers depends on factors such as the networking protocol (Ethernet or Fibre Channel), data rate, transmission distance, and compatibility with networking equipment.

Media Couplers/Converters

When you have two dissimilar types of network media, a *media converter* is used to allow them to connect. They are sometimes referred to as *couplers*. Depending on the conversion being done, the converter can be a small device barely larger than the connectors themselves or a large device within a sizable chassis.

Reasons for not using the same media throughout the network, and thus reasons for needing a converter, can range from cost (gradually moving from coax to fiber), disparate segments (connecting the office to the factory), or the need to run particular media in a setting (the need for fiber to reduce EMI problems in a small part of the building).

Figure 5.10 shows an example of a media converter. The one shown converts between 10/100/1000TX and SFP.

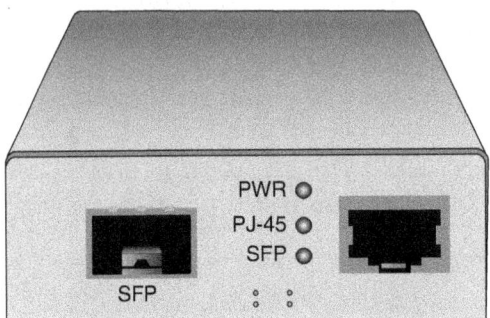

FIGURE 5.10   A Common Media Converter

The following converters are commonly implemented.

▶ Single-mode fiber to Ethernet

▶ Single-mode to multimode fiber

▶ Multimode fiber to Ethernet

▶ Fiber to coaxial

TIA/EIA 568A and 568B Wiring Standards

568A and 568B are telecommunications standards from TIA and EIA. These 568 standards specify the pin arrangements for the RJ-45 connectors on UTP or STP cables. The number 568 refers to the order in which the wires within the cable are terminated and attached to the connector.

The *TIA/EIA 568A* and *568B* standards (often referred to as *T568A* and *T568B* for termination standard) are similar; the difference is the order in which the pins are terminated. The signal is the same for both. Both are used for patch cords in an Ethernet network.

> **ExamAlert**
>
> The only notable difference between T568A and T568B is that pairs 2 and 3 (orange and green) are swapped.

Network media might not always come with connectors attached, or you might need to make custom-length cables. This is when you need to know something about how these standards actually work. Before you can crimp on the connectors, you need to know in which order the individual wires will be attached to the connector. Figure 5.11 shows the pin number assignments for the T568A and T568B standards.

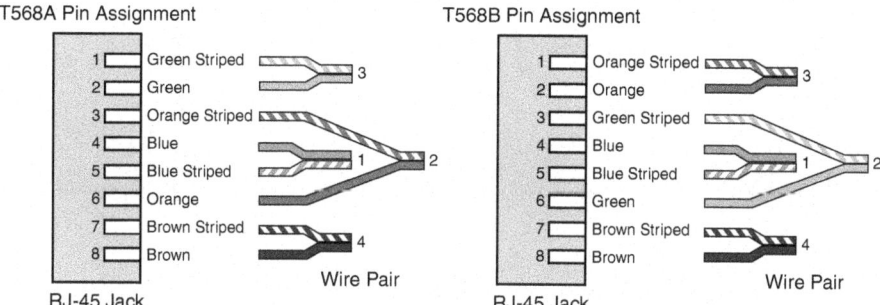

FIGURE 5.11   **Pin Assignments for the T568A and T568B Standards**

Straight-Through Versus Crossover Cables

Two types of cables are used to connect devices to hubs and switches: crossover cables and straight-through cables. The difference between the two types is that, in a crossover cable, two of the wires are crossed; in a straight-through cable, all the wires run straight through.

Specifically, in a crossover cable, wires 1 and 3 and wires 2 and 6 are crossed. Wire 1 at one end becomes wire 3 at the other end, wire 2 at one end becomes wire 6 at the other end, and vice versa in both cases. You can see the differences between the two cables in Figures 5.12 and 5.13. Figure 5.12 shows the pinouts for a straight-through cable, and Figure 5.13 shows the pinouts for a crossover cable.

ExamAlert

The crossover cable can be used to directly network two PCs without using a hub or switch. This is done because the cable performs the function of the switch.

Note

Auto MDI-X ports on newer interfaces detect whether the connection requires a crossover and automatically choose the MDI or MDI-X configuration to match the other end of the link.

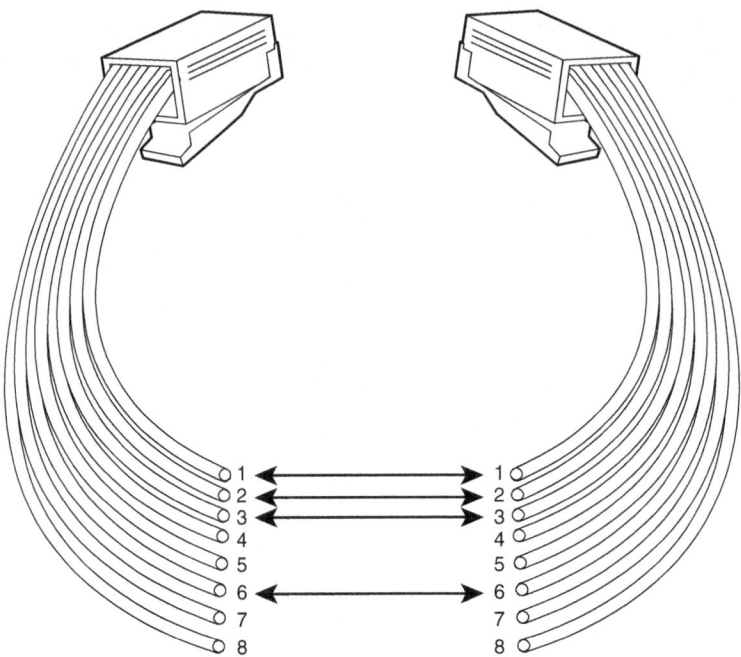

FIGURE 5.12   **Pinouts for a Straight-Through Twisted-Pair Cable**

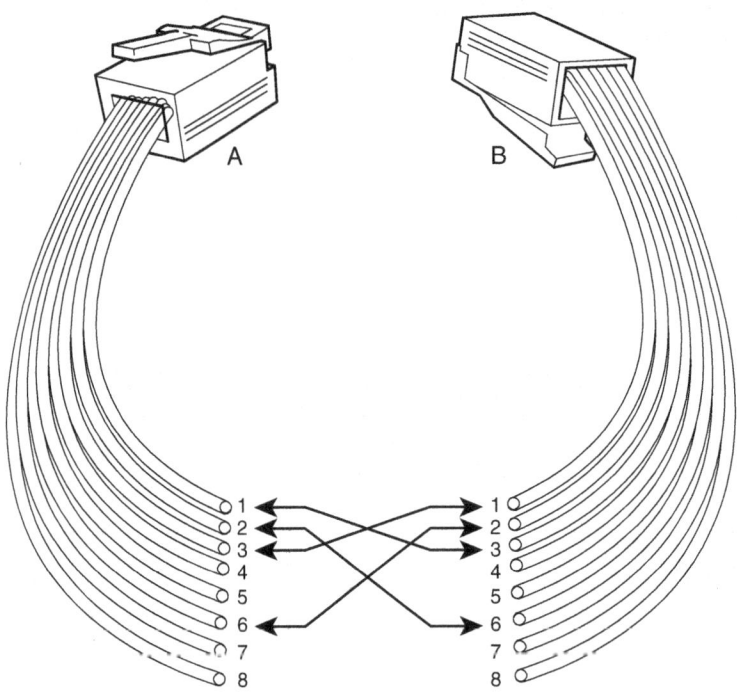

FIGURE 5.13   **Pinouts for a Crossover Twisted-Pair Cable**

To make a crossover Ethernet cable, you need to use both the 568A and 568B standards. One end of the cable can be wired according to the 568A standard and the other with the 568B standard.

A T1 crossover cable, the pinouts of which are shown in Figure 5.14, is used to connect two T1 CSU/DSU devices in a back-to-back configuration. RJ-45 connectors are used on both ends.

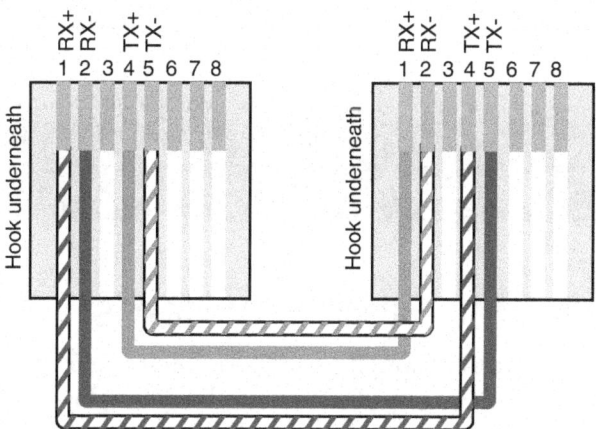

PRI (T1) Crossover/Loopback Cable

FIGURE 5.14   Pinouts for a T1 Crossover Cable

Rollover and Loopback Cables

The rollover cable is a Cisco proprietary cable used to connect a computer system to a router or switch console port. The rollover cable resembles an Ethernet UTP cable; however, it is not possible to use it on anything but Cisco equipment. Like UTP cable, the rollover cable has eight wires inside and an RJ-45 connector on each end that connects to the router and the computer port.

As far as pinouts are concerned, pin 1 on one end of the rollover cable connects to pin 8 at the other end of the cable. Similarly, pin 2 connects to pin 7, and so on. The ends are simply reversed. As soon as one end of the rollover cable is connected to the PC and the other to the Cisco terminal, the Cisco equipment can be accessed from the computer system using a program such as PuTTY.

> **ExamAlert**
>
> Remember that the rollover cable is a proprietary cable used to connect a PC to a Cisco router.

A loopback cable, also known as a *plug*, is used to test and isolate network problems. If made correctly, the loopback plug causes the link light on a device such as a *network interface card (NIC)* to come on. This is a quick and cheap way to test simple network cabling problems. The loopback plug redirects outgoing data signals to the system. The system then believes that it is both sending and receiving data.

The loopback cable is basically a troubleshooting tool used to test the device to see if it is sending and receiving properly. It uses UTP cable and RJ-45 connectors.

> **ExamAlert**
>
> Know that a loopback cable is a basic troubleshooting tool.

Ethernet Copper and Fiber Standards

A number of IEEE standards relate to networking and cover everything from implementation to security. The 802.3 standards relate to Ethernet deployment, and many of the early ones have become outdated. Make sure that you are familiar with the information that follows for the progression of popular standards from not that long ago to today.

10BASE-T

The 10BASE-T standard specifies an Ethernet network that commonly uses unshielded twisted-pair cable. In some implementations that require a greater resistance to interference and attenuation, STP can be used because it has extra shielding and is more able to combat interference.

The 10BASE-T standard uses broadband transmission and has a maximum physical segment length of 100 meters. Repeaters are sometimes used to extend the maximum segment length, although the repeating capability is now often built into networking devices used in twisted-pair networks. 10BASE-T specifies transmission speeds of 10 Mbps and can use several categories of UTP cable with RJ-45 connectors. The maximum number of computers supported on a 10BASE-T network is 1,024.

100BASE-TX

At one time, 10 Mbps networks were considered fast enough, but those days are long gone. Today, companies and home users alike demand more bandwidth than that, and 100BASE-TX transmits network data at speeds up to 100 Mbps. 100BASE-TX is most often implemented with UTP cable, but it can use STP; therefore, it suffers from the same 100-meter distance limitations as other UTP-based networks. 100BASE-TX uses Category 5, or higher, UTP cable, and it uses independent transmit and receive paths and therefore can support full-duplex operation. 100BASE-TX is the most common implementation of the Fast Ethernet (802.3u) standard.

> **Tip**
>
> Repeaters are sometimes needed when you connect segments that use 100BASE-TX or 100BASE-FX.

A counterpart to 100BASE-TX is 100BASE-FX, which is the IEEE standard for running Fast Ethernet over fiber-optic cable. Due to the expense of fiber implementations, 100BASE-FX is largely limited to use as a network backbone. 100BASE-FX can use two-strand multimode fiber or single-mode fiber media. The maximum segment length for half-duplex multimode fiber is 412 meters, but this maximum increases to an impressive 10,000 meters for full-duplex single-mode fiber. 100BASE-FX often uses SC or ST fiber connectors. Table 5.3 summarizes the characteristics of the 802.3u Fast Ethernet specifications.

TABLE 5.3 **Summary of 802.3u Fast Ethernet Characteristics**

Characteristic	100BASE-TX	100BASE-FX
Transmission method	Baseband	Baseband
Speed	100 Mbps	100 Mbps
Distance	100 meters	412 meters (multimode half duplex); 10,000 meters (single-mode full duplex)
Cable type	Category UTP, STP	Fiber-optic
Connector type	RJ-45	SC, ST

An additional fiber option, 100BASE-SX, is considered a lower-cost alternative to 100BASE-FX. It uses LEDs instead of lasers and can be used for shorter distances (up to 300 meters).

1000BASE-T

The Gigabit Ethernet standard 1000BASE-T, or 1000BASE-TX, is given the IEEE 802.3ab designation. The 802.3ab standard specifies Gigabit Ethernet over Category 5 or better UTP cable. The standard allows for full-duplex transmission using the four pairs of twisted cable. To reach speeds of 1000 Mbps over copper, a data transmission speed of 250 Mbps is achieved over each pair of twisted-pair cable. Table 5.4 summarizes the characteristics of 1000BASE-T.

TABLE 5.4 **Summary of 1000BASE-T Characteristics**

Characteristic	Description
Transmission method	Baseband
Speed	1000 Mbps
Total distance/segment	75 meters
Cable type	Category 5 or better
Connector type	RJ-45

10GBASE-T

The 802.3an standard brings 10-gigabit speed to regular copper cabling. Although transmission distances may not be those of fiber, they allow a potential upgrade from 1000 Mbps networking to 10 Gbps networking using the current wiring infrastructure.

The 10GBASE-T standard specifies 10 Gbps transmissions over UTP or STP twisted-pair cables. The standard calls for a cable specification of Category 6 or Category 6a. With Category 6, the maximum transmission range is 55 meters; with the augmented Category 6a cable, the transmission range increases to 100 meters. Category 6 and 6a cables are specifically designed to reduce attenuation and crosstalk, making 10 Gbps speeds possible. The 802.3an standard specifies regular RJ-45 networking connectors. Table 5.5 outlines the characteristics of this standard.

TABLE 5.5 **Summary of 802.3an Characteristics**

Characteristic	Descriptions
Transmission method	Baseband
Speed	10 Gbps
Total distance/segment	100 meters Category 6a cable; 55 meters Category 6 cable
Cable type	Category 6, 6a UTP or STP
Connector	RJ-45

40GBASE-T

The 40GBASE-T standard provides for 4-pair balanced (40 Gbps on 4-twisted pairs cable) twisted-pair Category 8 copper cabling up to 30 meters. It is defined in the IEEE 802.3bq standard, and it is expected to be used primarily within datacenters. Table 5.6 outlines the characteristics of the 802.3bq standard.

TABLE 5.6 **Summary of 802.3bq Characteristics**

Characteristic	Descriptions
Transmission method	Baseband
Speed	40 Gbps
Total distance/segment	30 meters Category 8 cable
Cable type	Category 8
Connector	RJ-45

1000BASE-LX and 1000BASE-SX

Both 1000BASE-LX and 1000BASE-SX are Gigabit Ethernet standards for fiber.

As a fiber standard for Gigabit Ethernet, 1000BASE-LX utilizes single-mode fiber. It can also run over multimode fiber with a maximum segment length of 550m.

The 1000BASE-SX standard is intended for use with multimode fiber and has a maximum length of 220 meters for default installations (550 meters is possible with the right optics and terminations). This standard is popular for intrabuilding links in office buildings.

10GBASE-LR and 10GBASE-SR

The 10GBASE-LR standard is easy to remember in that the *LR* stands for long range: the maximum fiber length is 10 kilometers, but it varies greatly depending on the type of single-mode fiber used. 10GBASE-SR is a multimode fiber intended for the short range (up to 400 meters): it is considered the lowest cost, lowest power, and smallest form factor optical option available at this speed.

Multiplexing Options

Virtual circuits establish a bidirectional communication link between devices and use it for their communication links. Multiplexing was discussed earlier in the "Broadband Versus Baseband Transmissions" section, but you should know that bidirectional wavelength division multiplexing (WDM) is the transmission of optical channels on a fiber propagating simultaneously in both directions.

Several types of WDM multiplexing can be employed during these links. One form of multiplexing optical signals is *dense wavelength-division multiplexing (DWDM)*. This method replaces SONET/SDH regenerators and can amplify the signal and enable it to travel a greater distance. The main components of a DWDM system include the following:

▶ Terminal multiplexer

▶ Line repeaters

▶ Terminal demultiplexer

An alternative to DWDM is *coarse wavelength-division multiplexing (CWDM)*. This method is commonly used with television cable networks. The main thing to know about it is that it has relaxed stabilization requirements; thus, you can have vastly different speeds for download than upload.

Cram Quiz

1. Which of the following connectors is commonly used with fiber cabling?

   ○ **A.** RJ-45

   ○ **B.** BNC

   ○ **C.** SC

   ○ **D.** RJ-11

2. What kind of cable would you associate with an F-type connector?

   ○ **A.** Fiber-optic

   ○ **B.** UTP

   ○ **C.** Coaxial

   ○ **D.** STP

3. Which of the following is not a type of fiber-optic connector used in network implementations?

   ○ **A.** MPO

   ○ **B.** SC

     ○ **C.** BNC

     ○ **D.** LC

4. Which of the following fiber connectors uses a twist-type connection method?

     ○ **A.** ST

     ○ **B.** SC

     ○ **C.** BNC

     ○ **D.** SA

5. Which of the following is a fiber standard for Gigabit Ethernet that utilizes single-mode fiber?

     ○ **A.** 1000BASE-SX

     ○ **B.** TIA/EIA 568a

     ○ **C.** RG-6

     ○ **D.** 1000BASE-LX

6. In a crossover cable, which wire is wire 1 crossed with?

     ○ **A.** 2

     ○ **B.** 3

     ○ **C.** 4

     ○ **D.** 5

7. Which of the following cables are specifically coated with a nonflammable material and do not give off toxic fumes if they catch fire?

     ○ **A.** SFP and GBIC

     ○ **B.** Cat 5e, Cat 6, Cat 6a

     ○ **C.** FDP

     ○ **D.** Plenum-rated

8. Which EIA/TIA category standard has a minimum of 250 MHz of bandwidth and specifies cable lengths up to 100 meters with 10/100/1000 Mbps transfer, along with 10 Gbps over shorter distances?

     ○ **A.** Category 5e

     ○ **B.** Category 6

     ○ **C.** Category 6a

     ○ **D.** Category 8

9. Which of the following are optical fiber connectors that utilize a push-on coupling mechanism, which allows for quick and easy connections without the need for screws or latches?

  ○ **A.** Receiver port (RX)

  ○ **B.** Transmitter port (TX)

  ○ **C.** MPO

  ○ **D.** EIA/TIA

Cram Quiz Answers

1. **C.** SC connectors are used with fiber-optic cable. RJ-45 connectors are used with UTP cable, BNC is used for thin coax cable, and RJ-11 is used for regular phone connectors.

2. **C.** F-type connectors are used with coaxial cables. They are not used with fiber-optic, unshielded twisted-pair (UTP), or shielded twisted-pair (STP) cabling.

3. **C.** BNC is a connector type used with coaxial cabling. It is not used as a connector for fiber-optic cabling. MPO, SC, and LC are all recognized types of fiber-optic connectors.

4. **A.** ST fiber connectors use a twist-type connection method. SC connectors use a push-type connection method. The other choices are not valid fiber connectors.

5. **D.** The 1000BASE-LX fiber standard for Gigabit Ethernet utilizes single-mode fiber. 1000BASE-SX is intended for use with multimode fiber and has a maximum length of 220 meters for default installations. TIA/EIA 568A and 568B are telecommunications standards that specify the pin arrangements for the RJ-45 connectors on UTP or STP cables. RG-6 is a common type of coaxial cable often used for cable TV and cable modems.

6. **B.** In a crossover cable, wires 1 and 3 and wires 2 and 6 are crossed.

7. **D.** Plenum-rated cables are coated with a nonflammable material, often Teflon or Kynar, and they do not give off toxic fumes if they catch fire. On routers, SFP modules and GBIC modules are often used to link a gigabit Ethernet port with a fiber network. Cat 5, Cat 5e, Cat 6, Cat 6a, and also Cat 7 and Cat 8 are general categories of twisted-pair cabling. FDP is an acronym for fiber distribution panel, which is a cabinet intended to provide space for termination, storage, and splicing of fiber connections.

8. **B.** Category 6 is high-performance UTP cable that can transmit data up to 10 Gbps. Category 6 has a minimum of 250 MHz of bandwidth and specifies cable lengths up to 100 meters with 10/100/1000 Mbps transfer, along with 10 Gbps over shorter distances. Category 6 cable typically is made up of four twisted pairs of copper wire, but its capabilities far exceed those of other cable types. Category 5e data-grade cable is used on networks that run at 10/100 Mbps and even up to 1000 Mbps. Category 5e cabling can be used up to 100 meters, depending on the implementation and standard used. Category 5e cable provides a minimum of 100 MHz of bandwidth. Category 6a, also called augmented 6, offers improvements over Category 6 by offering a minimum of 500 MHz of bandwidth.

It specifies transmission distances up to 100 meters with 10 Gbps networking speeds. The Category 8 standard was created for use where distances are short (such as between switches and servers in a datacenter). While it was not specifically intended for general office use (primarily due to cost), it will work great if used in a SOHO network because it can obtain speeds up to 40 Gbps at 2000 MHz and only for distances up to 30 meters (approximately 98 feet).

9. **C.** Multifiber push-on (MPO) connectors are optical fiber connectors widely used in high-density fiber-optic cabling systems, particularly in datacenters and telecommunications networks. As the name implies, MPO connectors utilize a push-on coupling mechanism, which allows for quick and easy connections without the need for screws or latches. Fiber transceivers are bidirectional and capable of operating in duplex mode. With either an SFP or GBIC, there is a receiver port (RX) and transmitter port (TX). These devices are static-sensitive as well as dust-sensitive, and dirty connectors can cause intermittent problems. The Electronic Industries Alliance/Telecommunications Industry Association (EIA/TIA) is an organization that focuses on developing standards for electronic components, electronic information, telecommunications, and Internet security. These standards are important to ensure uniformity of components and devices.

Troubleshooting Common Cable Connectivity Issues

▶ **5.2 Given a scenario, troubleshoot common cabling and physical interface issues.**

CramSaver

If you can correctly answer these questions before going through this section, save time by skimming the ExamAlerts in this section and then completing the Cram Quiz at the end of the section.

1. What is the correct term for delay time on a satellite-based network?

2. What are two types of crosstalk?

3. What tools are used to attach twisted-pair network cable to connectors within a patch panel?

4. What are the two parts of a toner probe?

5. True or false: Attenuation refers to the strengthening of data signals as they travel through a medium.

6. In the networking world, what refers to the rate of data delivery over a communication channel.

Answers

1. Latency is the time of the delay.

2. Two types of crosstalk are near-end crosstalk (NEXT) and far-end crosstalk (FEXT).

3. Punchdown tools are used to attach twisted-pair network cable to connectors within a patch panel.

4. A toner probe has two parts: the tone generator, or toner, and the tone locator, or probe.

5. False. *Attenuation* refers to the weakening of data signals as they travel through a medium.

6. In the networking world, *throughput* refers to the rate of data delivery over a communication channel.

ExamAlert

Remember that this objective begins with "Given a scenario." This means that you may receive a drag-and-drop, matching, or "live OS" scenario where you have to click through to complete a specific objective-based task.

When administering a wired network, you should be aware of a number of performance issues and common connectivity problems. Some of the topics lumped within this objective are more knowledge/definitions than actionable items, but make sure you are familiar with them all the same. We first look at some limitations, considerations, and issues. Following that, we discuss some of the common tools used by network technicians.

Limitations, Considerations, and Issues

Specifications and issues abound when trying to optimize a network and keep it up and running. It is very rare for resource demands to lessen over time, and they only seem to grow. Keeping up with that growth requires taking a lot into consideration and knowing the limitations of each technology. In the sections that follow, we look at some of the specifications/limitations and common issues associated with them.

Throughput, Speed, and Distance

There must be enough bandwidth to serve all users, and you need to be alert for bandwidth hogs. You want to look for top talkers (those that transmit the most) and top listeners (those that receive the most) and figure out why they are so popular.

In the networking world, *throughput* refers to the rate of data delivery over a communication channel. In this case, throughput testers test the rate of data delivery over a network. Throughput is measured in *bits per second (bps)*. Testing throughput is important for administrators to make them aware of exactly what the network is doing. With throughput testing, you can tell whether a high-speed network is functioning close to its expected throughput.

A throughput tester is designed to quickly gather information about network functionality—specifically, the average overall network throughput. Many software-based throughput testers are available online—some for free and some for a fee.

As you can see, throughput testers do not need to be complicated to be effective. A throughput tester tells you how long it takes to send data to a destination point and receive an acknowledgment that the data was received. To use the tester, enter the beginning point and then the destination point. The tester sends a predetermined number of data packets to the destination and then reports on the throughput level. The results typically display in *kilobits per second (Kbps)*, *megabits per second (Mbps)*, or *gigabits per second (Gbps)*. Table 5.7 shows the various data rate units.

TABLE 5.7 **Data Rate Units**

Data Transfer	Abbreviation	Speed
Kilobits per second	Kbps or Kbit/s	1,000 bits per second
Megabits per second	Mbps or Mbit/s	1,000,000 bits per second
Gigabits per second	Gbps or Gbit/s	1,000,000,000 bits per second
Kilobytes per second	KBps	1,000 bytes per second, or 8 kilobits per second
Megabytes per second	MBps	1,000,000 bytes per second, or 8 megabits per second
Gigabytes per second	GBps	1,000,000,000 bytes per second, or 8 gigabits per second

Administrators and techs can periodically conduct throughput tests and keep them on file to create a picture of network performance. If you suspect a problem with the network functioning, you can run a test to compare with past performance to see exactly what is happening.

One thing worth mentioning is the difference between throughput and bandwidth. These terms are often used interchangeably, but they have different meanings. When talking about measuring throughput, you measure the amount of data flow under real-world conditions—measuring with possible electromagnetic interference (EMI) influences, heavy traffic loads, improper wiring, and even network collisions. Take all this into account, take a measurement, and you have the network throughput. Bandwidth, in contrast, refers to the maximum amount of information that can be sent through a particular medium under ideal conditions.

> **Note**
>
> Be sure that you know the difference between throughput and bandwidth/speed.

Cabling Specifications/Limitations

Earlier in this chapter, we discussed the cabling types (Cat 7, Cat 8, and so on) that are available. Each of those categories of cables comes with its own limitations for throughput, speed, and distance—the three variables that a network administrator must so often juggle and balance.

Two types of websites that can be invaluable when it comes to networking are *speed test sites* and *looking-glass sites*. Speed test sites, as the name implies, are *bandwidth speed testers* that report the speed of the connection that you have to them and can be helpful in determining if you are getting the rate your ISP has promised.

Looking-glass sites are servers running *looking-glass (LG)* software that enables you to see routing information. The servers act as a read-only portal giving information about the backbone connection. Most of these servers will show **ping** information, trace (**tracert/traceroute**) information, and Border Gateway Protocol (BGP) information.

Cabling Considerations

When you're considering what cabling option to go with, money is almost always a factor. While the best option is always to use the best cabling and best devices, financial officers often insist that everything be done within budgetary constraints that prohibit always using the best. This means that you often have to work with what you have and try to keep it up and running cost-efficiently when things go wrong.

Damaged or bad wiring could be a patch cable (easy to replace) or the in-wall wiring (more difficult to replace). If you suspect wiring to be the faulty component, you can diagnose rather quickly by taking the device that is having trouble connecting to another location and/or bringing a working machine to this environment. You can use a multifunction cable tester to troubleshoot most wiring problems. You must check for cable continuity, as well as for shorts.

> **Note**
>
> Never assume that the cable you use is good until you test it and confirm that it is good. Sometimes cables break, and bad media can cause network problems.

Bent pins on a network cable or socket can result in very little or no contact being made on those connections. If the problem is with the cable, you can replace the cable. If the problem is with the client machine, fixing it can be difficult because most Ethernet ports are soldered directly to the motherboard. Often, the solution is to abandon that port and use a USB/Ethernet adapter to allow the client to continue to connect to the network.

Cabling Issues

Cabling can be used for many different scenarios; three common ones are as a crossover cable (used to connect any two devices of the same type), a rollover cable (used to connect a computer terminal to a router's console port), and a Power over Ethernet (PoE) cable.

An incorrect cable type—using a crossover cable instead of a standard cable, for instance, or single-mode cable in place of a multimode cable—can keep the host from being able to communicate on the network. Problems can also arise when a lower-category cable (Cat 5, for example) is used in a higher-speed (such as Cat 7) network, or an unshielded cable is used in an installment where a shielded cable is needed.

In many cases, a cable tester can be used to diagnose individual cabling issues, and the solution is to swap the incorrect cable with one suited for the purpose you are intending to use it for.

Signal Degradation

Attenuation refers to the weakening of data signals as they travel through a medium. Network media vary in their resistance to attenuation. Coaxial cable generally is more resistant than *unshielded twisted-pair (UTP)*; *shielded twisted-pair (STP)* is slightly more resistant than UTP; and fiber-optic cable does not suffer from attenuation. That's not to say that a signal does not weaken as it travels over fiber-optic cable, but the correct term for this weakening is *chromatic dispersion* rather than attenuation.

You must understand attenuation or chromatic dispersion and the maximum distances specified for network media. Exceeding a medium's distance without using repeaters can cause hard-to-troubleshoot network problems. A repeater is a network device that amplifies data signals as they pass, enabling them to travel farther. Most attenuation-related or chromatic dispersion-related difficulties on a network require using a network analyzer to detect them.

All media have recommended lengths at which the cable can be run. The reason is that data signals weaken as they travel farther from the point of origin. If the signal travels far enough, it can weaken so much that it becomes unusable. The weakening of data signals as they traverse the medium is called attenuation. The measurement of attenuation is done in decibels; thus, attenuation is also known as *dB loss*.

All copper-based cabling is particularity susceptible to attenuation. When cable lengths have to be run farther than the recommended lengths, signal repeaters can be used to boost the signal as it travels. If you work on a network with intermittent problems, and you notice that cable lengths are run too far, attenuation may be the problem.

> **ExamAlert**
>
> For the Network+ objective referencing cable problems associated with distance, think of attenuation.

Interference

Depending on where network cabling (commonly called *media*) is installed, *interference* can be a major consideration. Two types of media interference can adversely affect data transmissions over network media: *electromagnetic interference (EMI)* and crosstalk (discussed earlier).

EMI is a problem when cables are installed near electrical devices, such as air conditioners or fluorescent light fixtures. If a network medium is placed close enough to such a device, the signal within the cable might become corrupt. Network media vary in their resistance to the effects of EMI. Standard *unshielded twisted-pair (UTP)* cable is susceptible to EMI, whereas fiber cable, with its light transmissions, is resistant to EMI. When deciding on a particular medium, consider where it will run and the impact EMI can have on the installation.

EMI can reduce or corrupt signal strength. This can happen when cables are run too close to everyday office fixtures, such as computer monitors, fluorescent lights, elevators, microwaves, and anything else that creates an electromagnetic field. Again, the solution is to carefully run cables away from such devices. If they have to be run through EMI areas, shielded cabling or fiber cabling is needed.

Improper Termination

Improper termination of a network cable can have several negative effects on network performance. Improper termination can result in poor electrical contact between the cable and the connector, leading to signal loss or attenuation. This can degrade the quality of the transmitted signals and result in data errors or packet loss. Poorly terminated cables are more susceptible to electromagnetic interference (EMI) and radio frequency interference (RFI). This interference can distort the signals traveling through the cable, leading to data corruption and reduced network performance. Improper termination can cause crosstalk, where signals from adjacent wires interfere with each other. Crosstalk can result in signal degradation and data errors, especially in high-speed transmission environments.

Inadequate termination can cause signal reflections at the termination point, leading to impedance mismatches and signal distortion. This can affect signal integrity and increase the likelihood of transmission errors. Improperly terminated cables may be more prone to physical damage, such as broken or bent connector pins, loose connections, or cable fraying. Physical damage can further degrade signal quality and lead to network downtime or intermittent connectivity issues. Overall, improper termination can increase the error rates on the network, resulting in retransmissions, slower data transfer rates, and decreased network reliability.

> **ExamAlert**
>
> Improper termination of network cables can significantly impact network performance by causing signal loss, interference, crosstalk, reflections, physical damage, and increased error rates.

> **ExamAlert**
>
> While troubleshooting common cabling and physical interface problems, be sure you understand signal degradation issues including crosstalk, interference, and attenuation.

Incorrect Pinout

Most splits in a cable are intentional—enabling you to run the wiring in multiple directions with the use of a splitter. Depending on the type of cabling in question, it is not uncommon for each split to reduce the strength of the signal. It is also not uncommon for splitters to go bad. You should split the cable as few times as possible and check the splitter if a problem suddenly occurs in a run that is normally working.

If the split is unintentional, you are often dealing with an open/short, which is discussed later.

Bad Ports

On the router, the port configuration dictates what traffic is allowed to flow through. The router can be configured to enable individual port traffic in, out, or both and is referred to as *port forwarding*. If a port is blocked (such as 443 for HTTPS or 21 for FTP), the data will not be allowed through, and users will be affected.

> **ExamAlert**
>
> Think of port configuration and port forwarding as the same when it comes to the router.

A condition known as a *black hole* can occur when a router does not send back an expected message that the data has been received. It is known as a black hole from the view that data is being sent but is essentially being lost.

This condition occurs when the packet the router receives is larger than the configured size of the *maximum transmission unit (MTU)* and the Don't Fragment flag is configured on that packet. When this condition occurs, the router is supposed to send a "Destination Unreachable" message back to the host. If the packet is not received, the host does not know that the packet did not go through.

Although there are several solutions to this problem, the best is to verify whether a mismatch has occurred between the maximum size packet that clients can send and that the router can handle. You can use **ping** to check that packets of a particular size can move through the router by using the **-l** parameter to set a packet size and the **-f** parameter to set the Do Not Fragment bit.

Open/Short

In addition to the common issue of miswiring, other problems that can occur with cables (and that can be checked with a multifunction cable tester) include *open/short* faults. An open fault means that the cables are not making a full circuit; this can be due to a cut in the cable (across all or some of the wires). A short fault means that the data attempts to travel on wires other than those for which it is intended; this can be caused by miswiring or a twist in the cabling at a cut allowing the bare wires to touch.

> **ExamAlert**
>
> You should expect questions asking you what tool can be used to identify an open/short fault.

LED Status Indicators

Hubs and switches provide light-emitting diodes (LEDs) that provide information on the port status. For instance, by using the LEDs, you can determine whether there is a jabbering network card, whether there is a proper connection to the network device, and whether there are too many collisions on the network.

Duplexing Issues

When configuring a client for the network, you must be aware of two settings: port speed and duplex settings. They are adjusted in Windows in the Network Properties area. Speed and duplex mismatches can slow data rates to a crawl and prevent high-bandwidth applications (such as voice or streaming video) from being possible.

You have several choices for port speed and duplex settings, as Figure 5.15 illustrates. You can choose Auto Negotiation to detect the setting that the network uses. You also can choose one of the other settings to match the network configuration, such as 100 Mbps half duplex. If you work with a client system that is unable to log on to a network, you might need to ensure that the duplex setting and port speeds are correctly set for the network.

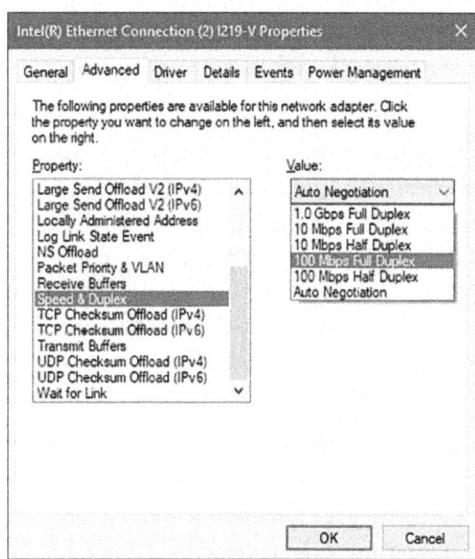

FIGURE 5.15 **Configuring Speed and Duplex Options**

TX/RX Reversed

Two primary types of cables can be used in an Ethernet network: a straight-through cable (as the name implies, all wires run straight through and are the same on both ends) and a crossover cable. In a crossover cable, two pairs of the wires are reversed; these are the TX and RX pairs (transmit and receive).

A crossover cable is intended to be used in specific applications only (such as to directly network two PCs without using a hub or switch) and will cause problems when used where a straight-through cable is called for (as a general rule, in all fixed wiring).

Dirty Optical Cables

Dirty fiber cables—or, more commonly, connectors—can cause a slowdown in traffic due to the need to be able to clearly transmit light. The "dirty" can be caused by exposure to liquids, dust, or other contaminants. Isopropyl alcohol can be used if wet cleaning is necessary (where it is not possible to simply blow away the dust).

Interface Issues

Interface issues are problems that occur at the interface between networking devices, often manifested as anomalies in the data transmitted over the network. Three common interface issues related to network cabling are runts, giants, and drops.

Runts are undersized Ethernet frames that do not meet the minimum frame length requirements specified by the Ethernet protocol (e.g., Ethernet II or IEEE 802.3) and typically occur when a collision or transmission error causes a frame to be prematurely terminated before it reaches the minimum length. Runts can result from issues such as network congestion, electrical interference, or faulty network equipment. They can cause performance degradation and may be indicative of underlying network problems that need to be addressed.

Just the opposite, *giants* are oversized Ethernet frames that exceed the maximum frame length allowed by the Ethernet protocol that often occur when a device incorrectly appends extra data to a frame or when fragmented frames are improperly reassembled. Giants can lead to network congestion, increased latency, and data corruption. Like runts, giants may indicate issues with network configuration, equipment malfunction, or network congestion.

Drops are instances where packets or frames are discarded or lost during transmission or processing. Drops can occur for various reasons, including network congestion, buffer overflows, misconfigured network devices, or hardware faults. Drops can result in data loss, retransmissions, and degraded network performance.

ExamAlert

For the exam, know the differences between runts, giants, and drops.

> **ExamAlert**
>
> Monitoring and troubleshooting dropped packets are essential for maintaining network reliability and performance.

A *cyclic redundancy check (CRC)* is a mathematical algorithm used to detect errors in transmitted data by calculating a checksum, which is appended to the data and verified by the receiving device. CRC errors occur when the data received at the destination device does not match the CRC value calculated from the received data. CRC errors often indicate issues such as signal interference, electrical noise, or physical damage to the cabling, causing data corruption during transmission.

Troubleshooting CRC errors typically involves inspecting and testing the network cabling infrastructure, including cables, connectors, and terminations, to identify any physical or environmental factors contributing to signal degradation. Using diagnostic tools such as network analyzers or cable testers can help pinpoint the location and cause of CRC errors. Remediation actions may include replacing damaged cables or connectors, improving cable routing and management practices, or implementing shielding to reduce electromagnetic interference.

Interface issues associated with network cabling, particularly those related to *port status* such as error disabled, administratively down, and suspended, indicate problems with the operational state of network interfaces. An interface in the *error disabled* state is effectively taken out of service by the network device due to persistent errors detected on the interface. Errors that can trigger this state include excessive CRC errors, input/output errors, late collisions, or other interface-related issues. When an interface is error disabled, it is effectively shut down, and traffic is no longer forwarded through that interface until the issue is resolved.

An interface in the *administratively down* state is intentionally deactivated by a network administrator through manual configuration or administrative commands. This status typically occurs when an administrator explicitly disables the interface using configuration commands on the network device. Interfaces may be administratively down for various reasons, such as maintenance, troubleshooting, or security-related changes.

An interface in the *suspended* state is temporarily disabled by the network device due to a detected issue or configuration mismatch. This status may occur when the device detects a violation of a configured policy, such as a port security violation, or when a port channel encounters operational issues. While suspended, the interface remains inactive, and traffic is not forwarded through it until the issue is resolved or the configuration is adjusted.

ExamAlert

Interface issues leading to a port status of error disabled, administratively down, or suspended can result from various factors, including physical layer problems such as faulty cabling, connector issues, or misconfigurations in network devices. The ability to identify and resolve the underlying causes, whether they are related to cabling, configuration, or device operation, is crucial for restoring normal network operation and minimizing downtime.

Hardware Issues

When it comes to networking, a lot of hardware can have issues that affect the ability to communicate. Two that the exam focuses on, though, are related to Power over Ethernet and transceiver related. Each is explored in the following sections.

Power over Ethernet Issues

When Power over Ethernet (PoE) is used, several common networking problems can arise, but common issues involve the power budget being exceeded and trying to work with an incorrect standard. The *power budget exceeded* problem occurs when the total power required by connected PoE devices exceeds the power budget allocated by the PoE switch or injector. Common causes include connecting high-power devices such as IP cameras, access points, or VoIP phones that draw more power than anticipated. When the power budget is exceeded, the PoE switch or injector may shut down PoE ports, resulting in a loss of power to connected devices and potentially disrupting network services. To mitigate this issue, network administrators can ensure that the PoE switch or injector has sufficient power capacity to support all connected devices, use PoE power management features to prioritize power allocation to critical devices, and/or replace or upgrade the PoE switch or injector to one with a higher power budget if necessary.

Another common issue arises when PoE devices are not compatible with the PoE standard supported by the network infrastructure (*incorrect standard*). PoE standards include IEEE 802.3af, IEEE 802.3at (PoE+), and IEEE 802.3bt (PoE++), each providing different power delivery capabilities. If a PoE device requires more power than the PoE standard supported by the switch or injector, it may not receive sufficient power or may not function correctly. Conversely, connecting a PoE device that only requires lower power to a high-power PoE+ or PoE++ port can result in overpowered operation, potentially damaging the device. To address this issue, network administrators should ensure that PoE devices are compatible with the PoE standard supported by the network infrastructure, verify the power requirements of PoE devices and select appropriate PoE switches or injectors that support the required power levels, and/or use power negotiation features supported by PoE standards to ensure optimal power delivery to connected devices.

Transceiver Issues

Transceiver mismatches occur when the transceivers used at either end of a network link are not compatible or do not support the same communication standards. When troubleshooting an SFP or GBIC, for example, you should make sure that you do not have a bad, or mismatched, transceiver. As simple as this advice may sound, it is important to verify that you are using a single-mode fiber with a single-mode interface and a multimode fiber cable for a multimode interface. Such a fiber type mismatch can cause the physical link to go completely down but does not always do so, thus making troubleshooting it difficult.

> **ExamAlert**
>
> To resolve transceiver mismatches, it is essential to ensure that both ends of the network link use compatible transceivers that support the same communication standards, data rates, and wavelengths.

Signal strength problems occur when the strength of the signals transmitted and received over a network link is inadequate or inconsistent: low signal strength can lead to degraded network performance, increased error rates, and intermittent connectivity issues. Signal strength problems may result from various factors, including distance, obstructions, and environmental factors. To address signal strength issues, network administrators can optimize the placement and orientation of network devices and antennas to maximize signal coverage and minimize signal attenuation; use signal amplifiers, repeaters, or boosters to strengthen signals over long-distance links or in areas with weak signal coverage; and/or employ signal quality monitoring tools to measure and analyze signal strength levels and identify areas requiring improvement or optimization.

Common Tools

A large part of network administration involves having the right tools for the job and knowing when and how to use them. Selecting the correct tool for a networking job sounds like an easy task, but network administrators can choose from a mind-boggling number of tools and utilities.

Given the diverse range of tools and utilities available, it is unlikely that you will encounter all the tools available—or even all those discussed in this chapter. For the Network+ exam, you are required to have general knowledge of the tools available and what they are designed to do.

Until networks become completely wireless, network administrators can expect to spend some of their time using a variety of media-related troubleshooting and installation tools. Some of these tools (such as the tone generator and locator) may be used to troubleshoot media connections, and others (such as wire crimpers and punchdown tools) are used to create network cables and connections.

The Basic Tools

Although many are costly, specialized networking tools and devices are available to network administrators and techs. The most widely used tools cost only a few dollars: the standard screwdrivers we use on almost a daily basis. As a network administrator or tech, you can expect with amazing regularity to take the case off a system to replace a network interface card (NIC) or perhaps remove the cover from a hub or switch to replace a fan. Advanced cable testers and other specialized tools will not help you when a screwdriver is needed.

Cable Crimpers, Strippers, and Snips/Cutters

Wire crimpers, also known as cable crimpers, are tools you might regularly use. Like many things, making your own cables can be fun at first, but the novelty soon wears off. Basically, a wire crimper is a tool that you use to attach media connectors to the ends of cables. For instance, you use one type of wire crimper to attach RJ-45 connectors on unshielded twisted-pair (UTP) cable. You use a different type of wire crimper to attach British Naval Connectors/Bayonet Neill–Concelman (BNCs) to coaxial cabling.

> **Tip**
>
> When making cables, always order more connectors than you need; a few mishaps will probably occur along the way.

In a sense, you can think of a wire crimper as a pair of special pliers. You insert the cable and connector separately into the crimper, making sure that the wires in the cable align with the appropriate connectors. Then, by squeezing the crimper's handles, you force metal connectors through the cable's wires, making the connection between the wire and the connector.

When you crimp your own cables, you need to be sure to test them before putting them on the network. It takes only a momentary lapse to make a mistake when creating a cable, and you can waste time later trying to isolate a problem in a faulty cable.

Two other commonly used wiring tools are strippers and snips/cutters. Wire strippers come in a variety of shapes and sizes. Some are specifically designed to strip the outer sheathing from coaxial cable, and others are designed to work best with UTP cable. All strippers are designed to cleanly remove the sheathing from wire to make sure a clean contact can be made.

Many administrators do not have specialized wire strippers unless they do a lot of work with copper-based wiring. However, standard wire strippers are good things to have on hand.

Wire snips, also known as wire cutters, are tools designed to cleanly cut the cable. Sometimes network administrators buy cable in bulk and use wire snips to cut the cable into desired lengths. The wire strippers are then used to prepare the cable for the attachment of the connectors.

Punchdown Tools

Punchdown tools are used to attach twisted-pair network cable to connectors within a patch panel. Specifically, they connect twisted-pair wires to the insulation displacement connector (IDC).

Tone Generator

A *toner probe* is a device that can save a network installer many hours of frustration. This device has two parts: the *tone generator*, or toner, and the *tone locator*, or probe. The toner sends the tone, and at the other end of the cable, the probe receives the toner's signal. This tool makes it easier to find the beginning and end of a cable. You might hear the tone generator and tone locator referred to as the *fox and hound*.

As you might expect, the purpose of the tone probe is to generate a signal that is transmitted on the wire you are attempting to locate. At the other end, you press the probe against individual wires. When it makes contact with the wire that has the signal on it, the locator emits an audible signal or tone.

The tone locator probe is a useful device, but it does have some drawbacks. First, it often takes two people to operate: one at each end of the cable. Of course, one person could just keep running back and forth, but if the cable is run over great distances, this can be a problem. Second, using the toner probe is time-consuming because it must be attached to each cable independently.

> **Note**
>
> Many problems that can be discovered with a tone generator are easy to prevent by taking the time to properly label cables. If the cables are labeled at both ends, you will not need to use such a tool to locate them.

> **Note**
>
> Toner probes are specifically used to locate cables hidden in floors, ceilings, or walls and to track cables from the patch panels to their destinations.

Loopback Adapter

A number of items fall under the loopback umbrella, and all of them serve the same purpose: they allow you to test a device/configuration/connectivity component using a dummy. The most popular loopback is the address used with **ping**, but Windows also includes a *loopback adapter*, which is a dummy network card (no hardware) used for testing a virtual network environment.

Various loopback adapters—actual hardware—can be purchased and used to test Ethernet jacks, fiber jacks, and so on.

TDR/OTDR

A *time-domain reflectometer (TDR)* is a device used to send a signal through a particular medium to check the cable's continuity. Good-quality TDRs can locate many types of cabling faults, such as a severed sheath, damaged conductors, faulty crimps, shorts, loose connectors, and more. Although network administrators will not need to use a tool such as this every day, it could significantly help in the troubleshooting process. TDRs help ensure that data sent across the network is not interrupted by poor cabling that may cause faults in data delivery.

> **Note**
>
> TDRs work at the physical layer of the OSI model, sending a signal through a length of cable, looking for cable faults.

Because the majority of network cabling is copper based, most tools designed to test cabling are designed for copper-based cabling. However, when you test fiber-optic cable, you need an optical tester.

An optical cable tester performs the same basic function as a wire media tester, but on optical media. The most common problem with an optical cable is a break in the cable that prevents the signal from reaching the other end. Due to the extended distances that can be covered with fiber-optic cables, degradation is rarely an issue in a fiber-optic LAN environment.

Ascertaining whether a signal reaches the other end of a fiber-optic cable is relatively easy, but when you determine that there is a break, the problem becomes locating the break. That's when you need a tool called an *optical time-domain reflectometer (OTDR)*. By using an OTDR, you can locate how far along in the cable the break occurs. The connection on the other end of the cable might be the source of the problem, or perhaps there is a break halfway along the cable. Either way, an OTDR can pinpoint the problem.

Unless you work extensively with fiber-optic cable, you are unlikely to have an OTDR or even a fiber-optic cable tester in your toolbox. Specialized cabling contractors will have them, though, so knowing they exist is important.

You can use a *light meter* to certify and troubleshoot fiber. A light source is placed on one end, and the light meter is used at the opposite end to measure loss.

Multimeter

One of the simplest cable-testing devices is a *multimeter*. By using the continuity setting, you can test for shorts in a length of coaxial cable. Or if you know the correct cable pinouts and have needlepoint probes, you can test twisted-pair cable.

A basic multimeter combines several electrical meters into a single unit that can measure voltage, current, and resistance. Advanced models can also measure temperature.

A multimeter has a display, terminals, probes, and a dial to select various measurement ranges. A digital multimeter has a numeric digital display, and an analog has a dial display. Inside a multimeter, the terminals are connected to different resistors, depending on the range selected.

Network multimeters can do much more than test electrical current:

▶ **Ping specific network devices:** A multimeter can ping and test response times of key networking equipment, such as routers, DNS servers, DHCP servers, and more.

▶ **Verify network cabling:** You can use a network multimeter to isolate cable shorts, split pairs, and other faults.

▶ **Locate and identify cable:** Quality network multimeters enable administrators to locate cables at patch panels and wall jacks using digital tones.

▶ **Provide documentation:** Multimeter results can be downloaded to a PC for inspection. Most network multimeters provide a means such as USB ports to link to a PC.

Cable Tester

A media tester, also called a *cable tester*, defines a range of tools designed to test whether a cable works properly. Any tool that facilitates the testing of a cable can be deemed a cable tester. However, a specific tool called a *media tester* enables administrators to test a segment of cable, looking for shorts, improperly attached connectors, or other cable faults. All media testers tell you whether the cable works correctly and where the problem in the cable might be.

Generically, the phrase *line tester* can be used for any device that tests a media line. Although available products may be Ethernet line testers, fiber line testers, and so on, most often a "line tester" is used to check telephone wiring and usually includes RJ-11 plugs as well as alligator clips.

A *cable certifier* is a type of tester that enables you to certify cabling by testing it for speed and performance to see that the implementation will live up to the ratings. Most stress and test the system based on noise and error testing. You need to know that the gigabit cable you think you have run is actually providing that speed to the network.

> **ExamAlert**
>
> For the exam, be sure you are familiar with cable issues and some of the tools that can help you diagnose them.

Wire Map

A *wire map* (sometimes called a *wiremap*, without the space between the words) is a test (when run, called wire mapping) to see that all Ethernet wiring is correct and there are no opens, shorts, or wires reversed on one end.

Tap

A *tap* is used to connect drop cables to a distribution cable much like a splitter. The difference between a tap and a splitter is that the splitter sends the incoming signal out to all paths equally, whereas a tap can apply a different amount of loss to each output path individually. This way, if you have one short path and one long path coming off the tap, the strength of the signal received by the host at the end of each path can be close to the same.

Fusion Splicer

A *fusion splicer* is an expensive tool used to join two optical cables. The splicing is typically done by an electric arc but could also be a laser or a flame. It is important that the splice be as undetectable as possible in order to keep from scattering or reflecting light as it passes through the splice and reducing the quality of the transmission.

Spectrum Analyzer

A *spectrum analyzer* measures the magnitude of an input signal versus frequency within the full frequency range of the instrument and can be used for a wide range of signals. Today, they are commonly used with Wi-Fi to reveal Wi-Fi hotspots and detect wireless network access with LED visual feedback. Such devices can be configured to scan specific frequencies. When working with 802.11b/g/n/ac/ax networks, you will most certainly require scanning for 2.4 GHz or 5 GHz RF signals.

Such devices can be used in the troubleshooting process to see where and how powerful RF signals are. Given the increase in wireless technologies, RF detectors are sure to continue to increase in popularity.

Fiber Light Meter

A *fiber light meter* measures the light moving through an optical fiber to look for problems with a cable. To use the meter, you connect one end of the fiber to a light source and put the meter on the other end. The meter reads the light it receives and determines the amount of signal loss, if any.

Cram Quiz

1. Which of the following describes the loss of signal strength as a signal travels through a particular medium?

   ○ **A.** Attenuation

   ○ **B.** Crosstalk

   ○ **C.** EMI

   ○ **D.** Chatter

2. A user calls to report periodic problems connecting to the network. Upon investigation, you find that the cable connecting the user's PC to the switch is close to a fluorescent light fixture. What condition is most likely causing the problem?

   ○ **A.** Crosstalk

   ○ **B.** EMI

   ○ **C.** Attenuation

   ○ **D.** Faulty cable

3. With a crossover cable, which two pairs are reversed?

   ○ **A.** RX and SX

   ○ **B.** TX and RX

   ○ **C.** SX and TX

   ○ **D.** SX and CX

4. While you were away, an air conditioning unit malfunctioned in a server room, and some equipment overheated. Which of the following would have alerted you to the problem?

   ○ **A.** Multimeter

   ○ **B.** Environmental monitor

   ○ **C.** TDR

   ○ **D.** OTDR

5. What tool would you use when working with an IDC?

   ○ **A.** Wire crimper

   ○ **B.** Media tester

   ○ **C.** OTDR

   ○ **D.** Punchdown tool

6. As a network administrator, you work in a wiring closet where none of the cables have been labeled. Which of the following tools are you most likely to use to locate the physical ends of the cable?

   ○ **A.** Toner probe

   ○ **B.** Wire crimper

   ○ **C.** Punchdown tool

   ○ **D. ping**

7. You are installing a new system into an existing star network, and you need a cable that is 45 feet long. Your local vendor does not stock cables of this length, so you are forced to make your own. Which of the following tools do you need to complete the task?

   ○ **A.** Optical tester

   ○ **B.** Punchdown tool

   ○ **C.** Crimper

   ○ **D.** UTP splicer

8. Which of the following does improper termination often result in? (Choose all that apply.)

   ○ **A.** Interference

   ○ **B.** Crosstalk

   ○ **C.** Reflections

   ○ **D.** Increased error rates

9. Which of the following is the space between the structural ceiling and a drop-down ceiling commonly used for heating, ventilation, and air conditioning systems and to run network cables?

   ○ **A.** DAC

   ○ **B.** Twinaxial

   ○ **C.** Plenum

   ○ **D.** SFTP+

Cram Quiz Answers

1. **A.** The term used to describe the loss of signal strength for media is *attenuation*. Crosstalk refers to the interference between two cables, EMI is electromagnetic interference, and chatter is not a valid media interference concern.

2. **B.** EMI is a type of interference that is often seen when cables run too close to electrical devices. Crosstalk occurs when two cables interfere with each other. Attenuation is a loss of signal strength. Answer D is incorrect also. It may be that a

faulty cable is causing the problem. However, the question asks for the most likely cause. Because the cable is running near fluorescent lights, the problem is more likely associated with EMI.

3. **B.** In a crossover cable, two pairs of the wires are reversed; these are the TX and RX pairs (transmit and receive).

4. **B.** Environmental monitors are used in server and network equipment rooms to ensure that the temperature does not fluctuate too greatly. In the case of a failed air conditioner, the administrator is alerted to the drastic changes in temperature. Multimeters, TDRs, and OTDRs are used to work with copper-based media.

5. **D.** You use a punchdown tool when working with an IDC. All the other tools are associated with making and troubleshooting cables; they are not associated with IDCs.

6. **A.** The toner probe tool, along with the tone locator, can be used to trace cables. Crimpers and punchdown tools are not used to locate a cable. The **ping** utility would be of no help in this situation.

7. **C.** When you're attaching RJ-45 connectors to UTP cables, the wire crimper is the tool you use. None of the other tools listed are used in the construction of UTP cable.

8. **A, B, C, D.** Improper termination of network cables can significantly impact network performance by causing signal loss, interference, crosstalk, reflections, physical damage, and increased error rates.

9. **C.** Plenum is the space between the structural ceiling and a drop-down ceiling. It is commonly used for heating, ventilation, and air-conditioning systems and to run network cables. Direct attach copper (DAC) cabling is a type of high-speed, short-distance cabling commonly used to connect networking equipment such as switches, routers, and servers. The cables consist of twinaxial copper cable assemblies with connectors, usually SFP+ (small form-factor pluggable) or QSFP (quad small form-factor pluggable), attached at both ends.

What's Next?

This chapter focused on wiring solutions. Chapter 6, "Wireless Solutions," looks at wireless solutions. Client systems communicate with a wireless access point using wireless LAN adapters. Such adapters are built into or can be added to laptops, handhelds, desktop computers, and even IoT devices. Wireless LAN adapters provide the communication point between the client system and the airwaves via an antenna.

CHAPTER 6

Wireless Solutions

This chapter covers the following official Network+ objectives:

▶ 2.3 Given a scenario, select and configure wireless devices and technologies.

This chapter covers CompTIA Network+ objective 2.3. For more information on the official Network+ exam topics, see the "About the Network+ Exam" section in the Introduction.

While once common, it is now rare to find a network that is wired-only. Networks of all shapes and sizes incorporate wireless segments into their networks. Home wireless networking has also grown significantly in the past few years with some home networks having speeds and capabilities surpassing those of small businesses.

Wireless networking enables users to connect to a network using radio waves instead of wires. Network users within range of a wireless *access point (AP)* can move around an office or any other location within range of a hotspot freely, without needing to plug into a wired infrastructure. The benefits of wireless networking clearly have led to its continued growth.

This chapter explores the many facets of wireless networking, starting with some of the concepts and technologies that make wireless networking possible.

Understanding Wireless Basics

▶ **2.3 Given a scenario, select and configure wireless devices and technologies.**

CramSaver

If you can correctly answer these questions before going through this section, save time by skimming the ExamAlerts in this section and then completing the Cram Quiz at the end of the section.

1. How many nonoverlapping channels are supported by 802.11a?

2. What are the ranges the 802.11b and 802.11g standards operate in?

3. True or false: Linux users can use the **iwconfig** command to view the state of their wireless network.

4. What does WPA3-Personal enable that replaces preshared key (PSK) in WPA2-Personal?

Answers

1. The 802.11a standard supports up to eight nonoverlapping channels.

2. The 802.11b and 802.11g standards operate in the 2.4 to 2.497 GHz range.

3. True. Linux users can use the **iwconfig** command to view the state of their wireless network.

4. For better password protection, WPA3-Personal uses Simultaneous Authentication of Equals (SAE), which replaces preshared key (PSK) in WPA2-Personal.

ExamAlert

Remember that this objective begins with "Given a scenario." This means that you may receive a drag-and-drop, matching, or "live OS" scenario where you have to click through to complete a specific objective-based task.

Wireless Channels and Frequencies

Radio frequency (RF) channels are an important part of wireless communication. A *channel* is the band of RF used for the wireless communication. Each IEEE wireless standard specifies the channels that can be used. *Channel width* is the width or bandwidth of the radio frequency channel used for transmitting data, and it determines the amount of spectrum available for data transmission and affects the data transfer rate and network performance. Channel width is typically measured in megahertz (MHz) and can vary depending on the wireless technology and the specific configuration of the network.

The 802.11a standard specifies radio frequency ranges between 5.15 and 5.875 GHz. In contrast, 802.11b and 802.11g standards operate in the 2.4 to 2.497 GHz range. 802.11n (known as *Wi-Fi 4*) can operate in either 2.4 GHz or 5 GHz ranges, and 802.11ac (known as *Wi-Fi 5*) operates in the 5 GHz range, while 802.11ax (known as *Wi-Fi 6*) can use 2.4 GHz or 5 GHz ranges. Wi-Fi 6E is an extension of the Wi-Fi 6 standard (802.11ax) and operates in the newly allocated 6 GHz frequency band: this additional spectrum provides more available channels and bandwidth, enabling Wi-Fi 6E devices to deliver higher performance, lower latency, and reduced interference compared to previous Wi-Fi generations.

Band steering is a technique used to optimize the utilization of different frequency bands, particularly in dual-band or tri-band Wi-Fi networks. It aims to improve network performance and client experience by directing devices to connect to the most suitable frequency band based on their capabilities and the network conditions. Many modern Wi-Fi routers and access points support dual-band (2.4 GHz and 5 GHz) or tri-band (2.4 GHz, 5 GHz, and 6 GHz) operation. Each frequency band has its advantages and limitations in terms of coverage, throughput, and interference. When a device attempts to connect to a Wi-Fi network, the access point or router employing band steering evaluates the capabilities of the client device. It determines whether the device supports both 2.4 GHz and 5 GHz bands and whether it has preferences or restrictions for certain bands. Based on the client's capabilities and the network conditions, band steering decides which frequency band offers the best performance and user experience for the device. Factors such as signal strength, congestion, interference, and available channels are taken into account in this decision-making process.

ExamAlert

While studying for the exam, know that frequency options include 2.4 GHz, 5 GHz, and 6 GHz. Know, as well, that band steering directs devices to connect to the most suitable frequency band based on their capabilities and network conditions.

Note

In general, band steering aims to steer capable devices to connect to the less congested and higher-performance 5 GHz band whenever possible.

There has been much discussion about Wi-Fi congestion and the need for regulatory intervention, particularly in the 2.4 GHz range, but the *regulatory impact* of such intervention has largely prevented attempts to limit the use of these channels.

Note

Standards such as IEEE *802.11h* specifically address regulatory requirements for operating WLANs in the 5 GHz frequency band, aiming to improve coexistence with other systems and reduce interference. One of the primary features introduced by 802.11h is Dynamic Frequency Selection (DFS), which enables WLAN devices to detect radar signals in the 5 GHz band and automatically switch to another channel to avoid interference. This Dynamic Frequency Selection helps ensure that WLANs operate without causing harmful interference to radar systems, which share the same frequency band. Another important aspect of 802.11h is Transmit Power Control (TPC), which allows WLAN devices to adjust their transmit power levels based on environmental conditions and regulatory requirements. Compliance with the 802.11h standard is essential for WLAN deployments in Europe, where regulatory agencies enforce specific requirements for operating WLANs in the 5 GHz band. Adhering to the DFS and TPC requirements outlined in 802.11h helps ensure that WLANs meet regulatory standards and operate legally within the designated frequency bands.

Note

Hertz (Hz) is the standard of measurement for radio frequency. Hertz is used to measure the frequency of vibrations and waves, such as sound waves and electromagnetic waves. One hertz is equal to one cycle per second. RF is measured in *kilohertz (KHz)*, one thousand cycles per second; *megahertz (MHz)*, one million cycles per second; or *gigahertz (GHz)*, one billion cycles per second.

As far as channels are concerned, 802.11a has a wider frequency band, enabling more channels and therefore more data throughput. As a result of the wider band, 802.11a supports up to eight nonoverlapping channels. 802.11b/g standards use the smaller band and support only up to three nonoverlapping channels.

It is recommended that nonoverlapping channels be used for communication. In the United States, 802.11b/g standards use 11 channels for data communication, as mentioned; three of these—channels 1, 6, and 11—are nonoverlapping. Most manufacturers set their default channel to one of the nonoverlapping channels to avoid transmission conflicts. With wireless devices, you can select which channel your WLAN operates on to avoid interference from other wireless devices that operate in the 2.4 GHz frequency range.

ExamAlert

While studying for the exam, know that 802.11b/g standards use 11 channels for data communication, of which three are nonoverlapping.

When troubleshooting a wireless network, be aware that overlapping channels can disrupt the wireless communications. For example, in many environments, APs are inadvertently placed close together—perhaps two APs in separate offices located next door to each other or between floors. Signal disruption results if channel overlap exists between the APs. The solution is to try to move the AP to avoid the overlap problem, or to change channels to one of the other nonoverlapping channels. For example, you could switch from channel 6 to channel 11.

Typically, you would change the channel of a wireless device only if it overlapped with another device. If a channel must be changed, it must be changed to another nonoverlapping channel. Table 6.1 shows the channel ranges for 802.11b/g wireless standards. Table 6.2 shows the channel ranges for 802.11a. 802.11n added the option of using both channels used by 802.11a and b/g and operating at 2.4 GHz/5 GHz. As such, you can think of 802.11n as an amendment that improved upon the previous 802.11 standards by adding *multiple input, multiple output (MIMO)* antennas and a huge increase in the data rate. 802.11n devices are still available, but they have largely been superseded today by 802.11ac, which became an approved standard in January 2014, and 802.11ax (which uses MU-MMO and is discussed later). Both 802.11ac and 802.11ax can be thought of as extensions of 802.11n.

ExamAlert

When troubleshooting a wireless problem in Windows, you can use the **ipconfig** command to see the status of IP configuration. Similarly, the **ip** command can be used in Linux. In addition, Linux users can use the **iwconfig** command to view the state of your wireless network. Using **iwconfig**, you can view such important information as the link quality, AP MAC address, data rate, and encryption keys, which can be helpful in ensuring that the parameters in the network are consistent.

Note

IEEE 802.11b/g wireless systems communicate with each other using radio frequency signals in the band between 2.4 GHz and 2.5 GHz. Neighboring channels are 5 MHz apart. Applying two channels that allow the maximum channel separation decreases the amount of channel crosstalk and provides a noticeable performance increase over networks with minimal channel separation.

Tables 6.1 and 6.2 outline the available wireless channels. When you're deploying a wireless network, it is recommended that you use channel 1, grow to use channel 6, and add channel 11 when necessary, because these three channels do not overlap.

ExamAlert

The 802.11n, 802.11ac, and 802.11ax standards are the most common today, and you will be hard-pressed to purchase (or even find) older technologies. It is, however, recommended that you know the older technologies for the exam.

TABLE 6.1 **RF Channels for 802.11b/g/n/ax**

Channel	Frequency Band
1	2412 MHz
2	2417 MHz
3	2422 MHz
4	2427 MHz
5	2432 MHz
6	2437 MHz
7	2442 MHz
8	2447 MHz
9	2452 MHz
10	2457 MHz
11	2462 MHz

Note

When looking at Table 6.1, remember that the RF channels listed (2412 for channel 1, 2417 for 2, and so on) are actually the center frequency that the transceiver within the radio and AP uses. There is only a 5 MHz separation between the center frequencies, and an 802.11b signal occupies approximately 30 MHz of the frequency spectrum. As a result, data signals fall within about 15 MHz of each side of the center frequency and overlap with several adjacent channel frequencies. This leaves you with only three channels (channels 1, 6, and 11 for the United States) that you can use without causing interference between APs.

TABLE 6.2 **RF Channels for 802.11a/ac/ax**

Channel	Frequency
36	5180 MHz
40	5200 MHz
44	5220 MHz
48	5240 MHz

Channel	Frequency
52	5260 MHz
56	5280 MHz
60	5300 MHz
64	5320 MHz

As mentioned, channels 1, 6, and 11 do not overlap. On a non-MIMO setup (such as with 802.11a, b, or g), always try to use one of these three channels. Similarly, if you use 802.11n/ac/ax with 20 MHz channels, stay with channels 1, 6, and 11 to be safe even though 802.11ac and ax channels can be 20 MHz, 40 MHz, 80 MHz, and 160 MHz wide.

ExamAlert

Understand the importance of channels 1, 6, and 11 as you study for the exam.

ExamAlert

For the exam, you should know the values in Table 6.2.

It is important to note that 802.11ac operates in the 5 GHz range only, while 802.11ax operates in both the 2.4 GHz and 5 GHz ranges (which no other standard had done since 802.11n) and is, thus, compatible with 802.11a/b/g/n/ac. Operating in both ranges creates more available channels (early chipsets, for example, support 8 channels in the 5 GHz and 4 channels in the 2.4 GHz range for a total of 12 available channels). With 802.11ac, MU-MIMO is limited to only downlink transmissions while 802.11ax creates MU-MIMO connections; consequently, a downlink MU-MIMO access point can transmit concurrently to multiple receivers, and an uplink MU-MIMO endpoint can simultaneously receive from multiple transmitters.

The 802.11ax standard supports up to eight MU-MIMO transmissions at a time (an increase from the four available with 802.11ac). Orthogonal frequency-division multiple access (OFDMA) is new with 802.11ax (and discussed in the upcoming section on channel bonding), as are several other technologies (including trigger-based random access, dynamic fragmentation, and spatial frequency reuse), enabling it to have a theoretical maximum speed of approximately 10 Gbps. New with 802.11ax is the use of 1024-QAM (quadrature amplitude modulation) to encode (modulate/demodulate) a larger number of data bits and increase throughput.

A subcategory of 802.11ax, known as Wi-Fi 6e (Wi-Fi 6 extended), will also work in the 6 GHz frequency: devices that are compatible will be able to operate on the 2.4, 5, and 6 GHz frequencies and benefit from less congested bands.

Table 6.3 offers a quick comparison of the wireless computing standards of relevance.

TABLE 6.3   **Wireless Computing Standards**

Standard	Time Period	Frequency	Maximum Data Range	Range	Examples
802.11a	Released in 1999	5 GHz	Up to 54 Mbps	35–75 meters	Older routers, access points
802.11b	Introduced in 1999	2.4 GHz	Up to 11 Mbps	35–100 meters	Early Wi-Fi devices
802.11g	Ratified in 2003	2.4 GHz	Up to 54 Mbps	38–140 meters	Residential Wi-Fi routers, laptops
802.11n	Finalized in 2009	2.4 GHz and 5 GHz	Up to 600 Mbps	70–250 meters	Modern Wi-Fi routers, smartphones
802.11ac	Introduced in 2013	5 GHz	Up to 1.3 Gbps	35–70 meters	Latest Wi-Fi routers, high-speed devices
802.11ax	Launched in 2019	2.4 GHz and 5 GHz with 6 GHz capability	Up to 9.6 Gbps	70–250 meters	Next-generation Wi-Fi routers, devices

Speed, Distance, and Bandwidth

When talking about wireless transmissions, you need to distinguish between *throughput* and *data rate*. From time to time, these terms are used interchangeably, but technically speaking, they are different. As shown later in this chapter, each wireless standard has an associated speed. For instance, 802.11n lists a theoretical speed of up to 600 Mbps, and 802.11ax has a theoretical maximum speed of a whopping 10 Gbps. This represents the speed at which devices using this standard can send and receive data. However, in network data transmissions, many factors prevent the actual speeds from reaching this end-to-end theoretical maximum. For instance, data packets include overhead such as routing information, checksums, and error recovery data. Although this might all be necessary, it can impact overall speed.

The number of clients on the network can also impact the data rate; the more clients, the more collisions. Depending on the network layout, collisions can have a significant impact on end-to-end transmission speeds. Wireless network signals degrade as they pass through obstructions such as walls or doors; the signal speed deteriorates with each obstruction.

All these factors leave you with the actual throughput of wireless data transmissions. Goodput represents the actual speed to expect from wireless transmissions (what is often thought of as throughput). In practical applications, wireless transmissions are approximately one-half or less of the data rate. Depending on the wireless setup, the transmission rate could be much less than its theoretical maximum.

> **ExamAlert**
>
> *Data rate* refers to the theoretical maximum of a wireless standard, such as the 600 Mbps for 802.11n or the 10 Gbps for 802.11ax. *Throughput* refers to the actual speeds achieved after all implementation and interference factors.

> **Note**
>
> Speed is always an important factor in the design of any network, and *high through-put (ht)* is a goal that has been around for a while. A number of the 802.11 standards offer a high throughput connection type, such as *802.11a-ht* and *802.11g-ht*. Although these implementations are better with the ht than without, it is true that today you will achieve better results with 802.11ax.

Channel Bonding

With channel bonding, you can use two channels at the same time. As you might guess, the ability to use two channels at once increases performance. Bonding can help increase wireless transmission rates with 802.11n from a maximum of 40 MHz up to 80 or even 160 MHz (for speed increases of 117 or 333 percent, respectively). 802.11n uses the orthogonal frequency-division multiplexing (OFDM) transmission strategy.

Whereas 802.11n stopped at four spatial streams, 802.11ac goes to eight (for another 100 percent speed increase). 802.11ax replaces OFDM with OFDMA, a multiuser version of OFDM that uses a digital modulation scheme. The multiple access is achieved by assigning subsets of subcarriers to individual users.

> **ExamAlert**
>
> If it seems as though 802.11ax keeps coming up in each discussion, the reason is that it is important to know this standard for the exam. Know that it works in both bands (2.4 and 5 GHz), utilizes multiuser MIMO (downlink and uplink), OFDMA (downlink and uplink), and higher data rates (thanks to 1024-QAM).

MIMO/MU-MIMO/Directional/Omnidirectional

A wireless antenna is an integral part of overall wireless communication. Antennas come in many shapes and sizes, with each one designed for a specific purpose. Selecting the right antenna for a particular network implementation is a critical consideration, and one that could ultimately decide how successful a wireless network will be. In addition, using the right antenna can save you money on networking costs because you need fewer antennas and APs.

> **ExamAlert**
>
> *Multiple input, multiple output (MIMO)* and *multiuser multiple input, multiple output (MU-MIMO)* are advanced antenna technologies that are key in wireless standards such as 802.11n, 802.11ac, 802.11ax, and LTE.

Many small home network adapters and APs come with a nonupgradeable antenna, but higher-grade wireless devices require you to choose an antenna. Determining which antenna to select takes careful planning and requires an understanding of what range and speed you need for a network. The antenna is designed to help wireless networks do the following:

▶ Work around obstacles

▶ Minimize the effects of interference

▶ Increase signal strength

▶ Focus the transmission, which can increase signal speed

The following sections explore some of the characteristics of wireless antennas.

Antenna Ratings

When a wireless signal is low and is affected by heavy interference, it might be possible to upgrade the antenna to create a more solid wireless connection.

To determine an antenna's strength, refer to its *gain value*. But how do you determine the gain value?

> **ExamAlert**
>
> For the exam, know that an antenna's strength is its gain value.

Suppose that a huge wireless tower is emanating circular waves in all directions. If you could see these waves, you would see them forming a sphere around the tower. The signals around the antenna flow equally in all directions, including up and down. An antenna that does this has a 0 dBi gain value and is called an *isotropic antenna*. The isotropic antenna rating provides a base point for measuring actual antenna strength.

> **Note**
>
> The *dB* in dBi stands for decibels, and the *i* stands for the hypothetical isotropic antenna.

An antenna's gain value represents the difference between the 0dBi isotropic and the antenna's power. For example, a wireless antenna advertised as 15dBi is 15 times stronger than the hypothetical isotropic antenna. The higher the decibel figure, the higher the gain.

When looking at wireless antennas, remember that a higher gain value means stronger send and receive signals. In terms of performance, the general rule is that every 3dB of gain added doubles an antenna's effective power output.

Antenna Coverage

When selecting an antenna for a particular wireless implementation, you need to determine the type of coverage the antenna uses. In a typical configuration, a wireless antenna can be either *omnidirectional* or *directional* (also called unidirectional). Which one you choose depends on the wireless environment.

An omnidirectional antenna is designed to provide a 360-degree dispersed wave pattern. This type of antenna is used when coverage in all directions from the antenna is required. Omnidirectional antennas are advantageous when a broad-based signal is required. For example, if you provide an even signal in all directions, clients can access the antenna and its associated AP from various locations. Because of the dispersed nature of omnidirectional antennas, the signal is weaker overall and therefore accommodates shorter signal distances.

Omnidirectional antennas are great in an environment that has a clear line of sight between the senders and receivers. The power is evenly spread to all points, making omnidirectional antennas well suited for home and small office applications.

Directional antennas are designed to focus the signal in a particular direction (which is why they are often referred to as unidirectional). This focused signal enables greater distances and a stronger signal between two points. The greater distances enabled by directional antennas give you a viable alternative for connecting locations, such as two offices, in a point-to-point configuration.

Directional antennas are also used when you need to tunnel or thread a signal through a series of obstacles. This arrangement concentrates the signal power in a specific direction and enables you to use less power for a greater distance than an omnidirectional antenna. Table 6.4 compares omnidirectional and directional wireless antennas.

TABLE 6.4 **Comparing Omnidirectional and Directional Antennas**

Characteristic	Omnidirectional	Directional	Advantage/Disadvantage
Wireless area coverage	General coverage area	Focused coverage area	Omnidirectional allows 360-degree coverage, giving it a wide coverage area. Directional provides a targeted path for signals to travel.
Wireless transmission range	Limited	Long point-to-point range	Omnidirectional antennas provide a 360-degree coverage pattern and, as a result, far less range. Directional antennas focus the wireless transmission; this focus enables greater range.
Wireless coverage shaping	Restricted	The directional wireless range can be increased and decreased.	Omnidirectional antennas are limited to their circular pattern range. Directional antennas can be adjusted to define a specific pattern, wider or more focused.

> **Note**
>
> In the wireless world, *polarization* refers to the direction in which the antenna radiates wavelengths. This direction can be vertical, horizontal, or circular. Today, vertical antennas are perhaps the most common. As far as the configuration is concerned, the sending and receiving antennas should be set to the same polarization.

ExamAlert

Omnidirectional antennas provide wide coverage but weaker signal strength in any one direction than a directional antenna.

Network Types

The difference between wireless network types lies in their architecture, purpose, and how devices communicate within the network. In a *mesh* network, devices communicate with each other dynamically and cooperatively to relay data across the network. Each node in the mesh network acts as a relay point, forwarding data to other nodes until it reaches its destination. Mesh networks are self-healing, meaning if one node fails or becomes unreachable, data can automatically reroute through alternate paths: and they are often used in scenarios where extended coverage, fault tolerance, and resilience are required, such as smart home systems, industrial IoT deployments, and outdoor wireless networks.

An *ad hoc* network, also known as a peer-to-peer network, is formed when devices connect directly to each other without the need for a central access point (AP). Devices in an ad hoc network communicate with each other directly, typically within a limited range, without relying on infrastructure components. Ad hoc networks are useful in situations where network infrastructure is unavailable or impractical, such as peer-to-peer file sharing, temporary network setups, and spontaneous collaboration among devices.

An *infrastructure* network is the most common type of wireless network, where devices connect to a central access point (AP) or multiple APs connected to a wired network. In an infrastructure network, the AP acts as a bridge between wireless clients and the wired network, facilitating communication between devices and providing Internet access. Devices in an infrastructure network communicate with each other through the AP, which manages the network and controls access to resources. Infrastructure networks are widely used in homes, offices, public hotspots, and enterprise environments for Internet access and local network connectivity.

A *point-to-point* network consists of two devices or nodes connected directly to each other to establish a dedicated communication link. Unlike other network types where multiple devices communicate with each other, point-to-point networks involve only two endpoints. Point-to-point links are commonly used for long-distance communication, such as building-to-building connections, backhaul links, and wireless bridges. They provide a high-speed, dedicated connection between two locations without the need for additional infrastructure or intermediate devices.

Table 6.5 compares wireless networking types.

TABLE 6.5 **Comparing Networking Types**

Type	Description	Characteristics	Examples
Mesh	A decentralized network architecture where each node relays data for the network	Redundant paths for data transmission, self-healing capabilities, scalability, suitable for large areas or environments with obstacles	Wireless sensor networks, smart home systems, municipal Wi-Fi networks
Ad hoc	A peer-to-peer network where wireless devices communicate with each other without the need for a central access point	No infrastructure required, devices can dynamically join or leave the network, suitable for temporary or mobile networks, limited range due to direct communication between devices	Mobile devices creating temporary networks, peer-to-peer file sharing, emergency response teams
Point-to-Point	A dedicated link established between two endpoints, typically used for connecting two distant locations	Direct communication between two points, high-speed and reliable connection, limited to two endpoints, suitable for long-distance communication	Wireless bridges connecting buildings, satellite communication links, point-to-point wireless backhaul in telecommunications
Infrastructure	A network architecture where devices communicate through a centralized access point or base station	Centralized management, wide coverage area, support for many devices, suitable for permanent installations or environments with high device density	Wi-Fi networks in homes, offices, public spaces, cellular networks using base stations, public Wi-Fi hotspots

> **Note**
>
> Each time *mesh* is mentioned, so is *self-healing*. This is the ability to automatically detect and recover from faults or disruptions without manual intervention. When a self-healing network detects an issue such as a failed component, congestion, or a broken link, it initiates predefined processes or algorithms to restore functionality and maintain continuous operation. This could involve rerouting traffic, activating backup resources, or dynamically adjusting configurations to bypass the problem area.

> **ExamAlert**
>
> Be sure you are familiar with the characteristics of the network types including mesh, ad hoc, point-to-point, and infrastructure.

Establishing Communications Between Wireless Devices

When you work with wireless networks, you must have a basic understanding of the communication that occurs between wireless devices. If you use an infrastructure wireless network design, the network has two key parts: the wireless client, also known as the *station (STA)*, and the AP. The AP acts as a *bridge* (or wireless bridge) between the STA and the wired network.

> **ExamAlert**
>
> When a single AP is connected to the wired network and to a set of wireless stations, it is called a *basic service set (BSS)*. An *extended service set (ESS)* describes the use of multiple BSSs that form a single subnetwork. Ad hoc mode is sometimes called an *independent basic service set (IBSS)*.

As with other forms of network communication, before transmissions between devices can occur, the wireless AP and the client must begin to talk to each other. In the wireless world, this is a two-step process involving *association* and *authentication*.

The association process occurs when a wireless adapter is turned on. The client adapter immediately begins scanning the wireless frequencies for wireless APs or, if using ad hoc mode, other wireless devices. When the wireless client is configured to operate in infrastructure mode, the user can choose a wireless AP with which to connect. This process may also be automatic, with the AP selection based on the SSID, signal strength, and frame error rate. Finally, the wireless adapter switches to the assigned channel of the selected wireless AP and negotiates the use of a port.

If at any point the signal between the devices drops below an acceptable level, or if the signal becomes unavailable for any reason, the wireless adapter initiates another scan, looking for an AP with stronger signals. When the new AP is located, the wireless adapter selects it and associates with it. This is known as *reassociation*.

> **ExamAlert**
>
> The 802.11 standards enable a wireless client to roam between multiple APs. An AP transmits a beacon signal every so many milliseconds. It includes a time stamp for client synchronization and an indication of supported data rates. A client system uses the beacon message to identify the strength of the existing connection to an AP. If the connection is too weak, the *roaming* client attempts to associate itself with a new AP. This association enables the client system to roam between distances and APs.

With the association process complete, the authentication process begins. After the devices associate, keyed security measures are applied before communication can take place. On many APs, authentication can be set to either *shared key authentication* or *open authentication*. The default setting for older APs typically is open authentication. Open authentication enables access with only the SSID and/or the correct WEP key for the AP. The problem with open authentication is that if you do not have other protection or authentication mechanisms in place, your wireless network is totally open to intruders. When set to shared key mode, the client must meet security requirements before communication with the AP can occur.

After security requirements are met, you have established IP-level communication. This means that wireless standard requirements have been met, and Ethernet networking takes over. There is basically a switch from 802.11 to 802.3 standards. The wireless standards create the physical link to the network, enabling regular networking standards and protocols to use the link. This is how the physical cable is replaced, but to the networking technologies there is no difference between regular cable media and wireless media.

Several components combine to enable wireless communications between devices. Each of these must be configured on both the client and the AP:

▶ **Service set identifier (SSID):** Whether your wireless network uses infrastructure mode or ad hoc mode, an SSID is required. The SSID is a configurable client identification that enables clients to communicate with a particular base station. Only client systems configured with the same SSID as the AP can communicate with it. SSIDs provide a simple password arrangement between base stations and clients in a basic service set identifier (BSSID) network. Extended service set identifiers (ESSIDs) are used for the ESS wireless network. BSSID is the MAC address assigned to a specific access point (AP) or wireless router, and each AP in a wireless network has a unique BSSID, which is used to identify and differentiate it from other APs within the same network. The ESSID is the name of the wireless network or the network's "broadcast" name: it is set by the network administrator and is shared among all APs in a network to create a single logical wireless network.

ExamAlert

ESSID is not unique to a single AP; multiple APs within the same network share the same ESSID. BSSID identifies individual access points, while ESSID identifies the entire wireless network or WLAN.

▶ **Wireless channel:** As stated earlier in the chapter, RF channels are an important part of wireless communications. A channel is the frequency band used for the wireless communication. Each standard specifies the channels that can be used. The 802.11a standard specifies radio frequency ranges between 5.15 GHz and 5.875 GHz. In contrast, the 802.11b and 802.11g standards operate in the 2.4 GHz to 2.497 GHz ranges. 802.11n and 802.11ax can operate in either the 2.4 GHz or 5 GHz range, and 802.11ac is at 5 GHz. Fourteen channels are defined in the IEEE 802.11 channel set, 11 of which are available in North America.

▶ **Security features:** IEEE 802.11 provides security using two methods: authentication and encryption. Authentication verifies the client system. In infrastructure mode, authentication is established between an AP and each station. Wireless encryption services must be the same on the client and the AP for communication to occur.

ExamAlert

Wireless devices ship with default SSIDs, security settings, channels, passwords, and usernames. To protect yourself, it is strongly recommended that you change these default settings. Today, many Internet sites list the default settings used by manufacturers with their wireless devices. This information is used by people who want to gain unauthorized access to your wireless devices.

Guest Networks

A *guest network* is a separate network SSID that is typically configured to provide Internet access to guests, visitors, or temporary users without compromising the security of the primary network. It allows guests to access the Internet while keeping them isolated from the main network, thereby protecting sensitive data and resources. Guest networks often have restrictions on the services and resources that guests can access. For example, they may allow Internet access only and block access to local network resources.

Administrators can implement bandwidth limits or Quality of Service (QoS) policies to ensure that guest network usage does not adversely affect the performance of the primary network. Depending on the configuration, guest networks may require authentication through a *captive portal*, where users must agree to terms of service or enter a password before gaining access to the Internet.

Captive portals are web pages that guests are redirected to when they attempt to connect to the guest network. They typically present users with a login page, terms of service agreement, acceptable use policy (AUP), or other authentication methods before granting access to the Internet. Captive portals may also provide branding opportunities for organizations and collect information about guest users for marketing or analytics purposes.

Configuring the Wireless Connection

Wireless connection configuration is fairly straightforward. Figure 6.1 shows an example of a simple wireless router. In addition to providing wireless access, it also includes a four-port wired switch.

FIGURE 6.1   **A Wireless Broadband Router for a Small Network**

Most of the broadband routers similar to the one shown in Figure 6.1 differ based on the following features:

▶ **Wireless bands:** The routers can provide only 2.4 GHz, only 5 GHz, or be either selectable (choosing one of the two) or simultaneous (using both).

▶ **Switch speed:** The ports on the switch can usually support either Fast Ethernet (10/100 Mbps) or Gigabit Ethernet (10/100/1000 Mbps).

▶ **Security supported:** The SSID, security mode, and passphrase may be configurable for each band, and some routers include a push-button feature for accessing setup. Some enable you to configure MAC address filtering and guest access, such as the one shown in Figure 6.2. MAC address filtering enables you to limit access to only those specified hosts. Guest access uses a different password and network name and enables visitors to use the Internet without having access to the rest of the network (thus avoiding your data and computers).

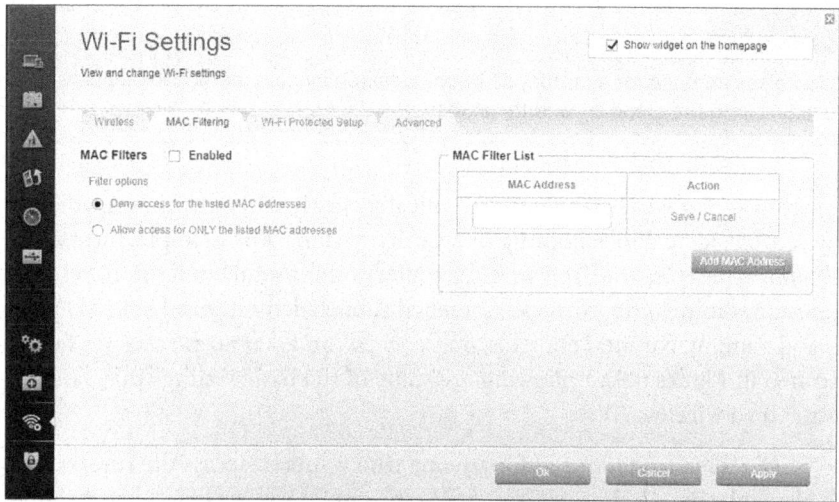

FIGURE 6.2 Configuring MAC Address Filtering on a SOHO Router

ExamAlert

Make sure that you understand the purpose of MAC address filtering.

▶ **Antenna:** The antenna may be a single external pole, two poles or even more, or be entirely internal. The model shown in Figure 6.1 uses an internal antenna, as shown in Figure 6.3.

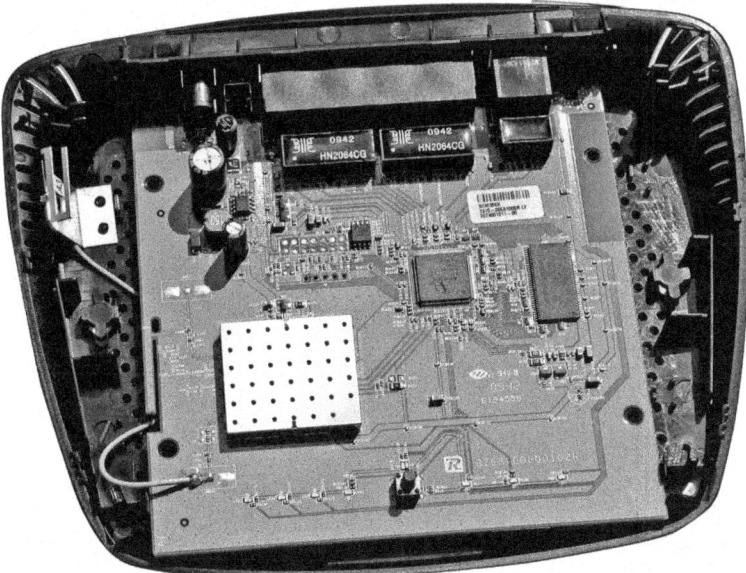

FIGURE 6.3 **The Antenna Is the Wire and Metal Component on the Left**

> **Note**
>
> The wireless antenna for a laptop, all-in-one desktop system, or mobile device is often built in to the areas around the screen.

The settings for a wireless router are typically clearly laid out. You can adjust many settings for troubleshooting or security reasons. For example, most newer *small office/home office (SOHO)* wireless routers offer useful configuration setup screens for administering firewall, screened subnet/demilitarized zone (DMZ), apps and gaming, parental controls, guest access, and diagnostic settings (as illustrated in Figure 6.4). Following are some of the basic settings that can be adjusted on a wireless AP:

▶ **SSID:** This name is used for anyone who wants to access the Internet through this wireless AP. The SSID is a configurable client identification that enables clients to communicate with a particular base station. In an application, only clients configured with the same SSID can communicate with base stations having the same SSID. SSID provides a simple password arrangement between base stations and clients.

As far as troubleshooting is concerned, if a client cannot access a base station, you need to ensure that both use the same SSID. Incompatible SSIDs are sometimes found when clients move computers, such as laptops or other mobile devices, between different wireless networks. They obtain an SSID from one network. If the system is not rebooted, the old SSID does not enable communication with a different base station.

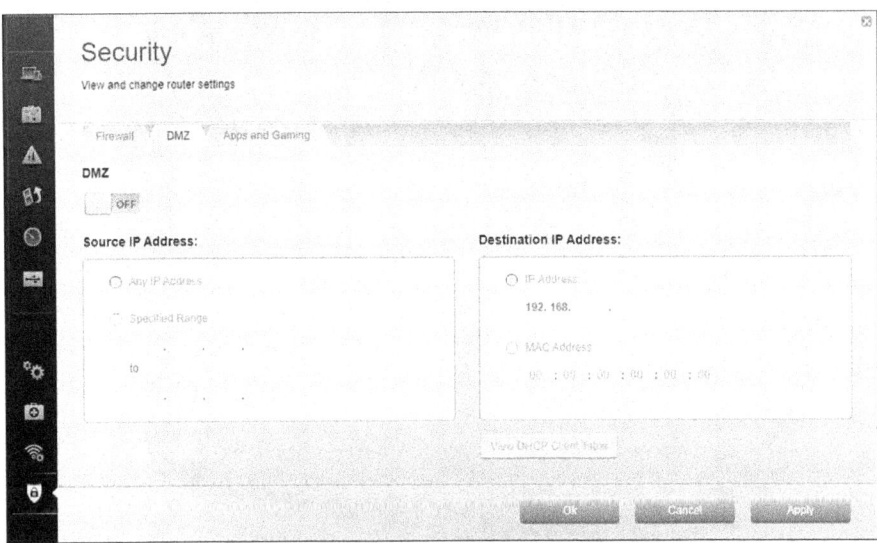

FIGURE 6.4 **Common Security Configuration Parameters for a Wireless Router**

▶ **Channel:** To access this network, all systems must use this channel. If needed, you can change the channel using the drop-down menu. The menu lists channels 1 through 11.

▶ **SSID broadcast:** In their default configuration, wireless APs typically broadcast the SSID name into the air at regular intervals. This feature is intended to allow clients to easily discover the network and roam between WLANs. The problem with SSID broadcasting is that it makes it a little easier to get around security. SSIDs are not encrypted or protected in any way. Anyone can snoop and get a look at the SSID and attempt to join the network if not secured.

> **Note**
>
> For SOHO use, *roaming* is not needed. This feature can be disabled for home use to improve the security of your WLAN. As soon as your wireless clients are manually configured with the right SSID, they no longer require these broadcast messages.

▶ **Authentication:** When configuring authentication security for the AP, you have several options depending on the age of the AP. At the lower (older) end, choices often include WEP-Open, WEP-Shared, and WPA-PSK. WEP-Open is the simplest of the authentication methods because it does not perform any type of client verification. It is a weak form of authentication because it requires no proof of identity. WEP-Shared requires that a WEP key be configured on both the client system and the AP. This makes authentication with WEP-Shared mandatory, so it is more secure for wireless transmission. To strengthen WEP encryption, a *Temporal Key Integrity Protocol (TKIP)* was employed. This protocol placed a 128-bit wrapper around the WEP encryption with a key that is based on things such as the MAC address of the destination device and the serial number of the packet. TKIP was designed as a backward-compatible replacement to WEP, and it could work with all existing hardware. Without the use of TKIP, WEP was considered weak. It is worth noting, however, that even TKIP has been broken.

Wi-Fi Protected Access with Pre-Shared Key (WPA-PSK) is a stronger form of encryption in which keys are automatically changed and authenticated between devices after a specified period of time, or after a specified number of packets have been transmitted.

On newer APs, the choices usually include WPA3, WPA Personal, WPA Enterprise, WPA2 Personal, and WPA2 Enterprise. Other choices can include WPA2/WPA Mixed Mode and RADIUS. Although WPA mandates the use of TKIP, WPA2 requires *Counter Mode with Cipher Block Chaining Message Authentication Code Protocol (CCMP)*. CCMP uses 128-bit AES encryption with a 48-bit initialization vector. With the larger initialization vector, it increases the difficulty in cracking and minimizes the risk of a replay attack. WPA3 (shown as an option in Figure 6.5) uses Simultaneous Authentication of Equals (SAE), which replaces preshared key (PSK) used in WPA2-Personal and is resistant to offline dictionary attacks. When given as a choice, WPA3-Personal adds more protection for individual users as a result of the password-based authentication even when the passwords that users choose are not all that complex.

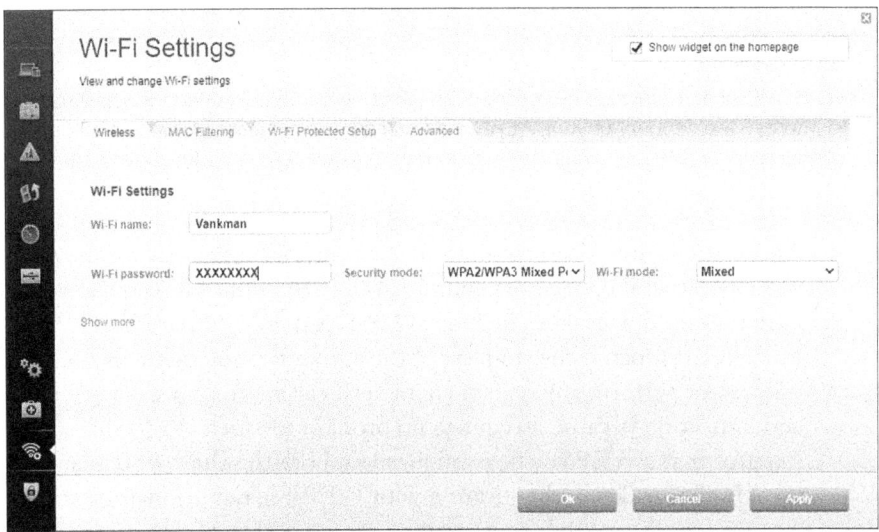

FIGURE 6.5   Newer Wireless Routers Offer WPA2/WPA3 Mixed Personal Mode as a Security Option

While PSK authentication relies on a shared passphrase, enterprise (802.1X or WPA-Enterprise) authentication leverages a centralized authentication server and user-specific credentials for authentication, making it suitable for larger, more secure networks. Enterprise authentication supports a variety of authentication methods, including Extensible Authentication Protocol (EAP) methods such as EAP-TLS, EAP-PEAP, and EAP-TTLS, which offer enhanced security features such as mutual authentication and certificate-based authentication.

> **ExamAlert**
>
> Know that two versions of WPA2 exist: WPA2-Personal (which uses preshared keys and protects unauthorized network access via a password) and WPA2-Enterprise (which verifies network users through a server). WPA3 also includes both a Personal and Enterprise version and helps prevent offline password attacks by using Simultaneous Authentication of Equals. SAE allows users to choose easier-to-remember passwords and, through forward secrecy, does not compromise traffic already transmitted even if the password becomes compromised.

▶ **Wireless mode:** To access the network, the client must use the same wireless mode as the AP. Today, most users configure the network for 802.11ac or 802.11ax for faster speeds.

▶ **DTIM period (seconds):** Wireless transmissions can broadcast to all systems—that is, they can send messages to all clients on the wireless network. Multiple broadcast messages are known as *multicast or broadcast traffic. Delivery Traffic Indication Message (DTIM)* is a feature used to ensure that when the multicast or broadcast traffic is sent, all systems are awake to hear the message. The DTIM setting specifies how often the DTIM is sent within the beacon frame. For example, if the DTIM setting by default is 1, this means that the DTIM is sent with every beacon. If the DTIM is set to 3, the DTIM is sent every three beacons as a DTIM wake-up call.

▶ **Maximum connection rate:** The transfer rate typically is set to Auto by default. This setting enables the maximum connection speed. However, it is possible to decrease the speed to increase the distance that the signal travels and boost signal strength caused by poor environmental conditions.

▶ **Network type:** This is where the network can be set to use the ad hoc or infrastructure network design.

It is easy to fall into the trap of thinking of wireless devices as being laptops connecting to the AP. Over the years, the number and the type of mobile devices that need to connect to the network have expanded tremendously. In addition to the laptops and tablets, gaming devices, media devices, cell phones, and IoT devices now all connect for wireless access. Although they might all seem different, they require the same information to connect.

Unfortunately, they all bring security concerns as well. *Bring-your-own-device (BYOD)* policies are highly recommended for every organization. Administrators can implement *mobile device management (MDM)* and *mobile application management (MAM)* products to help with the management and administration issues with these devices.

> **ExamAlert**
>
> Know that devices on networks today include such things as PCs, cell phones, laptops, tablets, gaming devices, media, and IoT devices.

Autonomous and Lightweight Access Points

An *autonomous AP* operates independently and does not require a centralized controller for management. Configuration and management of autonomous APs are performed directly on each individual AP, typically through a web-based interface or command-line interface (CLI). Autonomous APs are suitable for smaller deployments or environments where a centralized management system is not necessary. Each autonomous AP operates as a standalone device, handling tasks such as radio frequency (RF) management, client authentication, and access control independently.

A *lightweight AP*, also known as a controlled or managed AP, requires a centralized wireless LAN controller (WLC) for management and control. Lightweight APs are designed to offload many of the management functions to the WLC, including configuration, firmware updates, RF optimization, and client roaming. The WLC acts as a central point of control for all connected lightweight APs, providing centralized management and coordination of the wireless network. Lightweight APs are commonly used in larger deployments, such as enterprise networks, where centralized management, scalability, and seamless roaming are essential.

> **ExamAlert**
>
> The main difference between autonomous and lightweight access points is in their management architecture. Autonomous APs operate independently and are managed individually, while lightweight APs require a centralized wireless LAN controller for management and coordination. The choice between the two types of APs depends on factors such as the size of the deployment, management requirements, and scalability needs of the wireless network.

Cram Quiz

1. Which of the following wireless protocols can operate at 2.4 GHz? (Choose four.)

   ○ **A.** 802.11a

   ○ **B.** 802.11b

   ○ **C.** 802.11g

   ○ **D.** 802.11n

   ○ **E.** 802.11ac

   ○ **F.** 802.11ax

2. Under what circumstance would you change the default channel on an access point?

   ○ **A.** When channel overlap occurs between APs

   ○ **B.** To release and renew the SSID

   ○ **C.** To increase WPA2 security settings

   ○ **D.** To decrease WPA2 security settings

3. A client on your network has had no problems accessing the wireless network in the past, but recently they moved to a new office. Since the move, the client cannot access the network. Which of the following is most likely the cause of the problem?

   ○ **A.** The SSIDs on the client and the AP are different.

   ○ **B.** The SSID has been erased.

   ○ **C.** The client has incorrect broadcast settings.

   ○ **D.** The client system has moved too far from the AP.

4. You are installing a wireless network solution, and you require a standard that can operate using either 2.4 GHz or 5 GHz frequencies. Which of the following standards could you choose? (Choose two.)

   ○ **A.** 802.11a

   ○ **B.** 802.11b

   ○ **C.** 802.11g

   ○ **D.** 802.11n

   ○ **E.** 802.11ac

   ○ **F.** 802.11ax

5. You are installing a wireless network solution that uses a feature known as MU-MIMO. Which wireless networking standards are you possibly using? (Choose two.)

    ○ **A.** 802.11a

    ○ **B.** 802.11b

    ○ **C.** 802.11n

    ○ **D.** 802.11ac

    ○ **E.** 802.11ax

6. Which of the following aims to improve network performance and client experience by directing devices to connect to the most suitable frequency band based on their capabilities and the network conditions?

    ○ **A.** Band steering

    ○ **B.** Channel width

    ○ **C.** Radio frequency channels

    ○ **D.** Channel bonding

    ○ **E.** Point-to-point network

7. Which of the following does WPA3 use that replaces preshared key (PSK) used in WPA2-Personal and is resistant to offline dictionary attacks?

    ○ **A.** PSK

    ○ **B.** TKIP

    ○ **C.** SSID broadcasting

    ○ **D.** SAE

8. In which network architecture type do devices communicate through a centralized access point or base station?

    ○ **A.** Mesh

    ○ **B.** Infrastructure

    ○ **C.** Ad hoc

    ○ **D.** Point-to-point

9. Which of the following operates independently and does not require a centralized controller for management?

    ○ **A.** Lightweight AP

    ○ **B.** WLC

    ○ **C.** DTIM

    ○ **D.** Autonomous AP

10. Which of the following are web pages that guests are redirected to when they attempt to connect to the guest network?

   ○ **A.** BSSID

   ○ **B.** ESSID

   ○ **C.** Captive portals

   ○ **D.** Station (STA)

Cram Quiz Answers

1. B, C, D, and F. Wireless standards specify an RF range on which communications are sent. The 802.11b and 802.11g standards use the 2.4 GHz range. 802.11a and 802.11ac use the 5 GHz range. 802.11n can operate at 2.4 GHz and 5 GHz. 802.11ax operates in both the 2.4 GHz and 5 GHz ranges.

2. A. Ordinarily, the default channel used with a wireless device is adequate; however, you might need to change the channel if overlap occurs with another nearby AP. The channel should be changed to another, nonoverlapping channel. Changing the channel would not impact the WPA2 security settings.

3. D. An AP has a limited distance that it can send data transmissions. When a client system moves out of range, it cannot access the AP. Many strategies exist to increase transmission distances, including RF repeaters, amplifiers, and more powerful antennas or wireless APs. The problem is not likely related to the SSID or broadcast settings, because the client had access to the network before, and no settings were changed.

4. D and F. The IEEE standards 802.11n and 802.11ax can use either the 2.4 GHz or 5 GHz radio frequencies. Given a choice today, you should choose 802.11ax. 802.11a uses 5 GHz, and 802.11b and 802.11g use 2.4 GHz. 802.11ac operates at 5 GHz.

5. D and E. MU-MIMO is used by the 802.11ac and 802.11ax standards and makes multiuser MIMO possible (increasing the range and speed of wireless networking). MIMO, itself, enables the transmission of multiple data streams traveling on different antennas in the same channel at the same time.

6. A. Band steering is a technique used to optimize the utilization of different frequency bands, particularly in dual-band or tri-band Wi-Fi networks. It aims to improve network performance and client experience by directing devices to connect to the most suitable frequency band based on their capabilities and the network conditions. Channel width is the width or bandwidth of the radio frequency channel used for transmitting data, and it determines the amount of spectrum available for data transmission and affects the data transfer rate and network performance. Radio frequency (RF) channels are an important part of wireless communication. A channel is the band of RF used for the wireless communication. Each IEEE wireless standard specifies the channels that can be used. With channel bonding, you can use two channels at the same time. As you might guess, the ability to use two channels at once increases performance. A point-to-point network has a dedicated link established between two endpoints, typically used for connecting two distant locations.

7. **D.** WPA3 uses Simultaneous Authentication of Equals (SAE), which replaces preshared key (PSK) used in WPA2-Personal and is resistant to offline dictionary attacks. When given as a choice, WPA3-Personal adds more protection for individual users as a result of the password-based authentication even when the passwords that users choose are not all that complex. To strengthen WEP encryption, a Temporal Key Integrity Protocol (TKIP) was employed. Without the use of TKIP, WEP was considered weak. It is worth noting, however, that even TKIP has been broken. SSID broadcasting is intended to allow clients to easily discover the network and roam between WLANs. The problem with SSID broadcasting is that it makes it a little easier to get around security. SSIDs are not encrypted or protected in any way. Anyone can snoop and get a look at the SSID and attempt to join the network if not secured.

8. **B.** In the infrastructure network architecture, devices communicate through a centralized access point or base station. It offers centralized management, wide coverage area, and support for many devices, suitable for permanent installations or environments with high device density. A mesh network type is a decentralized network architecture where each node relays data for the network. An ad hoc network is a peer-to-peer network where wireless devices communicate with each other without the need for a central access point. In a point-to-point network, a dedicated link is established between two endpoints, typically used for connecting two distant locations.

9. **D.** An autonomous AP operates independently and does not require a centralized controller for management. Configuration and management of autonomous APs are performed directly on each individual AP, typically through a web-based interface or command-line interface (CLI). A lightweight AP, also known as a controlled or managed AP, requires a centralized wireless LAN controller (WLC) for management and control. Lightweight APs are designed to offload many of the management functions to the WLC, including configuration, firmware updates, RF optimization, and client roaming. Wireless transmissions can broadcast to all systems; that is, they can send messages to all clients on the wireless network. Multiple broadcast messages are known as multicast or broadcast traffic. The Delivery Traffic Indication Message (DTIM) is a feature used to ensure that when the multicast or broadcast traffic is sent, all systems are awake to hear the message. The DTIM setting specifies how often the DTIM is sent within the beacon frame.

10. **C.** Captive portals are web pages that guests are redirected to when they attempt to connect to the guest network. They typically present users with a login page, terms of service agreement, acceptable use policy (AUP), or other authentication methods before granting access to the Internet. BSSID is the MAC address assigned to a specific access point (AP) or wireless router, and each AP in a wireless network has a unique BSSID, which is used to identify and differentiate it from other APs within the same network. The ESSID is the name of the wireless network or the network's "broadcast" name: it is set by the network administrator and is shared among all APs in a network to create a single logical wireless network. Answer D is incorrect because if you use an infrastructure wireless network design, the network has two key parts: the wireless client, also known as the station (STA), and the AP.

What's Next?

Chapter 7, "Cloud Computing Concepts and Options," focuses on the definitions of cloud computing at the level you need to know for the Network+ exam.

CHAPTER 7

Cloud Computing Concepts and Options

> **This chapter covers the following official Network+ objectives:**
>
> ▶ 1.3 Summarize cloud concepts and connectivity options.

This chapter covers CompTIA Network+ objective 1.3. For more information on the official Network+ exam topics, see the "About the Network+ Exam" section in the Introduction.

At its core, *cloud computing* is a revolutionary technology that enables individuals and organizations to access and use computing resources (servers, storage, databases, networking, software, and more) over the Internet. Instead of relying on local hardware and infrastructure, cloud computing allows users to leverage the services of remote datacenters operated by cloud providers. This on-demand model offers numerous advantages, including scalability, cost-efficiency, flexibility, and accessibility, making it easier for businesses and individuals to deploy and manage IT resources without the need for extensive on-premises hardware and maintenance. In the past few years, cloud computing has become an integral part of modern technology ecosystems, supporting a wide range of applications, from web hosting and data storage to machine learning (ML) and artificial intelligence (AI), driving innovation and efficiency.

This chapter focuses on the definitions of cloud computing at the level you need to know for the Network+ exam. If you want to go further with the technology, consider the Cloud+ certification from CompTIA.

Cloud Concepts

▶ **1.3 Summarize cloud concepts and connectivity options.**

CramSaver

If you can correctly answer these questions before going through this section, save time by skimming the ExamAlerts in this section and then completing the Cram Quiz at the end of the section.

1. In which cloud delivery model are resources owned by the organization, and what organization acts as both the provider and the consumer?

2. With which cloud service model can consumers deploy but not manage or control any of the underlying cloud infrastructure (but they can have control over the deployed applications)?

3. What are some of the characteristics of cloud computing?

4. With what technology are functions such as firewalls, routers, switches, load balancers, and even entire network infrastructures replaced by software-based counterparts running on standard servers or cloud platforms?

Answers

1. In a private cloud model, the cloud is owned by the organization, and it acts as both the provider and the consumer.

2. With the Platform as a Service (PaaS) cloud service model, consumers can deploy but not manage or control any of the underlying cloud infrastructure (but they can have control over the deployed applications).

3. Regardless of the service model used, the characteristics include on-demand self-service, broad network access, resource pooling, rapid elasticity, and measured service.

4. With network functions virtualization (NFV), functions such as firewalls, routers, switches, load balancers, and even entire network infrastructures are replaced by software-based counterparts running on standard servers or cloud platforms.

The best way to think about this chapter is as an introduction to cloud computing and an agreement on the definition of what the terms associated with it really mean. The *National Institute of Standards and Technology (NIST)* defines three service models in Special Publication 800-145: Software as a Service (SaaS), Platform as a Service (PaaS), and Infrastructure as a Service (IaaS). It also defines possible delivery models that include private, public, and hybrid.

This chapter looks at each of these terms and what they mean as defined by the NIST and agreed upon by the computing community. Know that it is possible to mix and match the service models with the platform models so that you can have public IaaS, or private PaaS, and so on and that you utilize a Cloud Access Security Broker (CASB)—a software program—to sit between the cloud service users and cloud applications to monitor activity and enforce established security policies.

> **Note**
>
> The CASB can offer services beyond just monitoring users' actions but must always be able to enforce compliance with security policies. A service-level agreement (SLA) with a CASB is often used to spell out issues related to regulatory compliance and data protection.

Service Models

NIST defines three primary service models: SaaS, PaaS, and IaaS. Sometimes, others get tossed into the mix by marketing or tech companies (virtually anything can have *aaS* tacked to the end of it and its subscription referenced "as a Service"), but it is the primary models that you need to focus on for the exam. In the sections that follow, we walk through each of the three models and compare the key elements of each.

Software as a Service

According to NIST, Software as a Service (SaaS) is defined as follows:

> The capability provided to the consumer is to use the provider's applications running on a cloud infrastructure. The applications are accessible from various client devices through either a thin client interface, such as a web browser (for example, web-based email), or a program interface. The consumer does not manage or control the underlying cloud infrastructure, including network, servers, operating systems, storage, or even individual application capabilities, with the possible exception of limited user-specific application configuration settings.

The words used are significant, and the ones to focus on in this definition are that consumers can *use* the provider's applications and that they do not *manage or control* any of the underlying cloud infrastructure. Figure 7.1 depicts the responsibility of each party in the SaaS model.

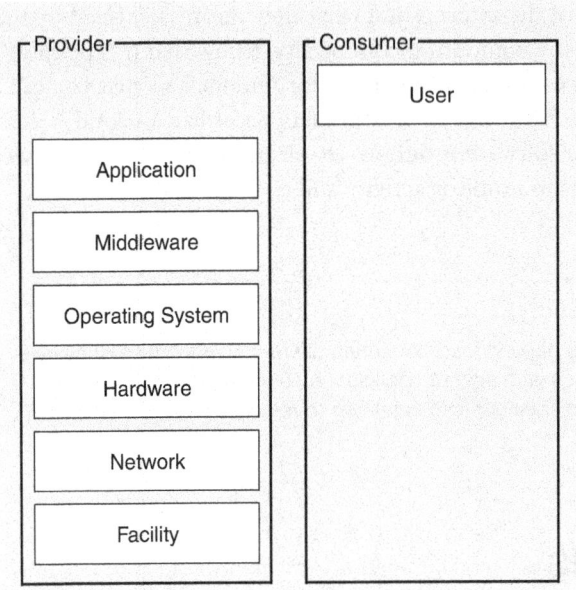

FIGURE 7.1 **The SaaS Service Model**

Platform as a Service

According to NIST, *Platform as a Service (PaaS)* is defined as follows:

> The capability provided to the consumer is to deploy onto the cloud infrastructure consumer-created or acquired applications created using programming languages, libraries, services and tools supported by the provider. The consumer does not manage or control the underlying cloud infrastructure including network, servers, operating systems, or storage, but has control over the deployed applications and possible configuration settings for the application-hosting environment.

The important words to focus on in this definition are that consumers can *deploy*, that they do not *manage or control* any of the underlying cloud infrastructure, but they can have *control over the deployed applications*. Figure 7.2 depicts the responsibility of each party in the PaaS model.

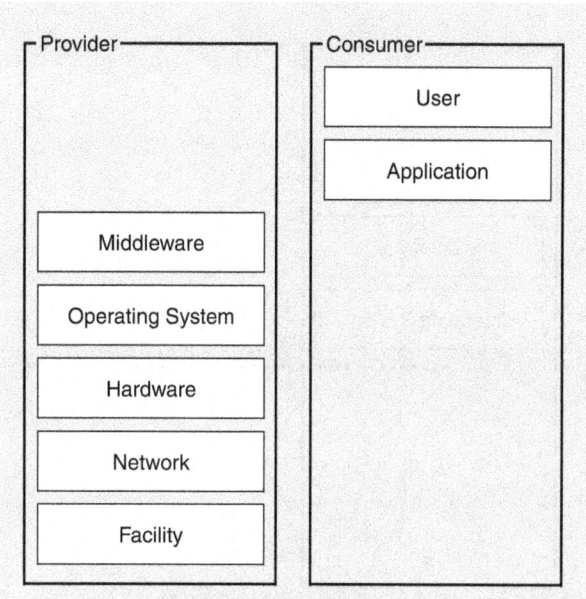

FIGURE 7.2   **The PaaS Service Model**

Infrastructure as a Service

According to NIST, *Infrastructure as a Service (IaaS)* is defined as follows:

> The capability provided to the consumer is to provision processing, storage, networks, and other fundamental computing resources where the consumer is able to deploy and run arbitrary software, which can include operating systems and applications. The consumer does not manage or control the underlying cloud infrastructure but has control over operating systems, storage, and deployed applications; and possible limited control of select networking components (e.g., host firewalls).

The words to focus on are that the consumer can *provision*, is able to *deploy and run*, but still does not *manage or control* the underlying cloud infrastructure, but now can be responsible for some aspects. Figure 7.3 depicts the responsibility of each party in the IaaS model.

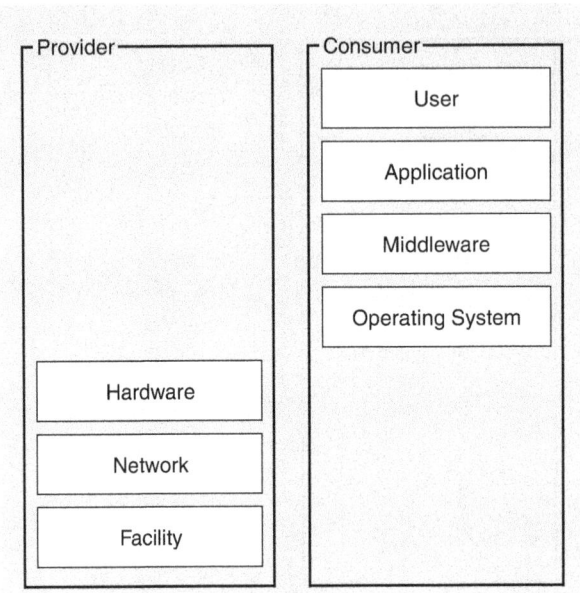

FIGURE 7.3   **The IaaS Service Model**

ExamAlert

Know that Infrastructure as a Service (IaaS) involves delivering the computer infrastructure in a hosted service model over the Internet.

Table 7.1 compares and contrasts the three as-a-service models you should know for the exam.

TABLE 7.1   **Cloud Service Models**

	SaaS	**IaaS**	**PaaS**
Definition	Provides software applications over the Internet	Offers virtualized computing resources over the Internet	Delivers a platform allowing customers to develop, run, and manage applications without dealing with the infrastructure
Use Case	Ideal for end users who need ready-to-use applications	Suited for businesses requiring scalable computing resources	Suitable for developers needing an environment to build, deploy, and manage applications

	SaaS	IaaS	PaaS
Management	Managed entirely by the service provider	Offers control over virtualized infrastructure, including server, storage, and networking	Managed platform, including infrastructure, runtime environment, and middleware
Scalability	Limited scalability options, as users rely on the provider's infrastructure	Highly scalable, allowing users to adjust resources according to demand	Scalable, with the platform automatically handling resource provisioning
Customization	Limited customization options, as users interact with pre-built applications	Provides flexibility to customize and configure virtual machines and networking components	Offers customization options primarily focused on application development and deployment
Cost Model	Typically based on a subscription or usage-based model	Often billed on resource usage, such as compute instances and storage	Cost may vary based on factors such as resource consumption and the number of applications deployed
Examples	Gmail, Salesforce, Office 365	Amazon Web Services (AWS), Microsoft Azure, Google Cloud Platform	Google App Engine, Microsoft Azure App Service

Note

Regardless of the service model used, the characteristics of each of them are that they include on-demand self-service, broad network access, resource pooling, rapid elasticity, and measured service.

After you have a service model selected, both CompTIA and NIST recognize the different delivery models, which are discussed next.

ExamAlert

For the exam, know that there are three possible cloud service models: IaaS, PaaS, and SaaS. Other models that are labeled with *aaS* are often done so for marketing purposes.

Deployment Models

Cloud deployment can be done in various ways. It is possible to isolate the cloud (a private model), make it widely available (a public model), or do something in between (a hybrid model). In the discussion that follows, we look at deployment models and move into discussing cloud-based features.

Private Cloud

A *private cloud* is defined as follows:

> The cloud infrastructure is provisioned for exclusive use by a single organization comprising multiple consumers (e.g., business units). It may be owned, managed, and operated by the organization, a third party, or some combination of them, and it may exist on or off premises.

Under most circumstances, a private cloud is owned by the organization, and it acts as both the provider and the consumer. It has a security-related advantage in not needing to put its data on the Internet.

Public Cloud

A *public cloud* is defined as follows:

> The cloud infrastructure is provisioned for open use by the general public. It may be owned, managed, and operated by a business, academic, or government organization, or some combination of them. It exists on the premises of the cloud provider.

Under most circumstances, a public cloud is owned by the cloud provider, and it uses a pay-as-you-go model. A good example of a public cloud is webmail or online document sharing/collaboration.

Hybrid Cloud

A *hybrid cloud* is defined as follows:

> The cloud infrastructure is a composition of two or more distinct cloud infrastructures (private, community, or public) that remain unique entities, but are bound together by standardized or proprietary technology that enables data and application portability (e.g., cloud bursting for load balancing between clouds).

In other words, a hybrid can be any combination of other delivery models.

Notice the word *community* that appears in the definition along with *private* and *public*. While this deployment model is no longer tested on, a community cloud is defined as follows:

> The cloud infrastructure is provisioned for exclusive use by a specific community of consumers from organizations that have shared concerns (e.g., mission, security requirements, policy, and compliance considerations). It may be owned, managed, and operated by one or more of the organizations in the community, a third party, or some combination of them, and it may exist on or off premises.

The key to distinguishing between a community cloud and other types of cloud delivery is that it serves a *similar* group. There must be joint interests and limited enrollment.

Note

A common reason for using cloud computing is to be able to offload traffic to resources from a cloud provider if your own servers become too busy. This is known as *cloud bursting*, and it requires load-balancing/prioritizing technologies such as *Quality of Service (QoS)* protocols to make it possible.

ExamAlert

For the exam, you should know that the most deployed cloud delivery models are private, public, and hybrid.

Table 7.2 compares and contrasts the three cloud deployment models you should know for the exam.

TABLE 7.2 **Cloud Deployment Models**

	Public	**Private**	**Hybrid**
Ownership	Owned and operated by a third-party provider	Owned and operated by a single organization	Combination of public and private cloud infrastructure
Use Case	Ideal for startups, small businesses, and organizations with variable workloads	Suited for organizations with sensitive data, regulatory compliance requirements, or specific security needs	Suitable for organizations seeking a balance between cost efficiency, scalability, and control over data and applications

	Public	Private	Hybrid
Accessibility	Accessible to the general public via the Internet	Accessible only to authorized users within the organization's network	Allows data and applications to be shared between public and private cloud environments
Security	Security managed by the cloud provider, offering standard security measures	Security managed internally, providing greater control and customization over security protocols	Security measures may vary based on the environment, but often require additional attention to data protection and compliance
Scalability	Offers unlimited scalability, allowing users to easily scale resources up or down based on demand	Scalability depends on the organization's infrastructure and resources, with less flexibility compared to public clouds	Provides scalability benefits of both public and private clouds, allowing organizations to scale resources dynamically while maintaining control over sensitive data
Cost Efficiency	Cost-effective for organizations with fluctuating resource requirements, as users only pay for the resources they use	Costs may be higher due to initial setup and maintenance costs, but can be more predictable over time	Offers cost efficiency by leveraging the benefits of both cloud types, allowing the organization to optimize resource allocation based on cost and performance requirements
Examples	Amazon Web Services (AWS), Microsoft Azure, Google Cloud Platform	Private cloud hosted on-premises or in a datacenter managed by the organization	A combination of public cloud services (AWS, Azure, etc.) and private cloud infrastructure

Multitenancy

One of the ways cloud computing is able to obtain cost efficiencies is by putting data from various clients on the same machines. This "multitenant" nature, known as multitenancy, means that workloads from different clients can be on the same system, and a flaw in implementation could compromise security. In theory, a security incident could originate with another customer at the cloud provider and bleed over into your data. Therefore, data needs to be protected from other cloud consumers and from the cloud provider as well.

Elasticity

According to NIST, one of the five essential characteristics of the cloud is not just elasticity, but rapid elasticity. This characteristic is defined as follows:

> Capabilities can be elastically provisioned and released, in some cases automatically, to scale rapidly outward and inward commensurate with demand. To the consumer, the capabilities available for provisioning often appear to be unlimited and can be appropriated in any quantity at any time.

The key words to focus on in this definition are *provisioned and released*, *scale*, and *appear to be unlimited*.

Scalability

Another feature that makes cloud computing valuable is scalability. According to NIST, "Performance can potentially be scaled to meet conditions of anticipated or real-world demand, within the parameters of a cloud service agreement." To be able to scale, elasticity is required, and autonomic autoscaling of resources is critical.

> **ExamAlert**
>
> For the exam, you should know the differences between *multitenancy*, *elasticity*, and *scalability* cloud concepts and be able to summarize them. Multitenancy refers to the ability of a cloud service provider to serve multiple tenants (users or organizations) using shared resources and infrastructure. Elasticity describes the cloud's ability to dynamically adjust resource allocation based on workload demands, ensuring optimal performance and efficiency. Scalability indicates the cloud's capacity to handle increasing workloads by adding resources such as compute, storage, or networking components. It gives a corporate IT department the ability to expand their cloud-hosted virtual machine (VM) environment with minimal effort and to quickly and easily scale up to meet increased/decreased demand.

Network Functions Virtualization (NFV)

Network Functions Virtualization (NFV) involves the virtualization of network functions and services that traditionally relied on dedicated, proprietary hardware appliances. With NFV, these functions, such as firewalls, routers, switches, load balancers, and even entire network infrastructures, are replaced by software-based counterparts running on standard servers or cloud platforms. The overall aim of NFV is to make networks more agile, scalable, and cost-effective by decoupling network functions from the physical hardware and allow for greater flexibility in managing and deploying network services, as they can be easily scaled up or down as needed, and new services can be introduced without the need for expensive hardware upgrades.

In essence, NFV leverages virtualization technology to abstract and virtualize network functions, making networks more adaptable to evolving demands and reducing operational complexity and costs for service providers and enterprises. It plays a crucial role in the evolution of modern, software-defined networking (SDN) and is instrumental in the transition to more dynamic and responsive network infrastructures.

> **Note**
>
> Software-defined networking is covered in detail in Chapter 4, "Network Implementations."

> **ExamAlert**
>
> For the exam, you should know that Network Functions Virtualization (NFV) is not an alternative to cloud computing, but an extension of it that adds in virtualization technology. It allows for the virtualization of a replica of a network's physical topology—and the way it behaves—without changing the logical topology and the way that devices are managed. NFV allows for the virtualization of network functions such as routers, firewalls, and switches, resulting in increased flexibility and scalability, thus making it a great technology for companies looking to virtualize an exact replica of their internal physical network

Cloud Connectivity Options

Most cloud providers offer a number of methods that clients can employ to connect to them. It is important, before making an investment in infrastructure, to check with your provider and see what methods it recommends and supports. One of the most common is to use an IPsec, hardware virtual private network (VPN) connection between your network(s) and the cloud provider. This method offers the capability to have a managed VPN endpoint that includes automated multidata center redundancy and failover.

A dedicated direct connection, known as a private-direct connection, is another, simpler, method. You can combine the dedicated network connection(s) with the hardware VPN to create a combination that offers a secure IPsec-encrypted private connection while also reducing network costs.

> **Note**
>
> IPsec was covered in detail in Chapter 1, "Networking Models, Ports, Protocols, and Services."

Amazon Web Services (AWS) is one of the most popular cloud providers on the market. It allows the two connectivity methods discussed (calling the dedicated connection "AWS Direct Connect") and a number of others that are variations, or combinations, of these two.

> **ExamAlert**
>
> Remember that common cloud interconnection methods include Internet, VPN, and Direct Connect. In other words, these methods are commonly used to connect a private cloud to a public cloud.

Virtual Private Cloud (VPC)

Virtual Private Cloud (VPC) technology is a fundamental component of cloud computing infrastructure, offered by cloud service providers like Amazon Web Services (AWS), mentioned earlier, as well as Microsoft Azure and Google Cloud Platform (GCP). VPC allows users to create isolated and logically segmented private networks within a public cloud environment via an endpoint that allows the VPC to be connected with other services without the need for additional technologies.

With VPC technology, users can define their own network architecture, including IP address ranges, subnets, routing tables, and access controls, while leveraging the underlying resources and scalability of the public cloud. This isolation provides a secure and dedicated network environment where organizations can deploy and manage their applications, databases, and services, ensuring data privacy and security.

Key features of VPC technology typically include

▶ **Network Isolation:** VPCs offer a secure and isolated network environment within the public cloud, ensuring that resources within one VPC cannot communicate directly with resources in another VPC unless explicitly allowed.

▶ **Customizable Network Configuration:** Users have the flexibility to design their VPC's network architecture, defining IP address ranges, subnets, and routing rules to suit their specific requirements.

▶ **Security Controls:** VPCs enable users to implement robust security measures, such as network access control lists (ACLs) and security groups, to control traffic flow and protect resources within the VPC.

▶ **Endpoint Usage:** A VPC endpoint allows the VPC to be connected with other services without the need for additional technologies like a virtual private network (VPN) connection or Internet gateway. Resources within the VPC must make any requests, as the connected services are not able to initiate requests via the VPC endpoint.

▶ **Scalability:** VPCs can be scaled up or down as needed, allowing organizations to adapt to changing workloads and resource demands.

▶ **Connectivity Options:** VPCs can be connected to on-premises datacenters or other VPCs, providing hybrid and multi-cloud networking capabilities.

Overall, Virtual Private Cloud technology empowers organizations to harness the benefits of cloud computing while maintaining control, security, and customization over their network infrastructure, making it a crucial component of cloud adoption for businesses of all sizes.

Cloud Gateways

Cloud gateway technology acts as a bridge between the internal network infrastructure and the cloud. It is a crucial component for businesses that want to harness the benefits of cloud computing while maintaining control and connectivity with their existing IT infrastructure. Two popular approaches are to utilize Internet gateways and/or Network Address Translation (NAT) gateways to intelligently route network traffic between different locations, optimizing the flow of data for performance and cost-effectiveness.

> **ExamAlert**
>
> Network Address Translation (NAT) is required to communicate over the Internet with private IP addresses. Although Internet routers are almost always required for routing, by default they won't route private IP addresses to public IP addresses.

An *Internet gateway* facilitates communication between resources within a VPC and the broader Internet. It acts as the entry and exit point for traffic flowing into and out of the VPC, allowing resources, such as virtual servers or applications, to connect to external services, websites, or users over the Internet.

It enables outbound Internet access for resources within a VPC, allows incoming traffic from the Internet to reach specific resources within the VPC, and facilitates secure and controlled communication between cloud resources and external services.

A NAT gateway, on the other hand, is used to manage outbound Internet connectivity for resources within a private subnet of a VPC. It enables private resources, which do not have public IP addresses, to initiate outbound connections to the Internet while keeping them protected and hidden behind the NAT gateway's public IP address. It provides private resources with a means to access the Internet while preserving their anonymity, offers enhanced security by not exposing private IP addresses to the Internet, and automatically manages Network Address Translation, routing, and high availability to ensure reliable connectivity.

Network Security, Groups, and Lists

Security is one of the most important issues to discuss with your cloud provider. Cloud computing holds great promise when it comes to scalability, cost savings, rapid deployment, and empowerment. As with any technology where so much is removed from your control, though, risks are involved. Each risk should be considered carefully to identify ways to help mitigate it. Naturally, the responsibilities of both the organization and the cloud provider vary depending on the service model chosen, but ultimately the organization is accountable for the security and privacy of the outsourced service.

Software and services not necessary for the implementation should be removed or at least disabled. Patches and firmware updates should be kept current, and log files should be carefully monitored. You should find the vulnerabilities in the implementation before others do and work with your service provider(s) to close any holes.

When it comes to data storage on the cloud, encryption is one of the best ways to protect it (keeping it from being of value to unauthorized parties), and VPN routing and forwarding can help. Backups should be performed regularly (and encrypted and stored in safe locations), and access control—implemented through the use of security groups and security lists—should be a priority.

> **Note**
>
> For a good discussion of cloud computing and data protection from the perspective of one vendor, visit https://cloud.google.com/transparency?hl=en.

Network security groups (NSGs) serve as a virtual firewall within cloud computing environments to control and regulate network traffic to and from resources like virtual machines, network interfaces, and subnets. They act as a crucial layer of defense, helping organizations secure their cloud infrastructure. Some ways in which NSGs function as a virtual firewall include

> **Note**
>
> A great reference detailing security groups can be found at https://docs.aws.amazon.com/vpc/latest/userguide/vpc-security-groups.html.

▶ **Rule-based Filtering:** NSGs operate based on a set of customizable rules that define what traffic is allowed or denied. These rules are analogous to the rules in a physical firewall and can be configured to meet specific security requirements.

▶ **Inbound and Outbound Control:** NSGs allow organizations to specify rules for both inbound and outbound traffic. Inbound rules control incoming connections to resources, while outbound rules govern outgoing traffic from resources. This granularity provides comprehensive security control.

▶ **Port-level Filtering:** NSGs enable filtering of traffic at the port level, allowing or blocking connections based on TCP and UDP ports. This feature is essential for permitting or restricting access to specific services or applications running on cloud resources.

▶ **Source and Destination IP Address Filtering:** NSGs allow administrators to define source and destination IP addresses or IP address ranges for traffic. This ensures that only authorized parties can communicate with cloud resources and helps prevent unauthorized access.

▶ **Priority and Rule Evaluation:** NSG rules have assigned priorities, and when multiple rules apply to a packet, the rule with the highest priority takes precedence. This allows for fine-grained control over traffic flow and security policy enforcement.

▶ **Logging and Monitoring:** NSGs often provide logging and monitoring capabilities, allowing administrators to track network traffic and security rule violations. This information is valuable for auditing, troubleshooting, and identifying potential threats.

▶ **Integration with Subnets and Resources:** NSGs can be associated with subnets, network interfaces, or individual virtual machines, providing granular control over network security. This allows organizations to tailor security policies to the specific needs of different resources.

▶ **Dynamic Updates:** NSGs support dynamic updates to rules, allowing organizations to adapt to changing security requirements without disrupting services. This flexibility is crucial for maintaining an effective security posture in dynamic cloud environments.

Network security lists preceded NSGs and similarly enable organizations to define a set of security rules that govern traffic flow and access control for their cloud computing resources. To use, you must associate a given security list with an entity—typically a subnet—and then virtual network interface cards (VNICs) that exist within the subnet are subject to the security lists (usually limited to five or fewer) associated with the subnet. The rules are essential for controlling access, protecting against unauthorized connections, and maintaining the security and integrity of the cloud-based assets.

Cram Quiz

1. With which cloud service model can consumers use the provider's applications but not manage or control any of the underlying cloud infrastructure?

   ○ **A.** SaaS

   ○ **B.** PaaS

   ○ **C.** IaaS

   ○ **D.** GaaS

2. Which of the following involves offloading traffic to resources from a cloud provider if your own servers become too busy?

   ○ **A.** Ballooning

   ○ **B.** Cloud bursting

   ○ **C.** Bridging

   ○ **D.** Harvesting

3. Which of the following does NIST define as a composition of two or more distinct cloud infrastructures?

   ○ **A.** Private cloud

   ○ **B.** Public cloud

   ○ **C.** Community cloud

   ○ **D.** Hybrid cloud

4. "To the consumer, the capabilities available for provisioning often appear to be unlimited and can be appropriated in any quantity at any time." Which cloud concept does this statement BEST describe?

   ○ **A.** Scalability

   ○ **B.** AWS Direct Connect

   ○ **C.** CASB

   ○ **D.** Elasticity

5. Which of the following means that workloads from different clients can be on the same system, and a flaw in implementation could compromise security?

   ○ **A.** VPC

   ○ **B.** IaC

   ○ **C.** Multitenancy

   ○ **D.** DaaS

6. Which of the following cloud service models primarily focuses on delivering a platform that allows developers and customers to develop, run, and manage applications without dealing with the infrastructure?

   ○ **A.** PaaS

   ○ **B.** SaaS

   ○ **C.** IaaS

   ○ **D.** Scalability as a Service

7. Which of the following can be used to connect a private cloud to a public cloud? (Choose all that apply.)

   ○ **A.** IPsec VPN

   ○ **B.** AWS Direct Connect

   ○ **C.** Internet

   ○ **D.** Dynamic Updates

8. Which of the following allows for the virtualization of network functions such as routers, firewalls, and switches, resulting in increased flexibility and scalability?

   ○ **A.** Cloud gateway

   ○ **B.** Hybrid cloud

   ○ **C.** Rapid elasticity

   ○ **D.** NFV

9. Which of the following technologies allows users to create isolated and logically segmented private networks within a public cloud environment via an endpoint? (Choose the BEST answer.)

   ○ **A.** VPC

   ○ **B.** Network isolation

   ○ **C.** Customizable network configuration

   ○ **D.** Security controls

Cram Quiz Answers

1. **A.** With the SaaS cloud service model, consumers are able to use the provider's applications, but they do not manage or control any of the underlying cloud infrastructure.

2. **B.** A common reason for using cloud computing is to be able to offload traffic to resources from a cloud provider if your own servers become too busy. This is known as cloud bursting.

3. **D.** The hybrid cloud delivery model is a composition of two or more distinct cloud infrastructures (public, private, and so on).

4. **D.** The statement describes elasticity. The key words to focus on in this definition are *provisioning* and *can be appropriated at any time*.

5. **C.** With multitenancy, a security incident could originate with another customer at the cloud provider and bleed over into your data. Therefore, data needs to be protected from other cloud consumers and from the cloud provider as well.

6. **A.** Platform as a Service (PaaS) delivers a platform allowing developers and customers to develop, run, and manage applications without dealing with the infrastructure. It is suitable for developers needing an environment to build, deploy, and manage applications. Software as a Service (SaaS) focuses on providing software applications such as Gmail, Salesforce, or Office 365 over the Internet. Infrastructure as a Service (IaaS) focuses on offering virtualized computing resources over the Internet. It is best suited for businesses requiring scalable computing resources. It offers control over virtualized infrastructure, including server, storage, and networking. Answer D also is incorrect. Scalability indicates the cloud's capacity to handle increasing workloads by adding resources such as compute, storage, or networking components.

7. **A, B, and C.** Common cloud interconnection methods include Internet, IPsec VPN, and AWS Direct Connect. In other words, these methods are commonly used to connect a private cloud to a public cloud. Answer D is incorrect because network security groups (NSGs) support dynamic updates to rules, allowing organizations to adapt to changing security requirements without disrupting services. This flexibility is crucial for maintaining an effective security posture in dynamic cloud environments.

8. **D.** NFV enables the virtualization of essential network functions like routers, firewalls, and switches, boosting flexibility and scalability. It's ideal for replicating internal physical networks virtually. Meanwhile, cloud gateway technology serves

as a vital link between internal network infrastructure and the cloud, crucial for maintaining control and connectivity. Answer B's flaw lies in misinterpreting the hybrid cloud, which encompasses various delivery models. Rapid elasticity, a crucial cloud characteristic according to NIST, ensures swift provisioning and release of resources, scaling to meet demand dynamically. The emphasis should be on provisioning, release, scalability, and the illusion of unlimited resources.

9. **A.** Virtual Private Cloud (VPC) technology allows users to create isolated and logically segmented private networks within a public cloud environment via an endpoint that allows the VPC to be connected with other services without the need for additional technologies. With VPC technology, users can define their own network architecture, including IP address ranges, subnets, routing tables, and access controls, while leveraging the underlying resources and scalability of the public cloud. Answers B, C, and D are incorrect because these are all key features of VPC technology.

What's Next?

Chapter 8, "Network Operations," focuses on several important topics: physical installations, network availability, organizational documents and policies, and disaster recovery technologies/techniques.

CHAPTER 8
Network Operations

This chapter covers the following official Network+ objectives:

▶ 2.4 Explain important factors of physical installations.

▶ 3.1 Explain the purpose of organizational processes and procedures.

▶ 3.2 Given a scenario, use network monitoring technologies.

▶ 3.3 Explain disaster recovery (DR) concepts.

▶ 3.5 Compare and contrast network access and management methods.

This chapter covers CompTIA Network+ objectives 2.4, 3.1, 3.2, 3.3, and 3.5. For more information on the official Network+ exam topics, see the "About the Network+ Exam" section in the Introduction.

This chapter examines two important parts of the role of a network administrator: documentation and the tools to use to monitor or optimize connectivity. Documentation, although not glamorous, is an essential part of the job. This chapter also looks at high availability, statistics, and disaster recovery concepts.

Physical Installation Factors

▶ **2.4 Explain important factors of physical installations.**

CramSaver

If you can correctly answer these questions before going through this section, save time by skimming the ExamAlerts in this section and then completing the Cram Quiz at the end of the section.

1. In a network installation with an intermediate distribution frame (IDF) and main distribution frame (MDF), what type of cabling is commonly used to connect network devices within the same IDF or MDF?

2. Which cabling standard is commonly used in network installations with intermediate distribution frames (IDFs) and main distribution frames (MDFs) to ensure compatibility and performance?

3. In a network installation, what is the primary function of a power distribution unit (PDU)?

4. True or false: Metered PDUs allow network administrators to monitor power usage and track energy consumption in real time.

Answers

1. Twisted-pair cable, particularly Cat 5e, Cat 6, or greater Ethernet cable, is commonly used for connecting network devices within the same IDF or MDF due to its cost-effectiveness, flexibility, and ability to support high-speed Ethernet connections.

2. The ANSI/TIA-568-C standard specifies the requirements for telecommunications cabling systems, including cabling types, connectors, and performance specifications, ensuring compatibility and performance in network installations with IDFs and MDFs.

3. The primary function of a PDU is to distribute power from a single source to multiple network devices, allowing efficient power management and connectivity for equipment such as servers, switches, and routers.

4. True. Metered PDUs allow network administrators to monitor power usage and track energy consumption in real time, providing valuable insights into power usage trends, optimizing power distribution, and identifying potential issues or inefficiencies in the network infrastructure.

The physical installation of a network is one of the most important factors in the success of the network for a plethora of reasons, including reliability, performance, security, scalability, and user experience. The physical installation includes components such as cables, switches, routers, and access points: a well-planned and properly installed physical infrastructure ensures reliable connectivity and minimizes the risk of hardware failures or malfunctions that can disrupt network operations. Optimal cable management and routing reduce signal degradation, interference, and crosstalk, leading to better network performance and higher data transfer speeds.

In the sections that follow, we walk through the most important factors relevant to the physical installation of the network.

Components of Wiring Distribution

The wiring closet, as the name implies, is the place in the network where you connect the cables and networking devices. These rooms have many names, including the wiring closet, the telecommunications room, and the network operations center (NOC). These telecommunications rooms contain the key network devices, such as the hubs, routers, switches, and servers. These rooms also contain the network media, such as patch cables that connect network devices to horizontal cables and the rest of the network.

Network Cross-Connects

The cable that runs throughout a network can be divided into two distinct sections:

▶ **Horizontal cabling:** Connects client systems to the network

▶ **Vertical (backbone) cabling:** Runs between floors to connect different locations on the network

Both of these cable types have to be consolidated and distributed from a location—a wiring closet.

Following are three types of cable distribution:

▶ **Vertical or main cross-connect:** The location where outside cables enter the building for distribution. This can include Internet and phone cabling.

▶ **Horizontal cross-connect:** The location where the vertical and horizontal connections meet.

▶ **Intermediate cross-connect:** A type typically used in larger networks. It provides an intermediate cross-connect between the main and horizontal cross-connects.

The term *cross-connect* refers to the point where the cables running throughout the network meet and are connected.

Horizontal Cabling

Within the telecommunications room, horizontal cabling connects the tele-communications room to the end user, as shown in Figure 8.1. Specifically, the horizontal cabling extends from the telecommunications outlet, or a network outlet with RJ-45 connectors, at the client end. It includes all cable from that outlet to the telecommunications room to the horizontal cross-connect—the distribution point for the horizontal cable. The horizontal cross-connect includes all connecting hardware, such as patch panels and patch cords. The horizontal cross-connect is the termination point for all network horizontal cables.

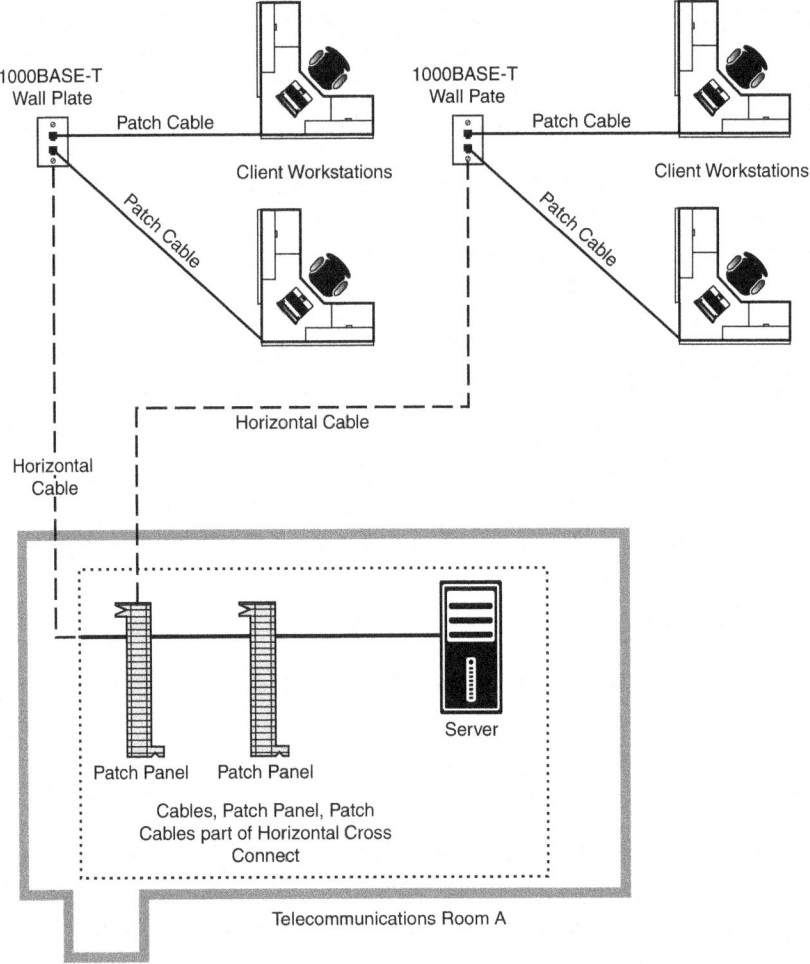

FIGURE 8.1 **Horizontal Cabling**

Horizontal cabling runs within walls and ceilings and therefore is called *permanent cable* or *structure cable*. The length of cable running from the horizontal connects and the telecommunication outlet on the client side should not exceed 90 meters. Patch cables used typically should not exceed 5 meters because of the 100-meter distance limitation of most UTP cable.

> **Note**
>
> Horizontal wiring includes all cabling run from the wall plate or network connection to the telecommunications closet. The outlets, cable, and cross-connects in the closet are all part of the horizontal wiring, which gets its name because the cable typically runs horizontally above ceilings or along the floor.

Vertical Cables

Vertical cable, or backbone cable, refers to the media used to connect telecommunications rooms, server rooms, and remote locations and offices. Vertical cable may be used to connect locations outside the local LAN that require high-speed connections. Therefore, vertical cable is often fiber-optic cable or high-speed UTP cable. Figure 8.2 shows the relationship between horizontal cable and vertical cable.

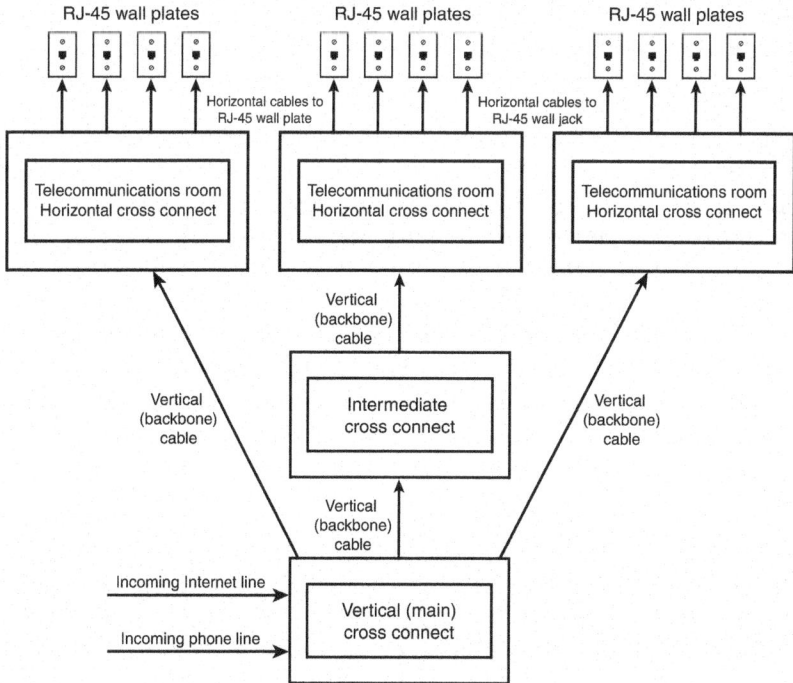

FIGURE 8.2 **Vertical and Horizontal Cabling**

Patch Panels

If you have ever looked in a telecommunications room, you have probably seen a distribution block, more commonly called a patch panel. A *patch panel* is a freestanding or wall-mounted unit with a number of RJ-45 port connections on the front. In a way, it looks like a wall-mounted hub without the *light-emitting diodes (LEDs)*. The patch panel provides a connection point between network equipment, such as hubs and switches, and the ports to which PCs are connected, which normally are distributed throughout a building.

> **Note**
>
> Not all environments use patch panels. In some environments, cables run directly between systems and a hub or switch. This is an acceptable method of connectivity, but it is not as easy to make tidy as a structured cabling system that uses a patch panel system and wall or floor sockets.

Also found in a wiring closet is the punchdown block. The wires from a telephony or UTP cable are attached to the punchdown block using a *punchdown tool*. To use the punchdown tool, you place the wires in the tip of the tool and push it into the connectors attached to the punchdown block. The wire insulation is stripped, and the wires are firmly embedded into the metal connector. Because the connector strips the insulation on the wire, it is known rather grandiosely as an *insulation displacement connector (IDC)*. Figure 8.3 shows a punchdown tool used for placing wires into a patch panel.

> **Note**
>
> Tools (hardware, software, and protocol-related) that can be used to solve networking issues are covered in Chapter 10, "Network Troubleshooting."

Using a punchdown tool is much faster than using wire strippers to prepare each individual wire and then twisting the wire around a connection pole or tightening a screw to hold the wire in place. In many environments, cable tasks are left to a specialized cable contractor. In others, the administrator is the one who must connect wires to a patch panel.

> **ExamAlert**
>
> Punchdown tools are used to attach twisted-pair network cable to connectors within a patch panel. Specifically, they connect twisted-pair wires to the IDC.

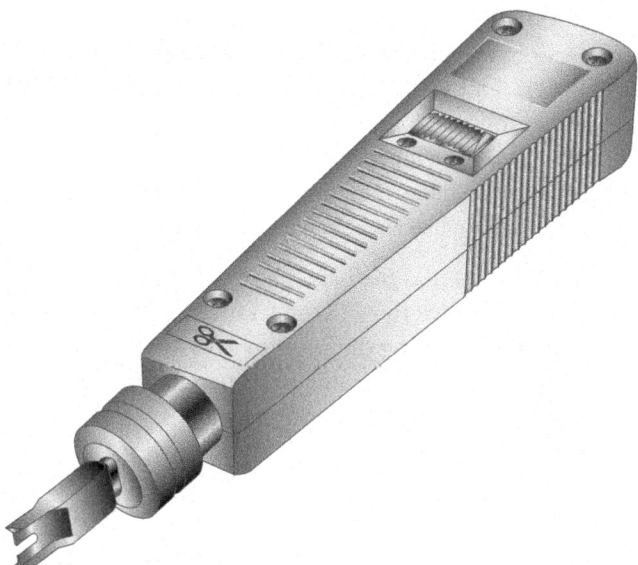

FIGURE 8.3 **Punchdown Tool**

Fiber Distribution Panels

Just as a patch panel is used to provide a connection point between network equipment, so too is a *fiber distribution panel (FDP)*. The difference between the two is that the FDP is a cabinet intended to provide space for termination, storage, and splicing of fiber connections.

> **ExamAlert**
>
> As you study for the exam, make sure you can identify the following exam objectives discussed here and the following: termination points, 66 block, 110 block, patch panel, and fiber distribution panel.

66 and 110 Blocks (T568A, T568B)

Two main types of punchdown blocks are used: type 66 and type 110. Type 66 is an older design used to connect wiring for telephone systems and other low-speed network systems and is not as widely used as type 110. The 66 block has 50 rows of IDC contacts to accommodate 25-pair twisted-pair cable. Block 66 was used primarily for voice communication. Although it was approved for Category 5 and greater, it is not really suitable for anything greater than 10BASE-T due to cross-talk problems. However, specialized certified blocks are available that do meet termination standards for Cat 5e, Cat 6, Cat 6a, Cat 7, or Cat 8.

In the network wiring closet, the 110 block is used to connect network cable to patch panels. The 110 connections can also be used at the other end of the network cable at the RJ-45 wall jack. The 110 blocks are preferred over the older 66 blocks because the 110 block improves on the 66 block by supporting higher frequencies and less crosstalk. Therefore, it supports higher-speed networks and higher-grade twisted-pair cable. The termination will be T568A or T568B, depending on which wiring standard is used.

In addition to 66 and 110 blocks, *Krone* and *Bix* blocks also exist. These two require different blades in the punchdown tools (Krone, for example, requires a separate scissor-like mechanism for trimming the wire) to work with them. Bix, which is short for Building Industry Cross-connect, is popular in older implementations, and Krone is more popular internationally.

MDF and IDF Wiring Closets

The preceding section looked at wiring closets. Two types of wiring closets are *main distribution frame (MDF)* and *intermediate distribution frame (IDF)*. The main wiring closet for a network typically holds the majority of the network gear, including routers, switches, wiring, servers, and more. This is also typically the wiring closet where outside lines run into the network. This main wiring closet is known as the MDF. One of the key components in the MDF is a primary patch panel. The network connector jacks attached to this patch panel lead out to the building for network connections.

In some networks, multiple wiring closets are used. When this is the case, the MDF connects to these secondary wiring closets, or IDFs, using a backbone cable. This backbone cable may be UTP, fiber, or even coaxial. In today's high-speed networks, UTP Gigabit Ethernet or high-speed fiber is the media of choice. Figure 8.4 shows the relationship between the MDF and the IDF.

> **Note**
>
> The MDF usually contains the termination point or demarc for telecommunications equipment. The demarcation point is where the telecom company's responsibility ends and the customer's private network responsibility begins.

A radio frequency identifier (RFID) reader and RFID badge card are also commonly used for access control to secure areas such as server rooms that must be secured. An MDF or IDF is often secured with mechanical access control that is accessed with a security device called a key fob. The MDF/IDF needs to be secured because it often contains high-level routers and switching equipment that cannot be accessed by general employees or visitors.

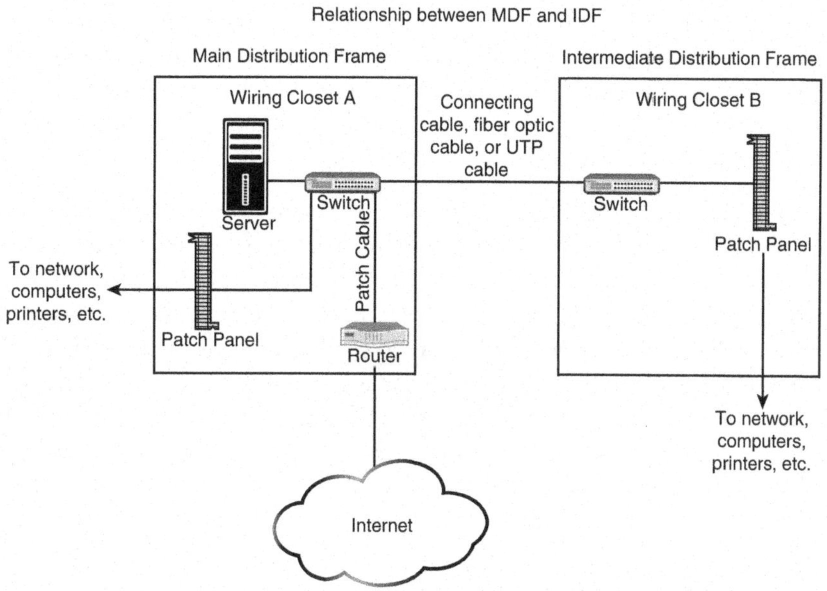

Relationship between MDF and IDF

FIGURE 8.4   **The Relationship Between MDFs and IDFs**

> **ExamAlert**
>
> Be prepared to identify the difference between an IDF and an MDF.

Rack Size

Rack size is a critical factor in network installation because the size of the rack determines the amount of space available for mounting networking equipment such as switches, routers, servers, and power distribution units (PDUs). Choosing an appropriately sized rack ensures that all equipment can be mounted securely and organized efficiently, minimizing clutter and maximizing airflow for proper equipment cooling.

Selecting racks that match the dimensions of the available space helps optimize space utilization and allows for future expansion without overcrowding or wasted space. Rack size considerations are essential for accommodating future growth and expansion of the network infrastructure. Choosing racks with sufficient capacity for additional equipment ensures scalability and flexibility to support evolving business needs and technological advancements over time.

Adequate rack size provides space for proper cable management, including routing, organizing, and securing network cables. Well-managed cables improve airflow, facilitate troubleshooting, and reduce the risk of cable damage or accidental disconnection, enhancing the overall reliability and maintenance of the network. Optimal rack sizing allows technicians and network administrators to access and work on equipment comfortably, minimizing downtime and disruptions during installation, upgrades, or repairs.

Lastly, rack size compatibility with standard equipment dimensions ensures interoperability and ease of integration with off-the-shelf networking hardware. Standard rack sizes, such as 19-inch or 23-inch widths, are commonly used in the industry and provide compatibility with a wide range of networking equipment from various vendors. A popular way to classify a rack is by its height, and thus racks are labeled with specifications such as "42U" or "7-foot." While the latter is easy to comprehend, in the case of the former, the "U" stands for "U-spaces" or "rack units," which are equal to 1.75 inches. The standard height is 48U, which is the same as 7-foot.

Port-side Exhaust/Intake

The direction in which airflow is directed by network equipment, such as switches or servers, through their ventilation ports should be *port-side exhaust/ intake* for maintaining optimal operating temperatures and preventing equipment overheating. In this configuration, the airflow is directed out of the equipment through its ventilation ports or exhaust fans located on the same side as the network ports (Ethernet ports, fiber-optic ports, and so on). The warm air generated by the operation of the equipment is expelled from the device's exhaust ports, usually toward the rear or side of the device.

Conversely, in this configuration, the airflow is drawn into the equipment through its ventilation ports or intake fans located on the same side as the network ports. Cool air from the environment is drawn into the device through its intake ports, usually located on the front or side of the device, to cool internal components such as processors, memory, and power supplies.

ExamAlert

The 2.4 exam objectives specifically call out rack size and port-side exhaust/intake. Be sure you are familiar with their meanings and important installation implications.

Lockable Environment

When it comes to keeping a network secure, it is imperative that the physical environment be just as secure—if not more so—than the online environment.

Not only should the datacenter be physically secured, but contents within it should also be protected. Lockable rackmount storage boxes, for example, are specialized enclosures designed to securely store and protect valuable equipment, tools, or accessories within a standard equipment rack. These storage boxes are typically mounted within the rack alongside other network equipment, such as servers, switches, and patch panels, providing a convenient and secure storage solution for items that require physical protection or restricted access.

Using Uninterruptible Power Supplies

No discussion of fault tolerance can be complete without a look at power-related issues and the mechanisms used to combat them. When you design a fault-tolerant system, your planning should definitely include *uninterruptible power supplies (UPSs)*. A UPS serves many functions and is a major part of server consideration and implementation.

On a basic level, a UPS, also known as a *battery backup*, is a box that holds a battery and built-in charging circuit. During times of good power, the battery is recharged; when the UPS is needed, it's ready to provide power to the server. Most often, the UPS is required to provide enough power to give the administrator time to shut down the server in an orderly fashion, preventing any potential data loss from a dirty power shutdown.

Why Use a UPS?

Organizations of all shapes and sizes need UPSs as part of their fault-tolerance strategies. A UPS is as important as any other fault-tolerance measure. Three key reasons make a UPS necessary:

▶ **Data availability:** The goal of any fault-tolerance measure is data availability. A UPS ensures access to the server if a power failure occurs—or at least as long as it takes to save a file.

▶ **Protection from data loss:** Fluctuations in power or a sudden power-down can damage the data on the server system. In addition, many servers take full advantage of caching, and a sudden loss of power could cause the loss of all information held in cache.

▶ **Protection from hardware damage:** Constant power fluctuations or sudden power-downs can damage hardware components within a computer. Damaged hardware can lead to reduced data availability while the hardware is repaired.

Power Loads, Threats, and Voltage

In addition to keeping a server functioning long enough to safely shut it down, a UPS safeguards a server from inconsistent power. This inconsistent power can take many forms. A UPS protects a system from the following power-related threats:

▶ **Blackout:** A total failure/power loss of the power supplied to the server.

▶ **Spike:** A short (usually less than 1 second) but intense increase in voltage. Spikes can do irreparable damage to any kind of equipment, especially computers.

▶ **Surge:** Compared to a spike, a surge is a considerably longer (sometimes many seconds) but usually less intense increase in power. Surges can also damage your computer equipment.

▶ **Sag:** A short-term voltage drop (the opposite of a spike). This type of voltage drop can cause a server to reboot.

▶ **Brownout/Under voltage event:** A drop in voltage that usually lasts more than a few minutes.

Many of these power-related threats can occur without your knowledge; if you don't have a UPS, you cannot prepare for them. For the cost, it is worth buying a UPS, if for no other reason than to sleep better at night.

Beyond the UPS

Power management is not limited only to the use of UPSs. In addition to these devices, you should employ *power generators* to be able to keep your systems up and running when the electrical provider is down for an extended period of time. *Redundant circuits* and *dual power supplies* should also be used for key equipment.

Any device fitted with multiple outputs that is specifically designed to distribute electric power is known as a *power distribution unit (PDU)*, and they are often plentiful in datacenters for supplying power to racks. The two main types of PDUs are Basic and Intelligent; the latter is any that is networked (allowing for remote management of power metering, toggling an outlet on/off, and so on).

> **ExamAlert**
>
> As you study for the exam, make sure you know the difference between redundant power supplies, uninterruptible power supplies, generators, and power distribution units as well as scenarios in which each would be the solution to a problem.

You want to make sure that power can stay up and running in the event of a crisis, so two other areas to pay attention to are the heating, ventilation, and air conditioning (HVAC) and the fire suppression system. It will do little good to keep the computers and servers running if you cannot keep the temperature within an operating range and provide safety in the event a fire occurs. Redundant systems should be considered for both of these crucial areas and regularly maintained.

Environmental Factors

Environmental concerns include considerations about temperature, humidity, electrical, and water/flood risks. Computer rooms should have fire and moisture detectors. Most office buildings have water pipes and other moisture-carrying systems in the ceiling. If a water pipe bursts (which is common in minor earthquakes), the computer room could become flooded. Water and electricity don't mix. Moisture monitors would automatically kill power in a computer room if moisture were detected, so the security professional should know where the water cut-offs are located.

Many computer systems require temperature and humidity control for reliable service. Humidity should be between 45 and 55 percent: maintaining relative humidity within this range helps prevent electrostatic discharge, equipment corrosion, and static electricity buildup, ensuring the reliability and longevity of networking hardware. High humidity levels can cause condensation to form inside networking equipment, leading to short circuits, corrosion, and other issues that can compromise equipment performance and reliability.

Large servers, communications equipment, and drive arrays generate considerable amounts of heat. An environmental system for this type of equipment is a significant expense beyond the actual computer system costs. Maintaining temperatures within the 20–25°C range (approximately 68–77°F in Fahrenheit) helps prevent overheating of networking equipment, reduces the risk of hardware failures, and prolongs the lifespan of components.

For fire suppression, halon gas systems were once widely installed due to their effectiveness in suppressing fires in enclosed spaces without leaving residue or causing equipment damage (making it suitable for protecting sensitive networking equipment and not damaging it the way a water-based system could). Such systems are now out of favor due to the climate harm they can cause, so alternatives are being used. Some examples of newly approved agents to be used in commercial buildings include carbon-based agents (such as perfluorohexane), carbon dioxide, and FM-200 (heptafluoropropane).

Cram Quiz

1. Which device is fitted with multiple outputs and is specifically designed to distribute electric power?

   ○ **A.** Demarcation points

   ○ **B.** FM-200 units

   ○ **C.** Port-side intakes

   ○ **D.** PDU

2. Which of the following is the optimal range for relative humidity in a datacenter or network installation environment to prevent equipment damage and corrosion?

   ○ **A.** 25–35 percent

   ○ **B.** 35–45 percent

   ○ **C.** 45–55 percent

   ○ **D.** 55–65 percent

3. Which of the following fire suppression systems was commonly used in datacenters and network installations due to its effectiveness in quickly extinguishing fires while minimizing equipment damage but now has largely been replaced by alternatives due to climate issues?

   ○ **A.** Halon gas system

   ○ **B.** Water sprinkler system

   ○ **C.** Foam-based system

   ○ **D.** Hydrogen peroxide (H_2O_2) system

4. What is the recommended temperature range for a datacenter or network installation environment to ensure optimal equipment performance and reliability?

   ○ **A.** 0–10°C

   ○ **B.** 20–25°C

   ○ **C.** 30–35°C

   ○ **D.** 40–55°C

5. Which of the following represents a temporary drop in voltage?

○ **A.** Total power failure

○ **B.** Power spike

○ **C.** Power surge

○ **D.** Undervoltage event

6. Which of the following is often secured with a mechanical locking device for access control because they often house important networking and telecommunications equipment?

○ **A.** Key fob

○ **B.** MDF/IDF

○ **C.** PDU

○ **D.** UPS

Cram Quiz Answers

1. **D.** A device fitted with multiple outputs that is specifically designed to distribute electric power is known as a power distribution unit (PDU), and these devices are often plentiful in datacenters for suppling power to racks.

2. **C.** The 45–55 percent range: maintaining relative humidity within this range helps prevent electrostatic discharge, equipment corrosion, and static electricity buildup, ensuring the reliability and longevity of networking hardware.

3. **A.** Halon gas is effective for suppressing fires in enclosed spaces without leaving residue or causing equipment damage, which is why it was used for protecting sensitive networking equipment. Climate issues, however, have created a need to seek alternative solutions.

4. **B.** The 20–25°C range: Maintaining temperatures within this range helps prevent overheating of networking equipment, reduces the risk of hardware failures, and prolongs the lifespan of components.

5. **D.** An undervoltage event, also known as a brownout, represents a temporary drop in voltage. A total failure, also known as a blackout, is a complete and lasting power loss of the power supplied to the server or device. A power spike is a short (usually less than 1 second) but intense increase in voltage. Spikes can do irreparable damage to any kind of equipment, especially computers. Compared to a spike, a surge is a considerably longer (sometimes many seconds) but usually less intense increase in power. Surges can also damage your computer equipment.

6. **B.** The main distribution frame (MDF) and intermediate distribution frame (IDF) are often secured with a mechanical locking device for access control because they often house important networking such as routers and switches and telecommunications equipment. A key fob is an actual access control device that controls entry, limiting someone's ability to enter or exit a room such as an MDF or IDF during certain times of the day. Any device fitted with multiple outputs that is

specifically designed to distribute electric power is known as a power distribution unit (PDU), and they are often plentiful in datacenters for supplying power to racks. A UPS, also known as a battery backup, is a box that holds a battery and built-in charging circuit. During times of good power, the battery is recharged; when the UPS is needed, it's ready to provide power to the server. Most often, the UPS is required to provide enough power to give the administrator time to shut down the server in an orderly fashion, preventing any potential data loss from a dirty power shutdown.

Organizational Processes and Procedures

▶ **3.1 Explain the purpose of organizational processes and procedures.**

CramSaver

If you can correctly answer these questions before going through this section, save time by skimming the ExamAlerts in this section and then completing the Cram Quiz at the end of the section.

1. Which network topology diagram documentation focuses on the direction in which data flows within the physical environment?

2. In computing, what are historical readings used as a measurement for future calculations referred to as?

3. True or false: Both logical and physical network diagrams provide an overview of the network layout and function.

4. True or false: Acceptable use policies define what controls are required to implement and maintain the security of systems, users, and networks.

Answers

1. The logical network topology diagram documentation refers to the direction in which data flows on the network within the physical topology. The logical diagram is not intended to focus on the network hardware but rather on how data flows through that hardware.

2. Baselines are historical readings used as a measurement for future calculations. An essential part of the administrator's role is keeping and reviewing baselines.

3. True. Both logical and physical network diagrams provide an overview of the network layout and function.

4. False. Security policies define what controls are required to implement and maintain the security of systems, users, and networks. Acceptable use policies (AUPs) describe how the employees in an organization can use company systems and resources: both software and hardware.

ExamAlert

Remember that this objective begins with "Explain the purpose." This means that you need to know and appreciate the role organizational documents and policies play in keeping a business up and running.

Administrators have several daily tasks, and new ones often crop up. In this environment, tasks such as documentation sometimes fall to the background. It's important that you understand why administrators need to spend valuable time writing and reviewing documentation. Having a well-documented network offers a number of advantages:

▶ **Troubleshooting:** When something goes wrong on the network, including the wiring, up-to-date documentation is a valuable reference to guide the troubleshooting effort. The documentation saves you money and time in isolating potential problems.

▶ **Training new administrators/technicians:** In many network environments, new administrators are hired, and old ones leave. In this scenario, documentation is critical. New administrators do not have the time to try to figure out where cabling is run, what cabling is used, potential trouble spots, and more. Up-to-date information helps new administrators quickly see the network layout.

▶ **Working with contractors and consultants:** Consultants, contractors, and even auditors occasionally may need to visit the network to make recommendations for the network or to add wiring or other components. In such cases, up-to-date documentation is needed. If documentation is missing, it would be much more difficult for these people to do their jobs, and more time and money would likely be required.

▶ **Compiling an asset inventory:** Knowing what you have, where you have it, and what you can turn to in the case of an emergency is both constructive and helpful. Asset inventory is a systematic process of identifying, labeling, tracking, documenting, and managing all hardware and software assets within the network infrastructure. This includes physical devices, such as servers, switches, routers, workstations, and printers, as well as software applications, licenses, and associated warranty support agreements. By maintaining accurate inventory records, organizations can effectively plan for upgrades, replacements, and expansions; minimize downtime; reduce costs; and ensure compliance with licensing and warranty obligations. Additionally, asset inventory serves as a valuable tool for risk management, security assessments, and disaster recovery planning within the network environment.

Quality network documentation does not happen by accident; rather, it requires careful planning. When creating network documentation, you must keep in mind who you are creating the documentation for and that it is a communication tool. Documentation is used to take technical information and present it in a manner that someone new to the network can understand. When planning network documentation, you must decide what you need to document.

All networks differ and so does the documentation required for each network. However, certain elements are always included in quality documentation:

▶ **Floor plan:** This diagram need not be complicated; it should simply show where everything is, including server rooms, wiring closets, MDF/ IDF, and so on. It is a layout of the area and what would be found in each location. A good way to think of this plan is that it would be a useful tool to hand to new junior administrators on their first day at work to familiarize them with where resources can be found.

▶ **Network topology:** Networks can be complicated. If someone new is looking over the network, it is critical to document the entire topology. This includes both the wired and wireless topologies used on the network. Network topology documentation typically consists of a diagram or series of diagrams labeling all critical components used to create the network. These diagrams utilize common symbols for components such as fire-walls, hubs, routers, and switches. Figure 8.5, for example, shows standard figures for, from left to right, a firewall, a hub, a router, and a switch.

FIGURE 8.5 **Diagram Symbols for a Firewall, a Hub, a Router, and a Switch**

▶ **Cable maps and rack diagrams:** Network wiring can be confusing. Much of it is hidden in walls and ceilings, making it hard to know where the wiring is and what kind is used on the network. Therefore, it is critical to keep documentation on network wiring up to date. *Cable maps* are graphical representations that depict the layout and connectivity of network cables within a specific area, such as a datacenter, server room, or network closet. Typically, they illustrate the paths of individual cables, including their endpoints, connections to network devices (such as switches, routers, servers, and patch panels), and the type of cables used (such as Ethernet, fiber optic, or coaxial), helping to identify cable runs, verify connections, and locate specific cables for maintenance or troubleshooting purposes. Similarly, *rack diagrams* elucidate what is on each rack and any unusual configurations that might be employed.

▶ **Network layer diagrams:** Layer 1, Layer 2, and Layer 3 network diagrams are visual representations that depict different aspects of a computer network's architecture and connectivity based on the OSI model layers. *Layer 1* network diagrams focus on the physical aspects of the network infrastructure and typically illustrate the layout of network devices, such as switches, routers, servers, workstations, and other hardware components within the physical space, such as a datacenter, server room, or network closet. *Layer 2* network diagrams focus on the data link layer of the OSI model, typically illustrating the logical connections between network devices, such as switches and bridges. *Layer 3* network diagrams focus on the network layer of the OSI model, typically illustrating the logical topology and interconnections between routers, subnets, and network segments at the network layer.

▶ **IDF/MDF documentation:** It is not enough to show that there is an intermediate distribution frame (IDF) and/or main distribution frame (MDF) in your building. You need to thoroughly document any and every free-standing or wall-mounted rack and the cables running between them and the end-user devices.

▶ **Server configuration:** A single network typically uses multiple servers spread over a large geographic area. Documentation must include schematic drawings of where servers are located on the network and the services each provides. This includes server function, server IP address, operating system (OS), software information, and more. Essentially, you need to document all the information you need to manage or administer the servers.

▶ **Network equipment:** The hardware used on a network is configured in a particular way—with protocols, security settings, permissions, and more. Trying to remember them all would be a difficult task. Having up-to-date documentation makes it easier to recover from a failure.

▶ **Network configuration, performance baselines, and key applications:** Documentation also includes information on all current network configurations, performance baselines taken, and key applications used on the network, such as up-to-date information on their updates, vendors, install dates, and more.

▶ **Detailed account of network services:** Network services are a key ingredient in all networks. Services such as Domain Name Service (DNS) and Dynamic Host Configuration Protocol (DHCP), and more, are an important part of documentation. You should describe in detail which server maintains these services, the backup servers for these services, maintenance schedules, how they are structured, and so on. *IP address management (IPAM)*, for example, benefits from effective documentation that outlines the overall strategy for assigning and managing IP addresses within the network. You should include details such as IP address ranges, subnetting schemes, allocation policies, and reserved address pools. This plan serves as a blueprint for IP address management and helps ensure consistency and efficiency in address allocation.

▶ **Wireless survey report:** A *site survey* is typically associated with wireless networking (thus also known as a *wireless survey*) and used to identify wireless access points and security settings. These surveys can be used to help you design and deploy an efficient network. During a wireless survey, trained technicians or network engineers use specialized equipment, such as wireless spectrum analyzers, Wi-Fi scanners, and signal strength meters, to measure and map the wireless signals emitted by access points (APs) and other wireless devices. The collected data is then analyzed to identify coverage gaps, dead zones, interference sources, and areas of poor signal quality. A subset of this, a *heat map* (also known as a signal coverage map or RF propagation map) is a graphical representation of wireless signal strength or coverage within a given area that uses color gradients or shading to visualize signal strength levels, with warmer colors (e.g., red or orange) indicating areas of stronger signal coverage and cooler colors (e.g., blue or green) indicating areas of weaker coverage or signal attenuation.

▶ **Audit and assessment report:** This report is used to see how well your operations/settings match what you intended them to. For more information, see "Network Device Logs" later in this chapter.

▶ **Standard operating procedures/work instructions:** Finally, documentation should include information on network policy and procedures. This information includes many elements, ranging from who can and cannot access the server room, to network firewalls, protocols, passwords, physical security, cloud computing use, mobile device use, and so on.

> **ExamAlert**
>
> Be sure that you know the types of information that should be included in network documentation. Know the differences between physical and logical diagrams as well as those showing racks, cabling, and network diagrams. Remember that network diagrams can include technology or devices that operate at Layers 1, 2, and 3 of the OSI reference model. Be sure you understand what devices operate at which layers of the OSI model.

Wiring and Port Locations

Network wiring schematics are an essential part of network documentation, particularly for midsize to large networks, where the cabling is certainly complex. For such networks, it becomes increasingly difficult to visualize network cabling and even harder to explain it to someone else. A number of software tools exist to help administrators clearly document network wiring in detail.

Several types of wiring schematics exist. They can be general, as shown in Figure 8.6, or they can be very specific, indicating the actual type of wiring used, the operating system on each machine, and so on. The more generalized they are, the less they need updating, whereas very specific schematics often need to be changed regularly. Table 8.1 represents another way of documenting data.

> **ExamAlert**
>
> For the exam, be familiar with the look of a general wiring schematic such as the one shown in Figure 8.6.

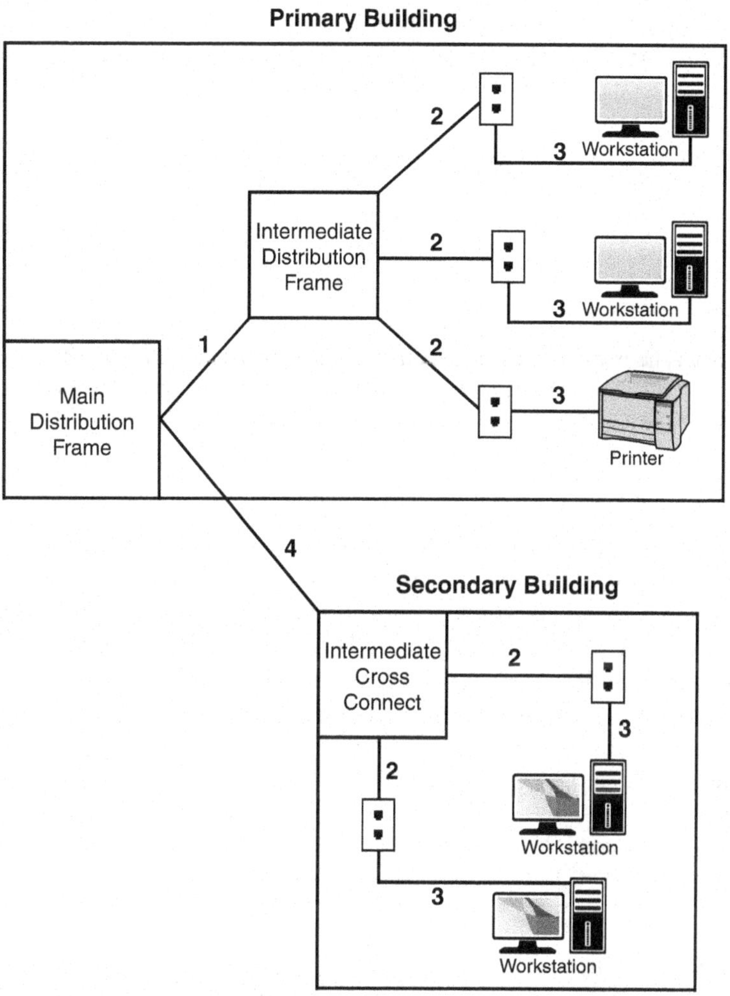

Primary Building

FIGURE 8.6 A General Wiring Schematic

TABLE 8.1 **Wiring Details**

Cable	Description	Installation Notes
1	Category 6 plenum-rated cable	Cable runs 50 feet from the MDF to IDF. Cable placed through the ceiling and through a mechanical room. Cable was installed 02/26/2021, upgrading a nonplenum Category 5e cable.
2	Category 6a plenum cable	Horizontal cable runs 45 feet to 55 feet from IDF to wall jack. Replaced Category 5 cable February 2021. Section of cable runs through the ceiling and over fluorescent lights.

Cable	Description	Installation Notes
3	Category 6 UTP cable	All patch cable connectors were attached in-house. Patch cable connecting the printer runs 45 feet due to printer placement.
4	8.3-micron core/125-micron cladding single mode	Connecting fiber cable runs 2 kilometers between the primary and secondary buildings.

Note

Due to the potential harm caused by PVC (plastics) in the UTP cable jacket when ignited, many municipalities mandate the installation of *plenum*-rated cable in areas inaccessible to sprinkler systems. A plenum refers to an enclosed space used for airflow.

Figure 8.6 provides a simplified look at network wiring schematics. Imagine how complicated these diagrams would look on a network with 1,000, 2,000, or even 6,000 computers. Quality network documentation software makes this easier; however, the task of network wiring can be a large one for administrators. Administrators need to ensure that someone can pick up the wiring documentation diagrams and have a good idea of the network wiring.

Caution

An important part of the administrator's role is reading schematics and determining where wiring runs are. Expect to see a schematic on your exam.

Port locations should be carefully recorded and included in the documentation as well. Simple Network Management Protocol (SNMP) can be used directly to map ports on switches and other devices; it is much easier, however, to use software applications that incorporate SNMP and use it to create ready-to-use documentation. A plethora of such programs are available; some are free and many are commercial products.

Note

SNMP was discussed in Chapter 1, "Networking Models, Ports, Protocols, and Services," and is covered in more detail later in this chapter.

Troubleshooting Using Wiring Schematics

Some network administrators do not take the time to maintain quality documentation. This failure to keep updated information will haunt them when it comes time to troubleshoot some random network problems. Without any network wiring schematics, the task will be frustrating and time consuming. The information shown in Figure 8.6 might be simplified, but you could use that documentation to evaluate the network and make recommendations.

> **Caution**
>
> When looking at a wiring schematic, pay close attention to where the cable is run and the type of cable used if the schematic indicates this information. If a correct cable is not used, a problem could occur.

> **Note**
>
> Network wiring schematics are a work in progress. Although changes to wiring do not happen daily, they do occur when the network expands or old cabling is replaced. It is imperative to remember that when changes are made to the network, the schematics and their corresponding references must be updated to reflect the changes. Out-of-date schematics can be very frustrating to work with.

Physical and Logical Network Diagrams

In addition to the wiring schematics, documentation should include diagrams of the physical and logical network design. Recall from Chapter 2, "Network Topologies, Architectures, and Types," that network topologies can be defined on a physical or a logical level. The *physical topology* refers to how a network is physically constructed—how it looks. The *logical topology* refers to how a network looks to the devices that use it—how it functions.

Network infrastructure documentation isn't reviewed daily; however, this documentation is essential for someone unfamiliar with the network to manage or troubleshoot the network. When it comes to documenting the network, you need to document all aspects of the infrastructure. This includes the physical hardware, physical structure, protocols, and software used.

> **ExamAlert**
>
> You should be able to identify physical and logical diagrams. You need to know the types of information that should be included in each diagram.

The physical documentation of the network should include the following elements:

▶ **Cabling information:** A visual description of all the physical communication links, including all cabling, cable grades, cable lengths, WAN cabling, and more.

▶ **Servers:** The server names and IP addresses, types of servers, and domain membership.

▶ **Network devices:** The location of the devices on the network. This information includes the printers, hubs, switches, routers, gateways, and more.

▶ **Wide-area network:** The location and devices of the WAN and components.

▶ **User information:** Some user information, including the number of local and remote users.

As you can see, many elements can be included in the physical network diagram. Figure 8.7 shows a physical segment of a network.

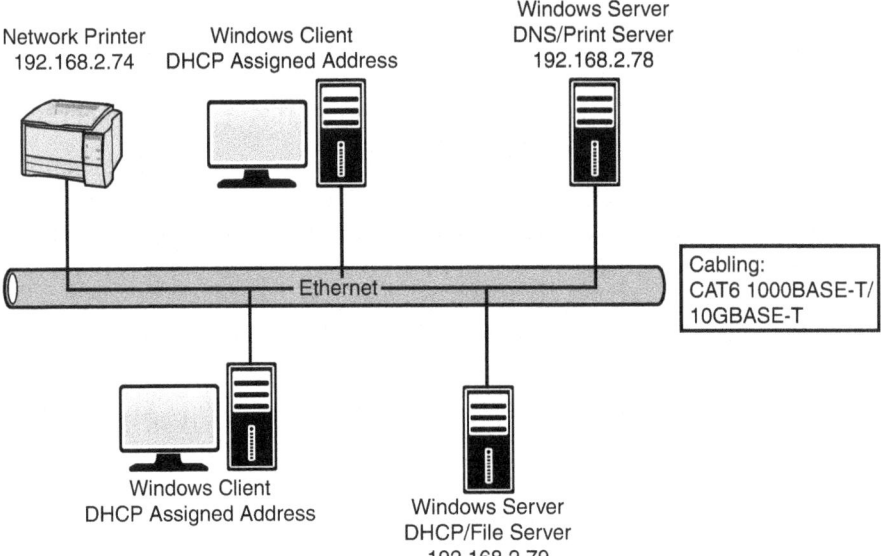

FIGURE 8.7 **A Physical Network Diagram**

> **Caution**
>
> You should recognize the importance of maintaining documentation that includes network diagrams, asset management, IP address utilization, vendor documentation, and internal operating procedures, policies, and standards.

Networks are dynamic, and changes can happen regularly, which is why the physical network diagrams also must be updated. Networks have different policies and procedures on how often updates should occur. Best practice is that the diagram should be updated whenever significant changes to the network occur, such as the addition of a switch or router, a change in protocols, or the addition of a new server. These changes impact how the network operates, and the documentation should reflect the changes.

> **Caution**
>
> There are no hard-and-fast rules about when to change or update network documentation. However, most administrators want to update whenever functional changes to the network occur.

The logical network refers to the direction in which data flows on the network within the physical topology. The logical diagram is not intended to focus on the network hardware but rather on how data flows through that hardware. In practice, the physical and logical topologies can be the same. In the case of the star/hub-and-spoke physical topology, data travels along the length of the cable from one computer to the next. So, the diagram for the physical and logical bus would be the same.

This is not always the case. For example, a topology can be in the physical shape of a star, but it is difficult to tell from looking at a physical diagram how data is flowing on the network.

In today's network environments, the star/hub-and-spoke topology is a common network implementation. Ethernet uses a physical star topology but a logical bus topology. In the center of the physical Ethernet star topology is a switch. It is what happens inside the switch that defines the logical bus topology. The switch passes data between ports as if they were on an Ethernet bus segment.

In addition to data flow, logical diagrams may include additional elements, such as the network domain architecture, server roles, protocols used, and more. Figure 8.8 shows how a logical topology may look in the form of network documentation.

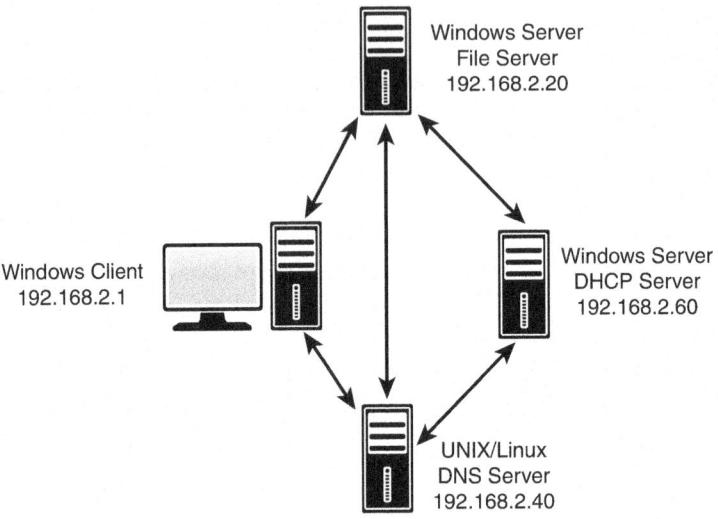

FIGURE 8.8   **A Logical Topology Diagram**

> **Caution**
>
> The logical topology of a network identifies the logical paths that data signals travel over the network.

Baseline/Golden Configurations

Baselines play an integral part in network documentation because they let you monitor the network's overall performance. In simple terms, a *baseline* is a measure of performance that indicates how hard the network is working and where network resources are spent. The purpose of a baseline is to provide a basis of comparison. For example, you can compare the network's performance results taken in March to results taken in June, or from one year to the next. More commonly, you would compare the baseline information at a time when the network is having a problem to information recorded when the network was operating with greater efficiency. Such comparisons help you determine whether there has been a problem with the network, how significant that problem is, and even where the problem lies.

To be of any use, baselining is not a one-time task; rather, baselines should be taken periodically to provide an accurate comparison. You should take an initial baseline after the network is set up and operational, and then again when major changes are made to the network. Even if no changes are made to the network, periodic baselining can prove useful as a means to determine whether the network is still operating correctly.

A *baseline/"golden"* configuration is a version used in configuration management as an ideal configuration against which configurations from similar devices can be compared. The "golden" configuration often implies that the baseline has been refined and validated to an even higher standard than just a baseline, typically representing the best-practice configuration for a given system or environment—thoroughly tested and proven to be reliable and secure.

All *network operating systems (NOSs)*, including Windows, macOS, UNIX, and Linux, have built-in support for network monitoring. In addition, many third-party software packages are available for detailed network monitoring. These system-monitoring tools provided in an NOS give you the means to take performance baselines, either of the entire network or for an individual segment within the network. Because of the different functions of these two baselines, they are called a system baseline and a component baseline.

To create a network baseline, network monitors provide a graphical display of network statistics. Network administrators can choose a variety of network measurements to track. They can use these statistics to perform routine troubleshooting tasks, such as locating a malfunctioning network card, a downed server, or a *denial-of-service (DoS)* attack.

> **Note**
>
> *Graphing*, and the process of seeing data visually, can be much more helpful in identifying trends than looking at raw data and log files.

Collecting network statistics is a process called *capturing*. Administrators can capture statistics on all elements of the network. For baseline purposes, one of the most common statistics to monitor is bandwidth usage. By reviewing bandwidth statistics, administrators can see where the bulk of network bandwidth is used. Then they can adapt the network for bandwidth use. If too much bandwidth is used by a particular application, administrators can actively control its bandwidth usage. Without comparing baselines, however, it is difficult to see what is normal network bandwidth usage and what is unusual.

> **Caution**
>
> Remember that baselines need to be taken periodically and under the same conditions to be effective. They are used to compare current performance with past performance to help determine whether the network is functioning properly or if troubleshooting is required.

Policies, Procedures, Configurations, and Regulations

Well-functioning networks are characterized by documented policies, procedures, configurations, and regulations. Because they are unique to every network, policies, procedures, configurations, and regulations should be clearly documented.

Policies

By definition, policies refer to an organization's documented rules about what is to be done, or not done, and why. Policies dictate who can and cannot access particular network resources, server rooms, backup media, and more.

Although networks might have different policies depending on their needs, some common policies include the following:

▶ **Network usage policy:** This policy defines who can use network resources such as PCs, printers, scanners, mobile devices, and remote connections. In addition, the usage policy dictates what can be done with these resources after they are accessed. No outside systems will be networked without permission from the network administrator.

▶ **Internet usage policy:** This policy specifies the rules for Internet use on the job. Typically, usage should be focused on business-related tasks. Incidental personal use is allowed during specified times.

▶ **Bring-your-own-device (BYOD) policy:** This policy defines what personally owned mobile devices (laptops, tablets, and smartphones) employees are allowed to bring to their workplace and use. *Mobile device management (MDM)* and *mobile application management (MAM)* systems can be used to help enterprises manage and secure the use of those mobile devices in the workplace and to interact with privileged company information and applications. Two things the policy needs to address are onboarding and offboarding. *Onboarding* the mobile device is the set of procedures to get it ready to go on the network (scanning for viruses, adding certain apps, and so forth). *Offboarding* is the process of removing company-owned resources when they are no longer needed (often done with a wipe or factory reset).

> **ExamAlert**
>
> For the exam, be familiar with onboarding (hiring) and offboarding (firing or leaving) processes.

▶ **Email usage policy:** Email must follow the same code of conduct as expected in any other form of written or face-to-face communication. All emails are company property and can be accessed by the company. Personal emails should be immediately deleted.

▶ **Personal software policy:** No outside software should be installed on network computer systems. All software installations must be approved by the network administrator. No software can be copied or removed from a site. Licensing restrictions must be adhered to.

▶ **Password policy:** This policy details how often passwords must be changed and the minimum level of security for each (number of characters, use of alphanumeric character set, longer phrases, and so on). Microsoft Windows, for example, uses local security policies that can define the password policy for the workstation. To access these, go to **Administrative Tools** (or go to **Run > secpol.msc**). Then, navigate to **Security Settings > Account Policies > Password Policy**.

▶ **Acceptable use policy (AUP):** This policy describes how the employees in an organization can use company systems and resources, both software and hardware. This policy should also outline the consequences for misuse. In addition, the policy (also known as a *use policy*) should address installation of personal software on company computers and the use of personal hardware, such as USB devices.

▶ **User account policy:** All users are responsible for keeping their password and account information secret. All staff are required to log off and sometimes lock their systems after they finish using them. Attempting to log on to the network with another user account is considered a serious violation.

▶ **International export controls:** A number of laws and regulations govern what can and cannot be exported to various countries when it comes to software and hardware. Employees should take every precaution to make sure they are adhering to the letter of the law.

▶ **Data loss prevention (DLP):** Losses from employees can quickly put a company in the red. All employees should understand that it is their responsibility to make sure all preventable losses are prevented.

▶ **Incident response plan:** When an incident occurs, all employees should understand it is their responsibility to be on the lookout for it and report it immediately to the appropriate party.

▶ **Disaster recovery (DR) plan:** Just as you should have a plan in place for responding to incidents, so, too, do you need one for disasters. This topic is explored in further detail in the section "Disaster Recovery and High-Availability" later in this chapter.

▶ **Business continuity plan (BCP):** When an incident occurs, it is too late to consider policies and procedures then; this must be done well ahead of time. *Business continuity* should always be of the utmost concern. Business continuity is primarily concerned with the processes, policies, and methods that an organization follows to minimize the impact of a system failure, network failure, or the failure of any key component needed for operation. *Business continuity planning (BCP)* is the process of implementing policies, controls, and procedures to counteract the effects of losses, outages, or failures of critical business processes. BCP is primarily a management tool that ensures that *critical business functions (CBFs)* can be performed when normal business operations are disrupted.

▶ **Nondisclosure agreement (NDA):** NDAs are the oxygen that many companies need to thrive. Employees should understand the importance of them to continued business operations and agree to follow them to the letter, and spirit, of the law. Many companies require new employees to sign NDAs and acceptable use policies (AUPs) when they are hired during the onboarding process.

▶ **Service-level agreement (SLA):** This is an agreement between you or your company and a service provider, typically a technical support provider or even cloud service provider. SLAs are also usually part of network availability and other agreements. They stipulate the performance you can expect or demand by outlining the expectations a vendor has agreed to meet. They define what is possible to deliver and provide the contract to make sure what is delivered is what was promised.

Note

Service-level agreements (SLAs) often address availability in terms of "nines," with five nines (taken to mean 99.999 percent) being the norm. This means the downtime allowed is less than 5 minutes and 15 seconds per year.

▶ **Memorandum of understanding (MOU):** This agreement between two or more parties indicates what the relationship is between the parties. It is sometimes a precursor to a contract, but is often used in place of it when a contract would not do (such as defining the relationship between departments of the same organization).

▶ **Safety procedures and policies:** Safety is everyone's business, and all employees should know how to do their job in the safest manner while also looking out for other employees and customers alike. A safety data sheet (SDS), or material safety data sheet (MSDS), is a standardized document that contains crucial occupational safety and health information and is mandated by the International Hazard Communication Standard (HCS).

▶ **Ownership policy:** The company owns all data, including users' email, voice mail, and Internet usage logs, and the company reserves the right to inspect them at any time. Some companies even go so far as controlling how much personal data can be stored on a workstation or mobile device.

This list is just a snapshot of the policies that guide the behavior for administrators and network users. Network policies should be clearly documented and available to network users. Often, these policies are reviewed with new staff members or new administrators. As they are updated, they are rereleased to network users. Policies are regularly reviewed and updated.

> **Note**
>
> You might be asked about network policies. Network policies dictate network rules and provide guidelines for network conduct. Policies are often updated and reviewed and are changed to reflect changes to the network and perhaps changes in business requirements.

Password-Related Policies

Although biometrics and smartcards have become more common, they still have a long way to go before they attain the level of popularity that username/ password combinations enjoy. Usernames and passwords do not require any additional equipment, which practically every other method of authentication does; the username and password process is familiar to users, easy to implement, and relatively secure. For that reason, these policies are worthy of more detailed coverage than the other authentication systems previously discussed.

> **Note**
>
> Biometrics are not as ubiquitous as username/password combinations, but they are coming up quickly. Many smartphones, for example, offer the ability to use a fingerprint scanner and/or gestures to access the system instead of username and password. Features such as these are expected to become even more common with future releases.

Passwords are a relatively simple form of authentication in that only a string of characters can be used to authenticate the user. However, how the string of characters is used and which policies you can put in place to govern them make usernames and passwords an excellent form of authentication.

Password Policies

All popular network operating systems include password policy systems that enable the network administrator to control how passwords are used on the system. The exact capabilities vary between network operating systems. However, generally they enable the following:

▶ **Minimum length of password:** Shorter passwords are easier to guess than longer ones. Setting a minimum password length does not prevent a user from creating a longer password than the minimum; however, each network operating system has a limit on how long a password can be.

▶ **Password expiration:** Also known as the maximum password age, password expiration defines how long the user can use the same password before having to change it. A general practice is that a password be changed every 30 days. In high-security environments, you might want to make this value shorter, but you should generally not make it any longer. Having passwords expire periodically is a crucial feature because it means that if a password is compromised, the unauthorized user will not indefinitely have access.

▶ **Prevention of password reuse:** Although a system might cause a password to expire and prompt the user to change it, many users are tempted to use the same password again. A process by which the system remembers the last 10 passwords, for example, is most secure because it forces the user to create completely new passwords. This feature is sometimes called enforcing password history.

▶ **Prevention of easy-to-guess passwords:** Some systems can evaluate the password provided by a user to determine whether it meets a required level of complexity. Enabling this function prevents users from having passwords such as *password*, *12345678*, their name, or their nickname.

Figure 8.9 shows an example of configuring a security policy in Windows for password complexity.

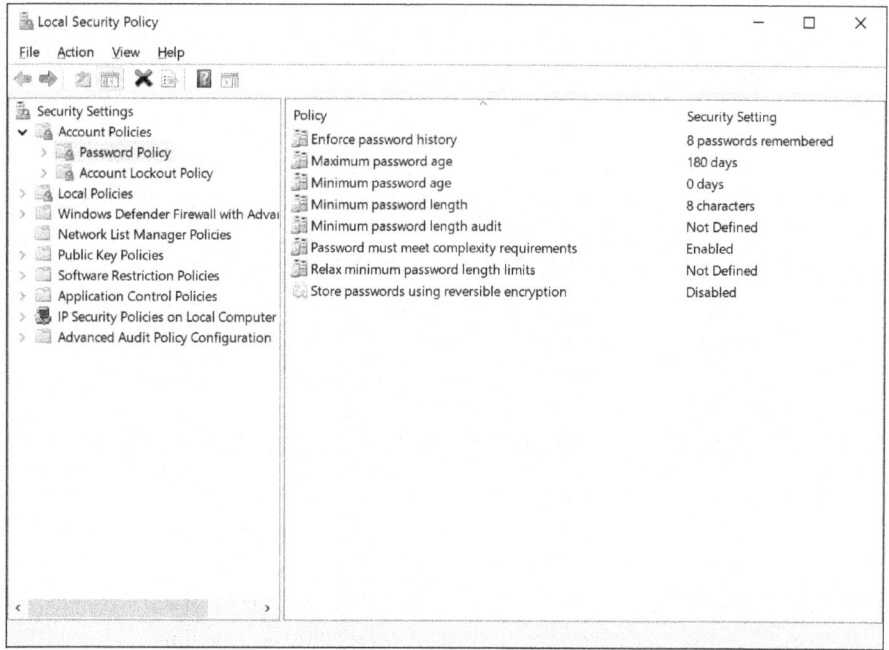

FIGURE 8.9   **Configuring a Password Security Policy.**

ExamAlert

You must identify an effective password policy. For example, a robust password policy would include forcing users to change their passwords on a regular basis.

Password Strength

No matter how good a company's password policy, it is only as effective as the passwords created within it. A password that is hard to guess, or strong, is more likely to protect the data on a system than one that is easy to guess, or weak.

If you are using only numbers and letters—and the OS is not case sensitive—36 possible combinations exist for each entry, and the total number of possibilities is 36^6. That might seem like a lot, but to a password-cracking program, it's not much security. A password that uses eight case-sensitive characters, with letters, numbers, and special characters, has so many possible combinations that a standard calculator cannot display the actual number.

There has always been a debate over how long a password should be. It should be sufficiently long that it is hard to break but sufficiently short that the user can easily remember it (and type it). In a normal working environment, passwords of 8 characters are sufficient. Certainly, they should be no fewer than 6 characters. In environments in which security is a concern, passwords should be 10 characters or more.

Users should be encouraged to use a password that is considered strong. A strong password has at least eight characters; has a combination of letters, numbers, and special characters; uses mixed case; and does not form a proper word. Examples are 3Ecc5T0h and e1oXPn3r. Such passwords might be secure, but users are likely to have problems remembering them. For that reason, a popular strategy is to use a combination of letters and numbers to form phrases or long words. Examples include d1eTc0La and tAb1eT0p. These passwords might not be quite as secure as the preceding examples, but they are still strong and a whole lot better than the name of the user's pet.

The National Institute of Standards and Technology (NIST) offers both requirements and recommendations when it comes to passwords. For NIST Digital Identity Guidelines, for example, see https://nvlpubs.nist.gov/nistpubs/SpecialPublications/NIST.SP.800-63b.pdf. These guidelines recommend that new passwords be screened against a list of known compromised passwords, password hints and knowledge-based security questions be skipped, and a limit placed on the number of failed authentication attempts that are allowed.

Procedures

Network procedures differ from policies in that they describe how tasks are to be performed. For example, each network administrator has backup procedures specifying the time of day backups are done, how often they are done, and where they are stored. A network is full of a number of procedures for practical reasons and, perhaps more important, for security reasons.

Administrators must be aware of several procedures when on the job. The number and exact type of procedure depend on the network. The network's overall goal is to ensure uniformity and ensure that network tasks follow a framework. Without this procedural framework, different administrators might approach tasks differently, which could lead to confusion on the network.

Network procedures might include the following:

▶ **Backup procedures:** These procedures specify when they are to be performed, how often a backup occurs, who does the backup, what data is to be backed up, and where and how it will be stored. Network administrators should carefully follow backup procedures.

▶ **Procedures for adding new users:** When new users are added to a network, administrators typically have to follow certain guidelines to ensure that the users have access to what they need to do their job, but no more. This is called the *principle of least privilege*.

▶ **Privileged user agreement:** Administrators and authorized users who have the ability to modify secure configurations and perform tasks such as account setup, account termination, account resetting, auditing, and so on need to be held to high standards.

▶ **Security procedures:** Some of the more critical procedures involve security. Security procedures are numerous but may include specifying what the administrator must do if security breaches occur, monitoring and reporting on weaknesses, and updating the OS and applications for potential security holes.

▶ **Network monitoring procedures:** The network needs to be constantly monitored. This includes tracking such things as bandwidth usage, remote access, user logons, and more.

▶ **Software procedures/system lifecycle:** All software must be periodically monitored and updated. Documented procedures dictate when, how often, why, and for whom these updates are done. When assets are disposed of, asset disposal procedures should be followed to properly document and log their removal. Two key variables worth noting are end-of-life (EOL) and end-of-support (EOS). *End-of-life (EOL)* refers to the stage in a product's lifecycle where the manufacturer discontinues production, sales, and support for the product. It means that the product is no longer being actively marketed or sold by the manufacturer; technical support and warranty coverage may no longer be available. This can pose security risks, compatibility issues, and operational challenges for equipment still in use. Similarly, *end-of-support (EOS)* refers to the date when the manufacturer discontinues providing software updates, security patches, and technical support for a product that has reached its end-of-life. Unsupported devices may become targets for cyberattacks, experience performance degradation, or fail to integrate with newer systems and protocols.

> **ExamAlert**
>
> For the exam, remember that end-of-life (EOL) refers to the stage in a product's lifecycle when the manufacturer discontinues production, sales, and support for the product; it means that the product is no longer being actively marketed or sold by the manufacturer. End-of-support (EOS) refers to the date when the manufacturer discontinues/terminates providing software updates, security patches, and technical support for a product that has reached its end-of-life.

Decommissioning is the process of retiring or removing hardware devices, software applications, or services from the network environment. It involves systematically shutting down, disabling, and removing components that are no longer needed or have reached the end of their useful life. After decommissioning is complete, it's important to verify that the components have been successfully removed from the network environment and that any associated data or configurations have been properly disposed of or migrated to alternative systems. This may involve conducting post-decommissioning checks, performing audits, and documenting the decommissioning activities for future reference.

Effective software management, including patch management, operating system management, and firmware management, is essential for maintaining the security, reliability, and performance of network infrastructure. By proactively managing software lifecycles and applying timely updates and patches, organizations can mitigate security risks, address software vulnerabilities, and ensure the continuous operation of their networks.

▶ **Procedures for reporting violations:** Users do not always follow outlined network policies. This is why documented procedures should exist to properly handle the violations. This might include a verbal warning upon the first offense, followed by written reports and account lockouts thereafter.

▶ **Remote-access policy and network admission procedures:** Many workers remotely access the network. This remote access is granted and maintained using a series of defined procedures. These procedures might dictate when remote users can access the network, how long they can access it, and what they can access. *Network admission control (NAC)*—also referred to as *network access control*—determines who can get on the network and is usually based on IEEE 802.1X guidelines. IEEE 802.1X authentication allows only authorized devices to connect to the network. The most secure form of IEEE 802.1X authentication is certificate-based authentication.

Change Management Documentation

Change management programs provide organizations with a formal process for identifying, requesting, approving, and implementing changes to configurations.

Change management procedures might include the following:

- ▶ **Documentation of reason for a change:** Before making any change at all, the first question to ask is *why*. A change requested by one user may be based on a misunderstanding of what technology can do, may be cost prohibitive, or may deliver a benefit not worth the undertaking.

- ▶ **Change request/service request:** An official request should be logged and tracked to verify what is to be done and what has been done. Within the realm of the change request should be the configuration procedures to be used, the rollback process that is in place, potential impact identified, and a list of those who need to be notified.

- ▶ **Approval process:** Changes should not be approved on the basis of who makes the most noise, but rather who has the most justified reasons. An official process should be in place to evaluate and approve changes prior to actions being undertaken. The approval can be done by a single administrator or a formal committee such as a change management board based on the size of your organization and the scope of the change being approved.

- ▶ **Maintenance window:** After a change has been approved, the next question to address is when it is to take place. Authorized downtime should be used to make changes to production environments.

- ▶ **Notification of change:** Those affected by a change should be notified after the change has taken place. The notification should not be just of the change but should include any and all impacts to them and identify whom they can turn to with questions.

- ▶ **Documentation:** One of the last steps of the change management process is always to document what has been done. This documentation should include information on network configurations, additions to the network, and physical location changes.

These procedures represent just a few of the ones that administrators must follow on the job. It is crucial that all these procedures are well documented, accessible, reviewed, and updated as needed to be effective.

Configuration Management/Documentation

One other critical form of documentation is configuration documentation, which plays an important role in *configuration management*. Many administrators believe they could never forget the configuration of a router, server, or switch, but it often happens. Although it is often a thankless, time-consuming task, documenting the network hardware and software configurations is critical for continued network functionality.

Production configuration represents the current operational state of configuration settings in a production environment, while the *backup configuration* serves as a safeguard by providing a copy of these settings that can be used for recovery and restoration purposes in case of emergencies or failures. The production configuration is optimized for performance, security, and functionality according to the requirements and objectives of the organization; thus, it undergoes rigorous testing and validation before being deployed to ensure reliability and stability in the production environment. The backup configuration, also known as a snapshot, refers to a copy or backup of the production configuration settings taken at a specific point in time.

Two primary types of network configuration documentation are required: software documentation and hardware documentation. Both include all configuration information so that if a computer or other hardware fails, both the hardware and software can be replaced and reconfigured as quickly as possible. The documentation is important because often the administrator who configured the software or hardware is unavailable, and someone else has to re-create the configuration using nothing but the documentation. To be effective in this case, the documentation must be as current as possible. Older configuration information might not help.

> **Note**
>
> Organizing and completing the initial set of network documentation are huge tasks, but they are just the beginning. Administrators must constantly update all documentation to keep it from becoming obsolete. Documentation is perhaps one of the less-glamorous aspects of the administrator's role, but it is one of the most important.

Regulations

The terms *regulation* and *policy* are often used interchangeably; however, there is a difference. As mentioned, policies are written by an organization for its employees. Regulations are actual legal restrictions with legal consequences.

These regulations are set not by the organizations but by applicable laws in the area. Improper use of networks and the Internet can certainly lead to legal violations and consequences. The following is an example of network regulation from an online company:

> Transmission, distribution, uploading, posting or storage of any material in violation of any applicable law or regulation is prohibited. This includes, without limitation, material protected by copyright, trademark, trade secret or other intellectual property right used without proper authorization, material kept in violation of state laws or industry regulations such as social security numbers or credit card numbers, and material that is obscene, defamatory, libelous, unlawful, harassing, abusive, threatening, harmful, vulgar, constitutes an illegal threat, violates export control laws, hate propaganda, fraudulent material or fraudulent activity, invasive of privacy or publicity rights, profane, indecent or otherwise objectionable material of any kind or nature. You may not transmit, distribute, or store material that contains a virus, 'Trojan Horse,' adware or spyware, corrupted data, or any software or information to promote or utilize software or any of Network Solutions services to deliver unsolicited email. You further agree not to transmit any material that encourages conduct that could constitute a criminal offense, gives rise to civil liability or otherwise violates any applicable local, state, national or international law or regulation.

Examples of regulations administrators regularly encounter include the General Data Protection Regulation (GDPR), which enhances and consolidates data protection measures for individuals and organizations operating within the European Union (EU). The Payment Card Industry Data Security Standard (PCI DSS) aims to minimize fraud and safeguard customer credit card data. The Health Insurance Portability and Accountability Act (HIPAA) of 1996 establishes nationwide guidelines for safeguarding health information. The Gramm-Leach-Bliley Act (GLBA) defines privacy regulations for the financial sector. Sarbanes-Oxley (SOX) regulates financial and accounting disclosure practices.

ExamAlert

For the exam and for real-life networking, remember that regulations often are enforceable by law.

Labeling

One of the biggest problems with documentation is the time needed to do it. To shorten this time, users, through human nature, take shortcuts and use code or shorthand when labeling devices, maps, reports, and the like. Although these shortcuts can save time initially, they can render the labels useless if a person other than the one who created the labels looks at them or if a long period of time has passed since they were created and the author cannot remember what the label now means.

To prevent this dilemma, it is highly recommended that each organization create standard labeling rules and enforce them at all levels.

Cram Quiz

You have been given a physical wiring schematic that shows the following:

Description	Installation Notes
Category 5E 350 MHz plenum-rated cable	Cable runs 50 feet from the MDF to the IDF.
	Cable placed through the ceiling and through a mechanical room.
	Cable was installed 01/15/2020, upgrading a nonplenum cable.
Category 5E 350 MHz nonplenum cable	Horizontal cable runs 45 feet to 55 feet from the IDF to a wall jack.
	Cable 6 replaced Category 5e cable February 2020.
	Section of cable run through ceiling and over fluorescent lights.
Category 6a UTP cable	Patch cable connecting printer runs 15 feet due to printer placement.
8.3-micron core/125-micron	Connecting fiber cable runs 2 kilometers cladding single mode between the primary and secondary buildings.

1. Given this information, what cable recommendation might you make, if any?

   ○ **A.** Nonplenum cable should be used between the IDF and MDF.

   ○ **B.** The horizontal cable run should use plenum cable.

   ○ **C.** The patch cable connecting the printer should be shorter.

   ○ **D.** Leave the network cabling as is.

2. You have been called in to inspect a network configuration. You are given only one network diagram, shown in the following figure. Using the diagram, what recommendation might you make?

Primary Building

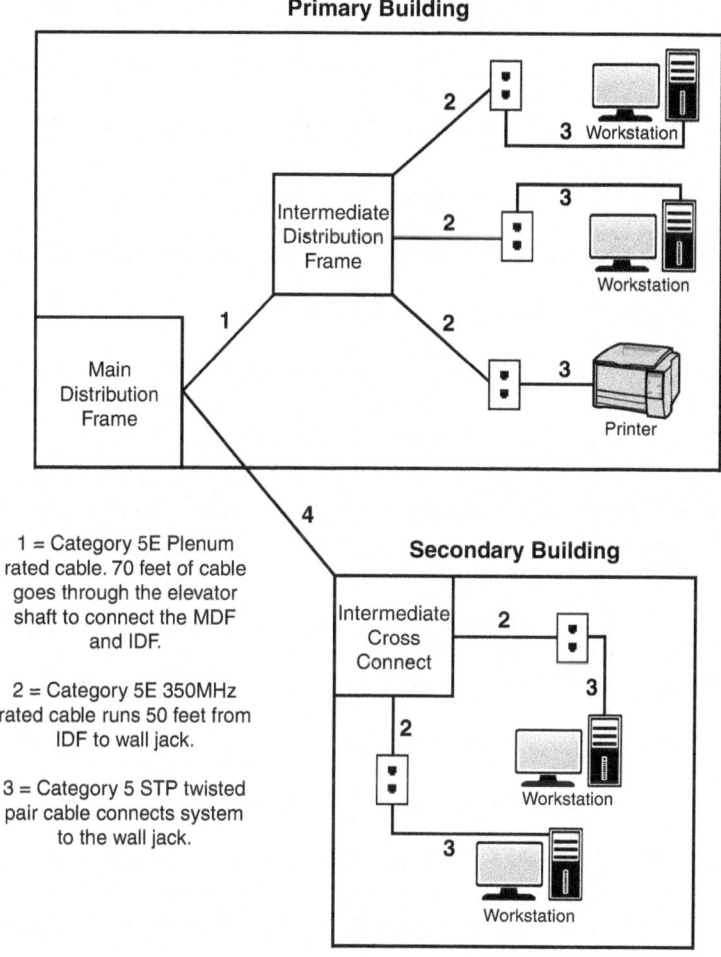

1 = Category 5E Plenum rated cable. 70 feet of cable goes through the elevator shaft to connect the MDF and IDF.

2 = Category 5E 350MHz rated cable runs 50 feet from IDF to wall jack.

3 = Category 5 STP twisted pair cable connects system to the wall jack.

Secondary Building

- ○ **A.** Cable 1 does not need to be plenum rated.
- ○ **B.** Cable 2 should be STP cable.
- ○ **C.** Cable 3 should be STP cable.
- ○ **D.** None. The network looks good.

3. The head of HR is complaining that the network cabling in their office is outdated and should be changed. What should they do to have the cabling evaluated and possibly changed?

- ○ **A.** Tell their supervisor that IT needs to get on the ball.
- ○ **B.** Tell your supervisor that IT needs to get on the ball.

○ **C.** Purchase new cabling at the local electronics store.

○ **D.** Complete a change request.

4. What stipulates the performance you can expect or demand by outlining the expectations a vendor has agreed to meet?

○ **A.** SLA

○ **B.** MOU

○ **C.** NDA

○ **D.** BCP

5. What stipulates that users should only be given the access rights and privileges required to what they need to do their job?

○ **A.** BYOD policy

○ **B.** MDM/MAM policies

○ **C.** Principle of least privilege

○ **D.** SDS/MSDS

6. Which of the following is a baseline configuration version used in configuration management as an ideal configuration against which configurations for similar devices can be compared?

○ **A.** GDPR

○ **B.** Golden configuration

○ **C.** EOL

○ **D.** EOS

Cram Quiz Answers

1. **B.** In this scenario, a section of horizontal cable runs through the ceiling and over fluorescent lights. This cable run might be a problem because such devices can cause electromagnetic interference (EMI). Alternatively, plenum cable is used in this scenario. Shielded twisted-pair (STP) may have worked as well.

2. **B.** In this diagram, Cable 1 is plenum rated and should be fine. Cable 3 is patch cable and does not need to be STP rated. Cable 2, however, goes through walls and ceilings. Therefore, it would be recommended to have a better grade of cable than regular UTP. STP provides greater resistance to EMI.

3. **D.** An official change request should be logged and tracked to verify what is to be done and what has been done. Within the realm of the change request should be the configuration procedures to be used, the rollback process that is in place, the potential impact identified, and a list of those that need to be notified.

4. **A.** A service-level agreement (SLA) is an agreement between you or your company and a service provider such as a cloud service provider or a technical support provider. SLAs are also usually part of network availability and other agreements. They stipulate the performance you can expect or demand by outlining the expectations a vendor has agreed to meet. Nondisclosure agreements (NDAs) are the oxygen that many companies need to thrive. Employees should understand the importance of them to continued business operations and agree to follow them to the letter, and spirit, of the law. Many companies require new employees to sign NDAs and acceptable use policies (AUPs) when they are hired during the onboarding process. Business continuity planning (BCP) is primarily a management tool that ensures that critical business functions (CBFs) can be performed when normal business operations are disrupted. A memorandum of understanding (MOU) is often used to define the relationship between two parties prior to, or in place of, a contract (such as between departments of the same organization).

5. **C.** The principle of least privilege dictates that users should only be given the access rights and privileges required to what they need to do their job, but no more. Bring-your-own-device (BYOD) policies define what personally owned mobile devices (laptops, tablets, and smartphones) employees are allowed to bring to their workplace and use. Mobile device management (MDM) and mobile application management (MAM) systems can be used to help enterprises manage and secure the use of those mobile devices in the workplace and to interact with privileged company information and applications. A safety data sheet (SDS), or material safety data sheet (MSDS), is a standardized document that contains crucial occupational safety and health information.

6. **B.** In configuration management, a baseline/golden configuration is an ideal configuration against which configurations for similar devices can be compared. The General Data Protection Regulation (GDPR) strengthens and unifies data privacy protection for individuals and companies within the European Union (EU). End-of-life (EOL) refers to the stage in a product's lifecycle when the manufacturer discontinues production, sales, and support for the product. It means that the product is no longer being actively marketed or sold by the manufacturer; technical support and warranty coverage may no longer be available. This can pose security risks, compatibility issues, and operational challenges for equipment still in use. Similarly, end-of-support (EOS) refers to the date when the manufacturer discontinues providing software updates, security patches, and technical support for a product that has reached its end-of-life.

Monitoring Network Performance

▶ **3.2 Given a scenario, use network monitoring technologies.**

CramSaver

If you can correctly answer these questions before going through this section, save time by skimming the ExamAlerts in this section and then completing the Cram Quiz at the end of the section.

1. What can be used to capture network data?

2. True or false: Port scanners detect open and often unsecured ports.

3. True or false: Interface monitoring tools can be used to create heat maps showing the quantity and quality of wireless network coverage in areas.

4. True or false: Always test updates on a lab machine before rolling out on production machines.

5. What is it known as when you roll a system back to a previous version of a driver or firmware?

Answers

1. Packet sniffers can be used by both administrators and hackers to capture network data.

2. True. Port scanners detect open and often unsecured ports.

3. False. Wireless survey tools can be used to create heat maps showing the quantity and quality of wireless network coverage in areas.

4. True. Always test updates on a lab machine before rolling out on production machines.

5. Rolling a system back to a previous version is known as *downgrading* and is often necessary when dealing with legacy systems and implementations.

ExamAlert

Remember that this objective begins with "Given a scenario." This means that you may receive a drag-and-drop, matching, or "live OS" scenario where you have to click through to complete a specific objective-based task.

When networks were smaller and few stretched beyond the confines of a single location, network management was a simple task. In today's complex, multisite, hybrid networks, however, the task of maintaining and monitoring network devices and servers has become a complicated but essential part of the network administrator's role. Nowadays, the role of network administrator often stretches beyond the physical boundary of the server room and reaches every node and component on the network. Whether an organization has 10 computers on a single segment or a multisite network with several thousand devices attached, the network administrator must monitor all network devices, protocols, and usage—preferably from a central location.

Given the sheer number and diversity of possible devices, software, and systems on any network, it is clear why network management is such a significant consideration. Although a robust network management strategy can improve administrator productivity, increase *uptime*, and reduce *downtime*, many companies choose to neglect network management because of the time involved in setting up the system or because of the associated costs. If these companies understood the potential savings, they would realize that neglecting network management provides false economies.

Network management and network monitoring are essentially methods to control, configure, and monitor devices on a network. Imagine a scenario in which you are a network administrator working out of your main office in Spokane, Washington, and you have satellite offices in New York, Dallas, Vancouver, and London. Network management allows you to access systems in the remote locations or have the systems notify you when something goes awry. In essence, network management is about seeing beyond your current boundaries and acting on what you see.

Network management is not one thing. Rather, it is a collection of tools, systems, and protocols that, when used together, enables you to perform tasks such as reconfiguring a network card in the next room or installing an application in the next state. With cloud computing, the need for administrators to monitor cloud-based systems and company data stored in the cloud increases the responsibilities. Thankfully, all major cloud providers (such as AWS, Microsoft, and Google), as well as many third parties, offer cloud-based network monitoring tools to make the task manageable.

Common Performance Metrics

The capabilities demanded from network management vary somewhat among organizations, but essentially, several key types of information and functionality are required, such as fault detection and performance monitoring. Some of the types of information and functions that network management tools can provide include the following:

- ▶ **Temperature:** You should make certain that devices are running within the acceptable range. Usually, the biggest problem is heat, so you need to get rid of it to keep systems from overheating or encountering chip creep.

- ▶ **Utilization:** Once upon a time, it was not uncommon for a network to have to limp by with scarce resources. Administrators would constantly have to trim logs and archive files to keep enough storage space available to service print jobs. Those days are gone, and any such hint of those conditions would be unacceptable today. For you to keep this from happening, one of the keys is to manage utilization and stay on top of problems before they escalate. Several areas of utilization to monitor are as follows:

  - ▶ **Bandwidth/throughput:** There must be enough bandwidth to serve all users, and you need to be alert for bandwidth hogs. You want to look for top talkers (those that transmit the most) and top listeners (those that receive the most) and figure out why they are so popular. *Flow data* can be used to ascertain this information and allow you to decide how to best respond. NetFlow is a network protocol analyzer developed by Cisco.

  - ▶ **Storage space:** Free space needs to be available for all users, and quotas may need to be implemented.

  - ▶ **Network device CPU:** Just as a local machine will slow when the processor is maxed out, so will the network.

  - ▶ **Network device memory:** It is next to impossible to have too much memory. You should balance loads to optimize the resources you have to work with.

  - ▶ **Wireless channel utilization:** Akin to bandwidth utilization is channel utilization in the wireless realm. As a general rule, a wireless network starts experiencing performance problems when channel utilization reaches 50 percent of the channel capacity.

- ▶ **Latency:** One of the biggest problems with satellite access is trouble with latency (the time lapse between sending or requesting information and the time it takes to return). Satellite communication experiences high latency due to the distance it has to travel as well as weather conditions. While latency is not restricted solely to satellites, it is one of the easiest forms of transmission to associate with it. In reality, latency can occur with almost any form of transmission.

▶ **Jitter:** Closely tied to latency, jitter differs in that the length of the delay between received packets differs. While the sender continues to transmit packets in a continuous stream and space them evenly apart, the delay between packets received varies instead of remaining constant. This issue can be caused by network congestion, improper queuing, or configuration errors.

▶ **Fault detection:** One of the most vital aspects of network management is knowing if anything is not working or is not working correctly. Network management tools can detect and report on a variety of faults on the network. Given the number of possible devices that constitute a typical network, determining faults without these tools could be an impossible task. In addition, network management tools not only might detect the faulty device but also shut it down. This means that if a network card is malfunctioning, you can remotely disable it. When a network spans a large area, fault detection becomes even more invaluable because it enables you to be alerted to network faults and to manage them, thereby reducing downtime.

ExamAlert

Most of this discussion involves your being alerted to some condition. Those alerts can generally be sent to you through email or Short Message Service (SMS) to any mobile device.

▶ **Performance monitoring:** Another feature of network management is the ability to monitor network performance. Performance monitoring is an essential consideration that gives you some crucial information. Specifically, performance monitoring can provide network usage statistics and user usage trends. This type of information is essential when you plan network capacity and growth. Monitoring performance also helps you determine whether there are any performance-related concerns, such as whether the network can adequately support the current user base.

▶ **Security monitoring:** Good server administrators have a touch of paranoia built in to their personality. A network management system (NMS) enables you to monitor who is on the network, what they are doing, and how long they have been doing it. More important, in an environment in which corporate networks are increasingly exposed to outside sources, the ability to identify and react to potential security threats is a priority. Reading log files to learn of an attack is a poor second to knowing that an attack is in progress and being able to react accordingly. One thing to look for is changes in raw data values; these changes can be identified through comparisons of *cyclic redundancy check (CRC)* values. Look for CRC errors, as well as *giants* (packets that are discarded because they exceed the medium's maximum packet size), *runts* (packets that are discarded because they are smaller than the medium's minimum packet size), and *encapsulation errors*.

▶ **Link state status:** You should regularly monitor link status to make sure that connections are up and functioning (or down, if expected to be). Breaks should be found and identified as quickly as possible to repair them or find workarounds. A number of link status monitors exist for the purpose of monitoring connectivity, and many can reroute (per a configured script file) when a down condition occurs.

▶ **Interface monitoring:** Just as you want to monitor for a link going down, you also need to know when an interface has problems. Particular problems to watch for include errors, utilization problems (unusually high, for example), discards, packet drops, resets, and problems with speed/duplex. An *interface monitoring tool* is invaluable for troubleshooting problems here.

▶ **Packet capture:** *Packet capture*, also known as packet sniffing or packet analysis, is a fundamental network monitoring technology used to capture, inspect, and analyze network traffic in real time or from stored packet captures. This can be a powerful tool for detecting anomalies and security threats within the network. Administrators can analyze packet contents to identify suspicious or malicious activity, such as unauthorized access attempts, network scanning, malware infections, or data exfiltration. Intrusion detection systems (IDS) and intrusion prevention systems (IPS) use packet capture to monitor network traffic for known attack signatures and abnormal behavior patterns, triggering alerts or blocking malicious traffic in real time.

▶ **Maintenance and configuration:** Want to reconfigure or shut down the server located in Australia? Reconfigure a local router? Change the settings on a client system? Remote management and configuration are key parts of the network management strategy, enabling you to centrally manage huge multisite locations.

▶ **Power monitoring:** A consistent flow of reliable energy is needed to keep a network up and running. A wide array of *power monitoring tools* is available to help identify and log problems that you can then begin to resolve.

▶ **Wireless monitoring:** As more networks go wireless, you need to pay special attention to issues associated with them. *Wireless survey tools* can be used to create heat maps showing the quantity and quality of wireless network coverage in areas. They can also allow you to see access points (including rogues) and security settings. These can be used to help you design and deploy an efficient network, and they can also be used (by you or others) to find weaknesses in your existing network (often marketed for this purpose as *wireless analyzers*).

Many tools are available to help monitor the network and ensure that it is properly functioning. Administrators—and those who want to obtain data that does not belong to them—can use tools such as a packet sniffer to monitor traffic. The following sections look at several monitoring tools.

SNMP

An SNMP management system is a computer running a special piece of software called a *network management system (NMS)*. These software applications can be free, or they can cost thousands of dollars. The difference between the free applications and those that cost a great deal of money normally boils down to functionality and support. All NMS applications, regardless of cost, offer the same basic functionality. Today, most NMS applications use graphical maps of the network to locate a device and then query it. The queries are built into the application and are triggered by pointing and clicking. You can actually issue SNMP requests from a command-line utility, but with so many tools available, this is unnecessary.

> **Note**
>
> Some people use the term *trap managers* for SNMP managers or network management systems (NMS). This reference is misleading, however, because an NMS can do more than just accept trap messages from agents.

Using SNMP and an NMS, you can monitor all the devices on a network, including switches, hubs, routers, servers, and printers, as well as any device that supports SNMP, from a single location. Using SNMP, you can see the amount of free disk space on a server in Jakarta or reset the interface on a router in Helsinki—all from the comfort of your desk in San Jose. Such power, though, brings with it some considerations. For example, because an NMS enables you to reconfigure network devices, or at least get information from them, it is common practice to implement an NMS on a secure workstation platform, such as a Linux or Windows server, and to place the NMS PC in a secure location.

Traps

SNMP *traps* are asynchronous notifications sent by network devices to a network management system or SNMP manager to alert it about significant events or conditions that occur within the network. They are used for proactive monitoring and management of network devices and infrastructure. They provide a way for network devices, such as routers, switches, and servers, to inform the SNMP manager about specific events or conditions without requiring the manager to continuously poll the devices for updates.

The traps are triggered by predefined events or conditions that occur within the network environment. These events could include system errors, interface status changes, hardware failures, security breaches, performance thresholds being exceeded, or other noteworthy occurrences. SNMP traps are typically delivered over UDP to the SNMP manager's IP address and port number. The SNMP manager listens for incoming trap messages and processes them accordingly. Upon receiving a trap, the manager can perform actions such as logging the event, generating alerts or notifications, updating network management databases, or triggering automated responses.

Management Information Base (MIB)

SNMP uses databases of information called MIBs to define what parameters are accessible, which of the parameters are read-only, and which can be set. MIBs are available for thousands of devices and services, covering every imaginable need. *Object identifiers (OIDs)* uniquely identify managed objects within an MIB hierarchy. Quite simply, an OID is an address used to identify each node in a tree structure. The addresses are integers separated by periods, corresponding to the path from the root through the series of ancestor nodes, to the node. Each node in the tree is controlled by an assigning authority who can create child nodes and delegate assigning authority for the child nodes.

To ensure that SNMP systems offer cross-platform compatibility, MIB creation is controlled by the *International Organization for Standardization (ISO)*. An organization that wants to create MIBs can apply to the ISO. The ISO then assigns the organization an ID under which it can create MIBs as it sees fit.

The assignment of numbers is structured within a conceptual model called the *hierarchical name tree*.

> **ExamAlert**
>
> When studying for the Network+ exam, be sure that you know SNMP and its use of traps, object identifiers (OIDs), and management information bases (MIBs).

Versions

There have been a few versions of SNMP over the years, and two you should be aware of are v2c and v3 because there are some differences between them. Following are some key areas of difference:

▶ **Security:** Version 2c does not provide robust security features out-of-the-box, and it relies primarily on community strings for authentication, which are essentially plaintext passwords used to control access to SNMP-enabled devices. This makes SNMPv2c vulnerable to security threats, such as eavesdropping, spoofing, and unauthorized access. Version 3 introduced significant improvements in security, offering authentication, encryption, and access control mechanisms to protect SNMP communications. SNMPv3 supports multiple security models, including SNMPv1-style community-based security (with community strings), as well as more secure User-based Security Model (USM) and View-based Access Control Model (VACM). SNMPv3 allows for the use of strong cryptographic algorithms, such as HMAC-SHA, HMAC-MD5, and AES, to ensure data integrity and confidentiality.

▶ **Authentication:** In SNMP version 2c, authentication is based solely on community strings, which are transmitted in plaintext. Community strings serve as simple passwords that grant access to SNMP-enabled devices. However, community strings do not provide robust authentication mechanisms and are susceptible to interception and brute-force attacks. SNMP version 3 supports stronger authentication methods, including Message Digest Algorithm 5 (MD5) and Secure Hash Algorithm (SHA), which are used in conjunction with usernames and passwords. SNMPv3 allows for the creation of user accounts with associated authentication and privacy (encryption) settings, providing enhanced security compared to SNMPv2c.

▶ **Privacy/encryption:** SNMP version 2c does not offer built-in encryption capabilities. As a result, SNMP communications, including community strings and management information, are transmitted in plaintext over the network. This lack of encryption exposes SNMPv2c to security risks, such as data interception and tampering. SNMP version 3 supports data encryption to protect sensitive information exchanged between SNMP managers and agents. SNMPv3 allows for the use of privacy protocols, such as Data Encryption Standard (DES), Triple DES (3DES), and Advanced Encryption Standard (AES), to encrypt SNMP payloads and ensure confidentiality of SNMP communications.

▶ **Message integrity:** SNMP version 2c does not provide strong message integrity features. While community strings can help verify the source of SNMP messages, they do not offer robust protection against message tampering or alteration. SNMP version 3 includes mechanisms for ensuring message integrity using Hash-based Message Authentication Code (HMAC) algorithms, such as HMAC-MD5 and HMAC-SHA. These algorithms generate cryptographic hashes of SNMP messages, allowing SNMP managers and agents to verify the authenticity and integrity of transmitted data.

ExamAlert

For the exam, know that SNMP version 3 offers significant improvements in security, authentication, encryption, and message integrity compared to SNMP version 2c, and thus it is the preferred choice for environments requiring strong security measures and compliance with regulatory standards. You must remember Simple Network Management Protocol (SNMP) is an application layer protocol whose purpose is to collect statistics from TCP/IP devices. SNMP is used to monitor the health of network equipment. SNMPv1 and SNMPv2 are considered insecure, and SNMPv3 is the current and recommended standard.

Community Strings

Community strings are plaintext passwords used for authentication and access control in Simple Network Management Protocol implementations. They serve as a basic form of security for SNMP-enabled devices, allowing administrators to control access to management information and SNMP operations. Since they are transmitted in plaintext over the network, they are vulnerable to interception and unauthorized access. It's recommended to regularly review and update community string configurations, enforce access controls, and monitor SNMP traffic for suspicious activity to enhance network security.

> **Note**
>
> SNMP is a powerful tool with many ways a network administrator can use it. Here are the top five things to know about it for exam preparation:
>
> 1. SNMP is an application layer protocol whose purpose is to collect statistics from TCP/IP devices. To monitor the health of all systems, you install SNMP agents on the machines and then monitor those agents from a central location.
>
> 2. Configuring SNMP traps on a network can help administrators determine the causes of jitter and latency.
>
> 3. To monitor the health of systems, you install SNMP agents on the devices and then monitor those agents from a central location.
>
> 4. SNMP agents listen on UDP/161. SNMP does not use TCP for messaging. SNMP sends traps on UDP/161.
>
> 5. If alarms and metrics regarding network availability have stopped or malfunctioned, modifying the MIBs on the SNMP monitoring system would be the first thing to do or check out.

Network Performance, Load, and Stress Testing

To test the network, administrators often perform three distinct types of tests:

▶ Performance tests

▶ Load tests

▶ Stress tests

These test names are sometimes used interchangeably. Although some overlap exists, they are different types of network tests, each with different goals.

Performance Tests

A *performance test* is, as the name suggests, all about measuring the network's current performance level. The goal is to take ongoing performance tests and evaluate and compare them, looking for potential bottlenecks. For performance tests to be effective, they need to be taken under the same type of network load each time, or the comparison is invalid. For example, a performance test taken at 3 a.m. will differ from one taken at 3 p.m.

> **Note**
>
> The goal of performance testing is to establish *baselines* for the comparison of network functioning. The results of a performance test are meaningless unless you can compare them to previously documented performance levels. As such, the focus is usually on anomaly detection (alerting and notification). Anomaly detection algorithms continuously monitor network metrics and compare current values against established baseline metrics and threshold levels: when a deviation or anomaly is detected, such as a sudden spike or drop in network traffic, an unusual pattern of user behavior, or an increase in error rates, the system triggers an alert or notification to alert administrators or automated response mechanisms.

Load Tests and Send/Receive Traffic

Load testing has some overlap with performance testing. Sometimes called *volume* or *endurance testing*, load tests involve artificially placing the network under a larger workload. For example, the network traffic might be increased throughout the entire network. After this is done, performance tests can be done on the network with the increased load. Load testing is sometimes done to see if bugs exist in the network that are not currently visible but that may become a problem as the network grows. For example, the mail server might work fine with current requirements. However, if the number of users in the network grew by 10 percent, you would want to determine whether the increased load would cause problems with the mail server. Load tests are all about finding a potential problem before it happens.

Performance tests and load tests are actually quite similar; however, the information outcomes are different. Performance tests identify the current level of network functioning for measurement and benchmarking purposes. Load tests are designed to give administrators a look into the future of their network load and to see whether the current network infrastructure can handle it.

> **Note**
>
> Performance tests are about network functioning today. Load tests look forward to see whether performance may be hindered in the future by growth or other changes to the network.

Stress Tests

Whereas load tests do not try to break the system under intense pressure, stress tests sometimes do. They push resources to the limit. Although these tests are not done often, they are necessary and—for administrators, at least—entertaining. Stress testing has two clear goals:

▶ It shows you exactly what the network can handle. Knowing a network's breaking point is useful information when you consider network expansion.

▶ It enables you to test your backup and recovery procedures. If a test knocks out network resources, you can verify that your recovery procedures work. Stress testing enables you to observe network hardware failure.

Stress tests assume that someday something will go wrong, and you will know exactly what to do when it happens.

Performance Metrics

Whether the testing being done is related to performance, load, or stress, you have to choose the metrics you want to monitor and focus on. Although a plethora of options are available, the most common four are the following:

▶ **Error rate:** This metric identifies the frequency of errors.

▶ **Utilization:** This metric shows the percentage of resources being utilized.

▶ **Packet drops:** This metric shows how many packets of data on the network fail to reach their destination.

▶ **Bandwidth/throughput:** This metric involves the capability to move data through a channel as related to the total capability of the system to identify bottlenecks, throttling, and other issues.

Network Device Logs

In a network environment, all NOSs and most firewalls, proxy servers, and other network components have logging features. These logging features are essential for network administrators to review and monitor. Many types of logs can be used. The following sections review some of the most common log file types.

On a Windows Server system, as with the other operating systems, events and occurrences are logged to files for later review. Windows Server and desktop systems use Event Viewer to view many of the key log files. The logs in Event Viewer can be used to find information on, for example, an error on the system or a security incident. Information is recorded into key log files; however, you will also see additional log files under certain conditions, such as if the system is a domain controller or is running a DHCP server application.

The term *event logs* refers generically to all log files used to track events on a system. Event logs are crucial for finding intrusions and diagnosing current system problems. In a Windows environment, for example, three primary event logs are used: security, application, and system.

> **Note**
>
> Be sure that you know the types of information included in the types of log files.

Security Logs

A system's security log contains events related to security incidents, such as successful and unsuccessful logon attempts and failed resource access. Security logs can be customized, meaning that administrators can fine-tune exactly what they want to monitor. Some administrators choose to track nearly every security event on the system. Although doing so might be prudent, it can often create huge log files that take up too much space. Figure 8.10 shows a security log from a Windows system.

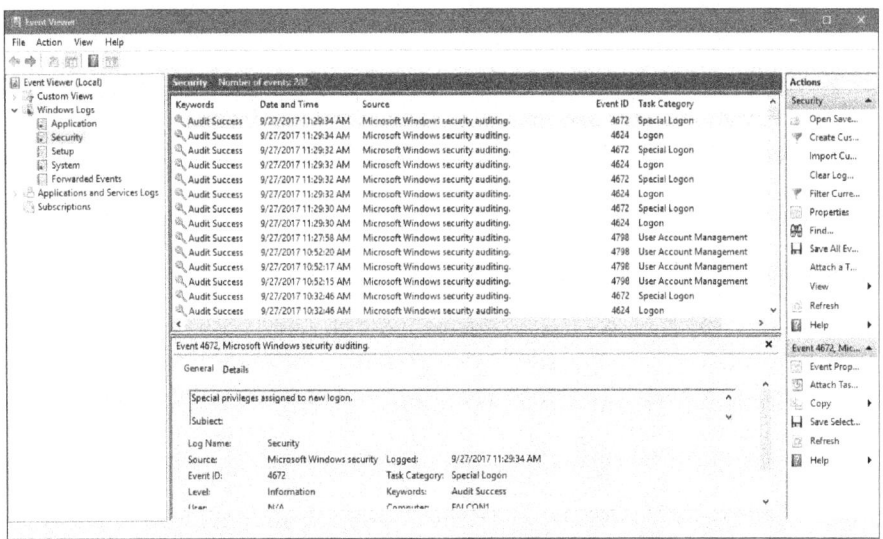

FIGURE 8.10 A Windows Security Log from Windows

Figure 8.10 shows that some successful logons and account changes have occurred. A potential security breach would show some audit failures for logon or logoff attempts. To save space and prevent the log files from growing too big, administrators might choose to audit only failed logon attempts and not successful ones.

Each event in a security log contains additional information to make it easy to get the details on the event:

- ▶ **Date:** The exact date the security event occurred.

- ▶ **Time:** The time the event occurred.

- ▶ **User:** The name of the user account that was tracked during the event.

- ▶ **Computer:** The name of the computer used when the event occurred.

- ▶ **Event ID:** The event ID telling you what event has occurred. You can use this ID to obtain additional information about the particular event. For example, you can take the ID number, enter it at the Microsoft support website, and gather information about the event. Without the ID, finding this information would be difficult.

To be effective, security logs should be regularly reviewed.

Application Log

An application log contains information logged by applications that run on a particular system rather than the operating system itself. Vendors of third-party applications can use the application log as a destination for error messages generated by their applications.

The application log works in much the same way as the security log. It tracks both successful events and failed events within applications. Figure 8.11 shows the details provided in an application log.

Figure 8.11 shows that two types of events occurred: general application information events and warning event events. Vigilant administrators would likely want to check the event ID of both the event and warning failures to isolate the cause.

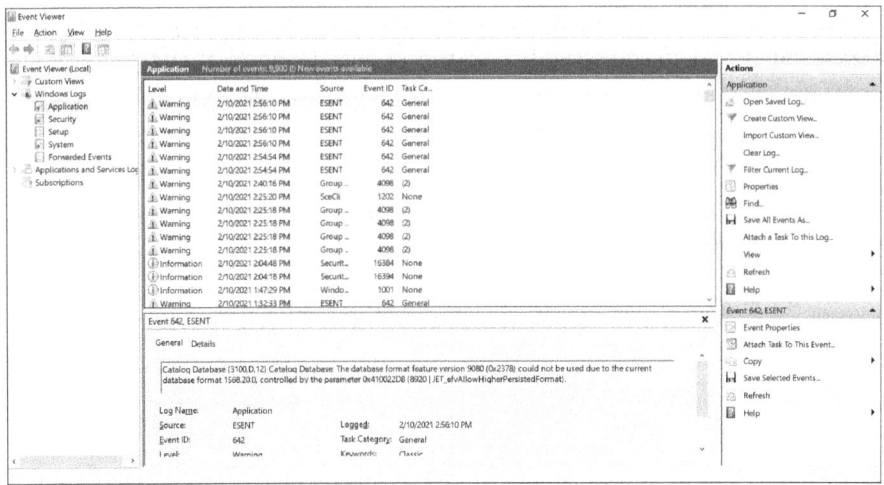

FIGURE 8.11   An Application Log in Windows

System Logs

System logs record information about components or drivers in the system, as shown in Figure 8.12. This is the place to look when you are troubleshooting a problem with a hardware device on your system or a problem with network connectivity. For example, messages related to the client element of *Dynamic Host Configuration Protocol (DHCP)* appear in this log. The system log is also the place to look for hardware device errors, time synchronization issues, or service startup problems.

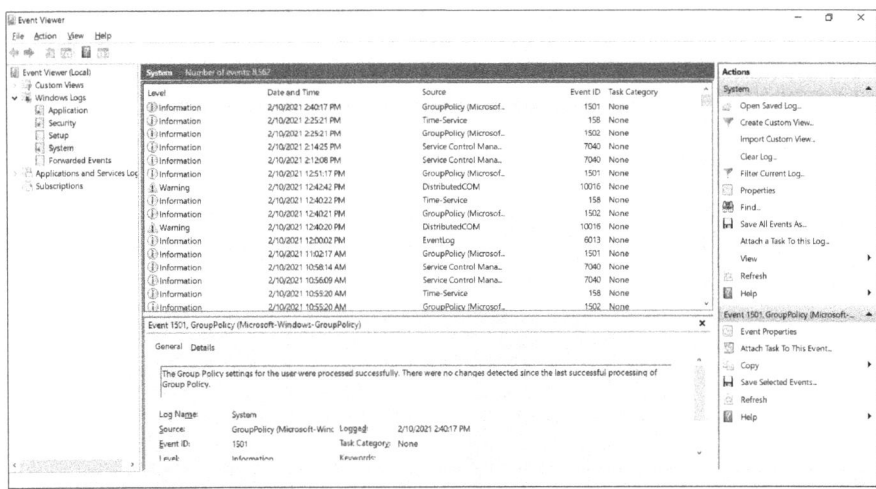

FIGURE 8.12   A System Log in Windows

History Logs

History logs are most often associated with the tracking of Internet surfing habits. They maintain a record of all sites that a user visits. Network administrators might review these for potential security or policy breaches, but generally these logs are not commonly reviewed.

Another form of history log is a compilation of events from other log files. For instance, one history log might contain all significant events over the past year from the security log on a server. History logs are critical because they provide a detailed account of alarm events that can be used to track trends and locate problem areas in the network. This information can help you revise maintenance schedules, determine equipment replacement plans, and anticipate and prevent future problems.

> **Note**
>
> Application logs and system logs can often be viewed by any user. Security logs can be viewed only by users who use accounts with administrative privileges.

Log Aggregation and Log Management

In a discussion of these logs, it becomes clear that monitoring them can be a huge issue. That is where *log aggregation* and *log management (LM)* come in. As the name implies, log aggregation is the process of collecting, consolidating, and centralizing log data from various sources within a network environment for analysis, monitoring, and troubleshooting purposes, and it is a subset of log management.

Log management describes the process of managing large volumes of system-generated computer log files. LM includes the collection, retention, and disposal of all system logs. Although LM can be a huge task, it is essential to ensure the proper functioning of the network and its applications. It also helps you keep an eye on network and system security.

Configuring systems to log all sorts of events is the easy part. Trying to find the time to review the logs is an entirely different matter. To assist with this process, third-party software packages are available to help with the organization and reviewing of log files. To find this type of software, you can enter **log management** into a web browser, and you will have many options to choose from. Some have trial versions of their software that may give you a better idea of how LM works.

Syslog is a message logging standard that has been around for many years. The tool used for creating log entries in UNIX/Linux-based systems is conveniently named syslog, but other tools can also be used. Syslog allows separation among three entities: the software that generates the message, the system that stores it, and the software used to analyze or report it. Every message is labeled with identifiers, such as a code indicating the software type generating the message and a severity level. A severity level of 0 is an emergency, 1 is an alert, 2 is critical, 3 is an error, 4 is a warning, 5 is a notice, 6 is information only, and 7 is for debugging.

> **ExamAlert**
>
> A syslog server (also known as the syslog collector or receiver) listens for and then logs data messages and events coming from the syslog client/device, which can help in identification and detailed investigation of security incidents. If events are not being logged or received, the network device may not be configured properly to log the proper severity level to the syslog server. Syslog messages are sent to the syslog server on port 514 with UDP.

Syslog is one tool while *security information and event management (SIEM)* is a comprehensive approach to security management that combines security information management (SIM) and security event management (SEM) functionalities into a unified platform. SIEM systems collect, aggregate, correlate, and analyze security-related data from various sources within an organization's IT infrastructure to provide real-time monitoring, threat detection, incident response, and compliance reporting capabilities.

A SIEM dashboard contains multiple views, which allow you to visualize and monitor patterns and trends. The display can include such elements as charts, sensors, graphs, and trends that can be easily explored by automatically filtering and exploring the data through interactive drilldowns.

> **Note**
>
> Log aggregation is the process by which security information and event management (SIEM) systems combine similar events to reduce event volume. Also note that SIEM event correlation is part of a SIEM solution. Event correlation aggregates log data and identifies malicious patterns across network devices and apps, making it possible to discover threats that might otherwise go unnoticed. If those patterns are a threat to security, then action can be taken.

API Integration

Application programming interface (API) integration with network monitoring involves connecting network monitoring tools or platforms with external systems, applications, or services via APIs to exchange data, automate workflows, and enhance monitoring capabilities. API integration allows network monitoring solutions to trigger automated actions or remediation workflows in response to detected events, anomalies, or performance issues. By interfacing with APIs of orchestration tools, ticketing systems, or configuration management platforms, network monitoring platforms can automate incident response processes, initiate corrective actions, or escalate alerts to relevant teams for resolution.

Port Mirroring

Port mirroring, also known as port monitoring, is a network monitoring technique used to copy and redirect network traffic from one or more network ports (or interfaces) to a designated monitoring port for analysis. It allows network administrators to observe and analyze the traffic passing through a specific switch port or set of ports without disrupting normal network operations.

> **Note**
>
> Port mirroring is a feature that can be used to enable monitoring of all packets arriving at an interface and can be used to troubleshoot network issues, such as latency or poor performance. It can also be used to direct network traffic to a monitoring system.

Patch Management

All applications, including productivity software, virus/malware checkers, and especially the operating system, release patches and updates often designed to address potential security weaknesses. Administrators must keep an eye out for these patches and install them when they are released. Manually tracking and applying patches across a large number of systems can be time consuming and error-prone, so patch management programs can automate the patch deployment process, streamlining administration tasks and reducing the workload on IT staff.

> **Note**
>
> The various types of updates discussed in this section apply to all systems and devices, including mobile devices and laptops, as well as servers and routers. Special server systems (and services) are typically used to deploy mass updates to clients in a large enterprise network.

Discussion items related to this topic include the following:

▶ **OS updates:** Most operating system updates relate to either functionality or security issues. For this reason, it is important to keep your systems up to date. Most current operating systems include the capability to automatically find updates and install them. By default, the automatic updates feature is usually turned on; you can change the settings if you do not want this feature enabled.

> **Note**
>
> Always test updates on a lab machine before rolling out on production machines.

▶ **Firmware updates:** Firmware updates keep the hardware interfaces working properly. Router manufacturers, for example, often issue patches when problems are discovered. Those patches need to be applied to the router to remove any security gaps that may exist. Figure 8.13 shows an example of checking a router's firmware.

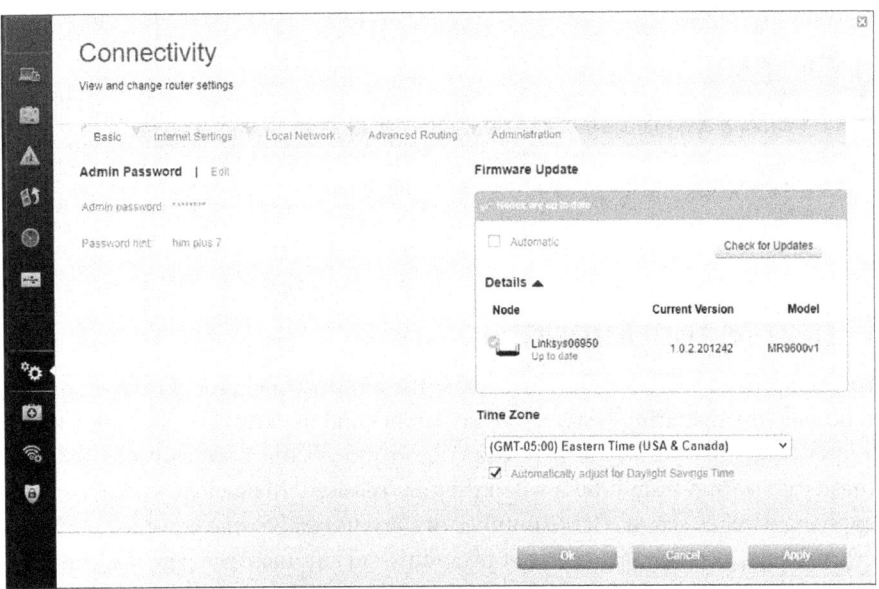

FIGURE 8.13   Checking a Router's Firmware

> **ExamAlert**
>
> Just as security holes can exist with operating systems and applications (and get closed through patches), they can also exist in firmware and be closed through updates.

▶ **Driver updates:** The main reason for updating drivers is that you have a piece of hardware that is not operating correctly. The failure to operate can be caused by the hardware interacting with software it was not intended to prior to shipping (such as OS updates). Because the problem can be from the vendor or the OS provider, updates can be automatically included (such as with Windows Update) or found on the vendor's site.

▶ **Feature changes/updates:** Not considered as critical as security or functionality updates, feature updates and changes can extend what you could previously do and extend your time using the hardware/software combination you have.

▶ **Major versus minor updates:** Most updates are classified as major (*must* be done) or minor (*can* be done). Depending on the vendor, the difference in the two may be telegraphed in the numbering: an update of 4.0.0 would be a major update, whereas 4.10.357 would be considered a minor one.

ExamAlert

As a general rule, the smaller the number of the update, the less significant it is.

▶ **Vulnerability patches:** Vulnerabilities are weaknesses, and patches related to them should be installed correctly with all expediency. After a vulnerability in an OS, a driver, or a piece of hardware has been identified, the fact that it can be exploited is often spread quickly: a *zero-day exploit* is any attack that begins the very day the vulnerability is discovered.

Note

If attackers learn of the weakness the same day as the developer, they have the ability to exploit it until a patch is released. There are times when the only thing that you, as a security administrator, can do between the discovery of the exploit and the release of the patch is to turn off the system or the service in question. Although this approach can be a costly undertaking in terms of productivity, it can be the only way to keep the network safe.

▶ **Upgrading versus downgrading:** Not all changes need to be upgraded. If, for example, a newly applied patch changes the functionality of a hardware component to the point that it will no longer operate as you need it to, you can consider reverting back to a previous state. This approach is known as *downgrading* and is often necessary when dealing with legacy systems and implementations.

> **ExamAlert**
>
> For the exam, know that removing patches and updates is considered downgrading.

Before you install or remove patches, it is important to do a *configuration backup*. Many vendors offer products that perform configuration backups across the network on a regular basis and allow you to roll back changes if needed. Free tools are often limited in the number of devices they can work with, and some of the more expensive ones include the capability to automatically analyze and identify the changes that could be causing any problems.

Network Discovery

Network discovery is a crucial aspect of network monitoring that involves identifying and cataloging devices, services, and resources within an IT infrastructure. Discovery forms the foundation of network monitoring by providing visibility into the network's components and their interconnections. It allows monitoring tools to identify active hosts, network devices, servers, endpoints, applications, and services. By discovering network assets and their attributes, such as IP addresses, MAC addresses, device types, operating systems, and installed software, network monitoring solutions can create an inventory of resources to monitor and analyze.

In *ad hoc* discovery, administrators may use tools like network scanners, ping sweeps, or port scanners to identify devices or conduct targeted scans to investigate specific network segments or address issues. Ad hoc discovery enables administrators to perform targeted scans or investigations in response to specific needs or events, such as troubleshooting network issues or verifying configurations.

Scheduled network discovery involves automated, recurring processes that periodically scan the network to update asset inventories, detect changes, and ensure accurate representation of the network topology. Scheduled discovery ensures that the network inventory remains up to date and accurate, facilitating proactive monitoring and management.

Traffic Analysis

Traffic analysis is a core component of network monitoring, providing insights into the flow of data packets within a network environment. It involves the examination, interpretation, and visualization of network traffic patterns, behaviors, and characteristics to gain actionable intelligence about the performance, security, and usage of the network. By analyzing traffic patterns and

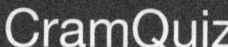
trends, administrators can identify congestion points, bottlenecks, or resource limitations that may impact network performance.

Availability Monitoring

Availability monitoring is focused on ensuring that network resources, services, and applications are accessible and operational for users. It involves continuously monitoring the availability and uptime of network devices, servers, applications, and services to detect and address potential outages, downtime, or service disruptions promptly. By proactively monitoring availability metrics, organizations can minimize downtime, maintain service continuity, and deliver a seamless user experience across their IT infrastructure.

Configuration Monitoring

Configuration monitoring was previously discussed in this chapter in terms of change management and involves overseeing the configuration settings and parameters of network devices, systems, and services. It involves tracking changes to configurations, verifying compliance with standards and policies, and ensuring consistency and stability across the network infrastructure. Configuration monitoring supports compliance with industry regulations, security standards, and organizational policies by ensuring that network configurations adhere to predefined guidelines and best practices. By comparing current configurations against baseline templates or security benchmarks, administrators can assess compliance status, identify noncompliant settings, and remediate security vulnerabilities or configuration drifts.

> **ExamAlert**
>
> Exam objective 3.2 calls out network discovery, ad hoc discovery, scheduled network discovery, availability monitoring, and configuration monitoring. That is a lot of discovery and monitoring! For the exam, know each of these well.

Cram Quiz

1. Which of the following involves pushing the network beyond its limits, often taking down the network to test its limits and recovery procedures?

   ○ **A.** Crash and burn

   ○ **B.** Stress test

   ○ **C.** Recovery test

   ○ **D.** Load test

2. You suspect that an intruder has gained access to your network. You want to see how many failed logon attempts were made in one day to help determine how the person got in. Which of the following might you do?

- ○ **A.** Review the history logs.
- ○ **B.** Review the security logs.
- ○ **C.** Review the logon logs.
- ○ **D.** Review the performance logs.

3. Which utility can be used to write syslog entries on a Linux-based operating system?

- ○ **A.** memo
- ○ **B.** record
- ○ **C.** logger
- ○ **D.** trace

4. Which of the following is not a standard component of an entry in a Windows-based security log?

- ○ **A.** Event ID
- ○ **B.** Date
- ○ **C.** Computer
- ○ **D.** Domain
- ○ **E.** User

5. You have just used a port scanner for the first time. On one port, it reports that there is not a process listening, and access to this port will likely be denied. Which state is the port most likely considered to be in?

- ○ **A.** Listening
- ○ **B.** Closed
- ○ **C.** Filtered
- ○ **D.** Blocked

6. You are required to monitor discards, packet drops, resets, and problems with speed/duplex. Which of the following types of monitoring would assist you?

- ○ **A.** Interface
- ○ **B.** Power
- ○ **C.** Environmental
- ○ **D.** Application

7. By default, the automatic update feature on most modern operating systems is

  ○ **A.** Disabled

  ○ **B.** Turned on

  ○ **C.** Set to manual

  ○ **D.** Ineffective

8. What should you do if a weakness is discovered that affects network security and no patch has yet been released?

  ○ **A.** Post information about the weakness on the vendor's site.

  ○ **B.** Call the press to put pressure on the vendor.

  ○ **C.** Ignore the problem and wait for the patch.

  ○ **D.** Take the at-risk system offline.

9. Which of the following statements are true regarding SNMP? (Choose all that apply.)

  ○ **A.** Simple Network Management Protocol (SNMP) is an application layer protocol whose purpose is to collect statistics from TCP/IP devices.

  ○ **B.** SNMPv1 and SNMPv2 are considered insecure, and SNMPv3 is the current and recommended standard.

  ○ **C.** SNMP agents listen on UDP/161. SNMP does not use TCP for messaging. SNMP sends traps on UDP/161.

  ○ **D.** Configuring SNMP traps on a network can help administrators determine the causes of jitter and latency.

10. Which is a specific network monitoring technique used to copy and redirect network traffic from one or more network ports (or interfaces) to a designated monitoring port for analysis?

  ○ **A.** Performance testing

  ○ **B.** Network discovery

  ○ **C.** Traffic analysis

  ○ **D.** Port mirroring

11. A network administrator wants to correlate security events and analyze them for suspected intrusions. Which should the administrator use?

  ○ **A.** SIEM

  ○ **B.** Port aggregation

  ○ **C.** Firmware updates

  ○ **D.** Vulnerability patches

Cram Quiz Answers

1. **B.** Whereas load tests do not try to break the system under intense pressure, stress tests sometimes do. Stress testing has two goals. The first is to see exactly what the network can handle. It is useful to know the network's breaking point in case the network ever needs to be expanded. Second, stress testing allows you to test your backup and recovery procedures.

2. **B.** The security logs can be configured to show failed or successful logon attempts as well as object access attempts. In this case, you can review the security logs and failed logon attempts to get the desired information. The failed logons will show the date and time when the failed attempts occurred.

3. **C.** The syslog feature exists in most UNIX/Linux-based distributions, and entries can be written using logger. The other options are not possibilities for writing syslog entries.

4. **D.** The standard components of an entry in a Windows-based security log include the date, time, user, computer, and event ID. The domain is not a standard component of a log entry.

5. **B.** When a port is closed, no process is listening on that port, and access to this port will likely be denied. When the port is open and listening, the host sends a reply indicating that a service is listening on the port. When the port is filtered or blocked, there is no reply from the host, meaning that the port is not listening or the port is secured and filtered.

6. **A.** An interface monitoring tool is invaluable for troubleshooting problems and errors that include utilization problems, discards, packet drops, resets, and problems with speed/duplex.

7. **B.** By default, the automatic update feature is usually turned on.

8. **D.** Often, the only thing that you, as a security administrator, can do between the discovery of the exploit and the release of the patch is to turn off the service. Although this approach can be a costly undertaking in terms of productivity, it can be the only way to keep the network safe.

9. **A, B, C, and D.** Yup! All these statements are true regarding SNMP. Simple Network Management Protocol (SNMP) is an application layer protocol whose purpose is to collect statistics from TCP/IP devices. To monitor the health of all systems, you install SNMP agents on the machines/devices and then monitor those agents (TCP/IP devices) from a central location. SNMPv1 and SNMPv2 are considered insecure, and SNMPv3 is the current and recommended secure version. Configuring SNMP traps on a network can help administrators determine the causes of jitter and latency.

10. **D.** Port mirroring is a feature that can be used to enable monitoring of all packets arriving at an interface and can be used to troubleshoot network issues, such as latency or poor performance. It can also be used to direct network traffic to a monitoring system. Performance testing is, as the name suggests, all about measuring the network's current performance level. The goal is to take ongoing performance tests and evaluate and compare them, looking for potential bottlenecks. For performance tests to be effective, they need to be taken under the same type of network load each time, or the comparison is invalid. Network discovery

is a crucial aspect of network monitoring that involves identifying and cataloging devices, services, and resources within an IT infrastructure. Traffic analysis is a core component of network monitoring, providing insights into the flow of data packets within a network environment. It involves the examination, interpretation, and visualization of network traffic patterns, behaviors, and characteristics to gain actionable intelligence about the performance, security, and usage of the network.

11. **A.** SIEM systems collect, aggregate, correlate, and analyze security-related data from various sources within an organization's IT infrastructure to provide real-time monitoring, threat detection, incident response, and compliance reporting capabilities. Port aggregation is the combining of multiple ports on a switch; it can be done one of three ways: auto, desirable, or on. Firmware updates keep the hardware interfaces working properly. Router manufacturers, for example, often issue patches when problems are discovered. Those patches need to be applied to the router to remove any security gaps that may exist. Vulnerability patches related to weaknesses and vulnerabilities should be installed correctly with all expediency. After a vulnerability in an OS, a driver, or a piece of hardware has been identified, the fact that it can be exploited is often spread quickly: a *zero-day exploit* is any attack that begins the very day the vulnerability is discovered.

Disaster Recovery and High Availability

▶ **3.3 Explain disaster recovery (DR) concepts.**

CramSaver

If you can correctly answer these questions before going through this section, save time by skimming the ExamAlerts in this section and then completing the Cram Quiz at the end of the section.

1. What is the difference between an incremental backup and a differential backup?

2. True or false: A brownout/undervoltage event is total failure of the power supplied to the server.

3. What are hot, warm, and cold sites used for?

4. True or false: The MTTR is the measurement of the anticipated or predicted incidence of failure of a system or component between inherent failures, whereas the MTBF is the measurement of how long it takes to repair a system or component after a failure occurs.

5. What is the concept of simultaneous management and utilization of multiple available paths for the transmission of streams of data called?

Answers

1. With incremental backups, all data that has changed since the last full or incremental backup is backed up. The restore procedure requires several backup iterations: the media used in the latest full backup and all media used for incremental backups since the last full backup. An incremental backup uses the archive bit and clears it after a file is saved to disk. With a differential backup, all data changed since the last full backup is backed up. The restore procedure requires the latest full backup media and the latest differential backup media. A differential backup uses the archive bit to determine which files must be backed up but does not clear it.

2. False. A total failure of the power supplied to the server is called a blackout.

3. Hot, warm, and cold sites are designed to provide alternative disaster recovery (DR) locations for network operations if a disaster occurs.

4. False. The MTBF is the measurement of the anticipated or predicted incidence of failure of a system or component between inherent failures, whereas the MTTR is the measurement of how long it takes to repair a system or component after a failure occurs.

5. Multipathing is the concept of simultaneous management and utilization of multiple available paths for the transmission of streams of data.

Even the most fault-tolerant networks can fail, which is an unfortunate fact. When those costly and carefully implemented fault-tolerance strategies fail, you are left with disaster recovery.

Disaster recovery (DR) can take many forms. In addition to disasters such as fire, flood, and theft, many other potential business disruptions can fall under the banner of disaster recovery. For example, the failure of the electrical supply to your city block might interrupt the business functions. Such an event, although not a disaster per se, might invoke the disaster recovery methods.

The cornerstone of every disaster recovery strategy is the preservation and recoverability of data. When talking about preservation and recoverability, you must talk about backups. Implementing a regular backup schedule can save you a lot of grief when fault tolerance fails or when you need to recover a file that has been accidentally deleted. When it's time to design a backup schedule, you can use three key types of backups: full, differential, and incremental.

Backups

Backups are equivalent to an insurance policy for a server, workstation, or any other data-containing device. Most of the time, they are nothing but time- and resource-consuming entities that make you wonder why you are wasting your time doing them. When disaster happens, however, you realize immediately their worth and kick yourself for not doing them with even more frequency.

Full Backups

The preferred method of backup is the *full backup* method, which copies all files and directories from the hard disk to the backup media. There are a few reasons why doing a full backup is not always possible. First among them is likely the time involved in performing a full backup.

> **ExamAlert**
>
> During a recovery operation, a full backup is the fastest way to restore data of all the methods discussed here, because only one set of media is required for a full restore.

Depending on the amount of data to be backed up, however, full backups can take an extremely long time when you are backing up and can use extensive system resources. Depending on the configuration of the backup hardware, completing this backup can considerably slow down the network. In addition, some environments have more data than can fit on a single medium. Therefore, doing a full backup is awkward because someone might need to be there to change the media.

The main advantage of full backups is that a single set of media holds all the data you need to restore. If a failure occurs, that single set of media should be all that is needed to get all data and system information back. The upshot of all this is that any disruption to the network is greatly reduced.

Unfortunately, its strength can also be its weakness. A single set of media holding an organization's data can be a security risk. If the media were to fall into the wrong hands, all the data could be restored on another computer. Using passwords on backups and using a secure offsite and onsite location can minimize the security risk.

Differential Backups

Companies that don't have enough time to complete a full backup daily can use the *differential backup*. Differential backups are faster than a full backup because they back up only the data that has changed since the last full backup. This means that if you do a full backup on a Saturday and a differential backup on the following Wednesday, only the data that has changed since Saturday is backed up. Restoring the differential backup requires the last full backup and the latest differential backup.

Differential backups know what files have changed since the last full backup because they use a setting called the archive bit. The archive bit flags files that have changed or have been created and identifies them as ones that need to be backed up. Full backups do not concern themselves with the archive bit because all files are backed up, regardless of date. A full backup, however, does clear the archive bit after data has been backed up to avoid future confusion. Differential backups notice the archive bit and use it to determine which files have changed. The differential backup does not reset the archive bit information.

Incremental Backups

Some companies have a finite amount of time they can allocate to backup procedures. Such organizations are likely to use *incremental backups* in their backup strategy. Incremental backups save only the files that have changed since the last full or incremental backup. Like differential backups, incremental backups use the archive bit to determine which files have changed since the last full or incremental backup. Unlike differentials, however, incremental backups clear the archive bit, so files that have not changed are not backed up

ExamAlert

Both full and incremental backups clear the archive bit after files have been backed up.

The faster backup time of incremental backups comes at a price—the amount of time required to restore. Recovering from a failure with incremental backups requires numerous sets of media—all the incremental backup media sets and the one for the most recent full backup. For example, if you have a full backup from Sunday and an incremental for Monday, Tuesday, and Wednesday, you need four sets of media to restore the data. Each set in the rotation is an additional step in the restore process and an additional failure point. One damaged incremental media set means that you cannot restore the data. Table 8.2 summarizes the various backup strategies.

TABLE 8.2 **Backup Strategies**

Backup Type	Advantage	Disadvantage	Data Backed Up	Archive Bit
Full	Backs up all data on a single media set. Restoring data requires the fewest media sets.	Depending on the amount of data, full backups can take a long time.	All files and directories are backed up.	Does not use the archive bit but resets it after data has been backed up.
Differential	Faster backups than a full backup.	The restore process takes longer than just a full backup. A differential backup uses more media sets than a full backup.	All files and directories that have changed since the last full backup.	Uses the archive bit to determine the files that have changed but does not reset the archive bit.
Incremental	Faster backup times	An incremental backup requires multiple disks; restoring data takes more time than the other backup methods.	The files and directories that have changed since the last full or incremental backup.	Uses the archive bit to determine the files that have changed and resets the archive bit.

> **Note**
>
> Although some of this information may not be tested on the exam, it is very important for any network administrator to be familiar with backup types and backup rotation schemes.

When thinking of media employed in backups and data to be backed up, never forget the cloud. Cloud backups provide an additional layer of redundancy for data. In case of hardware failures, natural disasters, or other catastrophic events, having data stored offsite ensures that it can be recovered without relying solely on local backups. Cloud backup solutions can scale easily to accommodate growing data storage needs: administrators can adjust storage capacity as needed without the hassle of procuring and configuring additional hardware. Many cloud backup services offer automated backup scheduling, reducing the burden on network administrators to manually manage backup routines. This automation helps ensure that backups are performed regularly and reliably.

Leading cloud backup providers typically implement security measures to protect data, including encryption, access controls, and compliance certifications. This helps mitigate the risk of data breaches or unauthorized access. Cloud backups often operate on a pay-as-you-go model, eliminating the need for large upfront investments in hardware and infrastructure. This can be particularly advantageous for organizations with limited IT budgets. Pay particular attention to versioning and retention policies with cloud providers. These policies allow administrators to restore previous versions of files and set rules for how long data should be retained.

Snapshots

In addition to the three types of backups previously discussed, there are also *snapshots*. Whereas a backup can take a long time to complete, the advantage of a snapshot—an image of the state of a system at a particular point in time—is that it is an instantaneous copy of the system. This snapshot is often accomplished by splitting a mirrored set of disks or by creating a copy of a disk block when it is written to preserve the original and keep it available.

Snapshots are popular with virtual machine implementations. You can take as many snapshots as you want (provided you have enough storage space) to be able to revert a machine to a "saved" state. Snapshots contain a copy of the virtual machine settings (hardware configuration), information on all virtual disks attached, and the memory state of the machine at the time of the snapshot. This makes the snapshots additionally useful for virtual machine cloning, allowing the machine to be copied once—or multiple times—for testing.

> **ExamAlert**
>
> Think of a snapshot as a photograph, which is where the name came from, of a moment in time of any system.

Backup Best Practices

Many details go into making a backup strategy a success. The following are issues to consider as part of your backup plan:

▶ **Using offsite storage:** Consider storing backup media sets offsite, such as in the cloud, so that if a disaster occurs in a building, a current set of media is available offsite. The offsite media should be as current as any that is onsite and should be secure.

▶ **Labeling media:** The goal is to restore the data as quickly as possible. Trying to find the media you need can prove difficult if it is not marked. Furthermore, labeling can prevent you from recording over something you need to keep.

▶ **Verifying backups:** Never assume that the backup was successful. Seasoned administrators know that checking backup logs and performing periodic test restores are part of the backup process.

▶ **Cleaning:** You need to occasionally clean the backup drive. If the inside gets dirty, backups can fail.

> **ExamAlert**
>
> A backup strategy must include offsite storage to account for theft, fire, flood, or other disasters.

Cold, Warm, Hot, and Cloud Sites

A disaster recovery plan might include the provision for a recovery site that can be quickly brought into play. These sites fall into three categories: hot, warm, and cold. The need for each of these types of sites depends largely on the business you are in and the funds available. Disaster recovery sites represent the ultimate in precautions for organizations that need them. As a result, they do not come cheaply.

The basic concept of a disaster recovery site is that it can provide a base from which the company can be operated during a disaster. The disaster recovery site normally is not intended to provide a desk for every employee. It's intended more as a means to allow key personnel to continue the core business functions.

In general, a cold recovery site is one that can be up and operational in a relatively short amount of time, such as a day or two. Provision of services, such as telephone lines and power, is taken care of, and the basic office furniture might be in place. But there is unlikely to be any computer equipment, even though the building might have a network infrastructure and a room ready to act as a server room. In most cases, cold sites provide the physical location and basic services.

Cold sites are useful if you have some forewarning of a potential problem. Generally, cold sites are used by organizations that can weather the storm for a day or two before they get back up and running. If you are the regional office of a major company, it might be possible to have one of the other divisions take care of business until you are ready to go. But if you are the only office in the company, you might need something a little hotter.

For organizations with the dollars and the desire, hot recovery sites represent the ultimate in fault-tolerance strategies. Like cold recovery sites, hot sites are designed to provide only enough facilities to continue the core business function, but hot recovery sites are set up to be ready to go at a moment's notice.

A hot recovery site includes phone systems with connected phone lines. Data networks also are in place, with any necessary routers and switches plugged in and turned on. Desks have desktop PCs installed and waiting, and server areas are equipped with the necessary hardware to support business-critical functions. In other words, within a few hours, the hot site can become a fully functioning element of an organization.

The issue that confronts potential hot-recovery site users is that of cost. Office space is expensive in the best of times, but having space sitting idle 99.9 percent of the time can seem like a tremendously poor use of money. A popular strategy to get around this problem is to use space provided in a disaster recovery facility, which is basically a building, maintained by a third-party company, in which various businesses rent space. Space is usually apportioned according to how much each company pays.

Sitting between the hot and cold recovery sites is the warm site. A warm site typically has computers but is not configured ready to go. This means that data might need to be upgraded or other manual interventions might need to be performed before the network is again operational. The time it takes to get a warm site operational lands right in the middle of the other two options, as does the cost.

One of the newer types of sites being marketed is a *cloud site*. Similar to a warm site, a cloud site is available when needed. The difference between the cloud site and the warm site is that the warm site is often dedicated to the company while a cloud site is controlled by a provider who may market availability of it to many different companies (like an insurance policy) knowing that the odds are good that only one will need it at a time.

High-Availability Approaches and Recovery Concepts

Critical business functions refer to those processes or systems that must be made operational immediately when an outage occurs. The business can't function without them, and many are information intensive and require access to both technology and data. When you evaluate your business's sustainability, realize that disasters do indeed happen. If possible, build infrastructures that don't have a *single point of failure (SPOF)* or connection. If you're the administrator for a small company, it is not uncommon for the SPOF to be a router or gateway, but you must identify all *critical nodes* and *critical assets*. The best way to remove an SPOF from your environment is to add in redundancy.

Know that every piece of equipment can be rated in terms of mean time between failures (MTBF) and mean time to recovery (MTTR). The MTBF is the measurement of the anticipated or predicted incidence of failure of a system or component between inherent failures, whereas the MTTR is the measurement of how long it takes to repair a system or component after a failure occurs.

The *recovery time objective (RTO)* is the maximum amount of time that a process or service is allowed to be down and the consequences still considered acceptable. Beyond this time, the break in business continuity is considered to affect the business negatively. The *recovery point objective (RPO)* is the maximum time in which transactions could be lost from a major incident—how much you are willing to walk away from in order to get everything up and running again. Both RTO and RPO have to be balanced in coming up with a policy for how to deal with incidents.

> **ExamAlert**
>
> For the exam, make sure you know the differences between MTBF, MTTR, RTO, and RPO. Know what the acronyms stand for and what they mean.

Some technologies that can help with availability are the following:

▶ **Fault tolerance** is the capability to withstand a fault (failure) without losing data. This can be accomplished through the use of RAID, backups, and similar technologies. Popular fault-tolerant RAID implementations include RAID 1, RAID 5, and RAID 10.

▶ **Load balancing** is a technique in which the workload is distributed among several servers. This feature can take networks to the next level; it increases network performance, reliability, and availability. A load balancer can be either a hardware device or software specially configured to balance the load.

> **ExamAlert**
>
> Remember that load balancing increases redundancy and therefore data availability. Also, load balancing increases performance by distributing the workload.

▶ **Multipathing** is the concept of simultaneous management and utilization of multiple available paths for the transmission of streams of data. By increasing the available paths that data can take, often by introducing redundancy, it is possible to decrease the likelihood of a path's failure bringing operations down.

▶ **Redundant hardware** is a key component of making sure systems have a chance of staying up in the event of a failure of any one component. Redundancy can apply to switches, routers, firewalls, and literally any other piece of hardware that network operations are dependent upon.

▶ **Network interface card (NIC) teaming** is the process of combining multiple network cards for performance and redundancy (fault tolerance) reasons. This can also be called bonding, balancing, or aggregation.

▶ **Port aggregation** is the combining of multiple ports on a switch; it can be done one of three ways: auto, desirable, or on.

▶ **Clustering** is a method of balancing loads and providing fault tolerance.

Use *vulnerability scanning* and *penetration testing* to find the weaknesses in your systems before others do. Make sure that *end-user awareness and training* are priorities when it comes to identifying problems and that you stress *adherence to standards and policies.* Those policies should include both network policies and security policies, both of which were discussed previously in this chapter.

> **ExamAlert**
>
> As you study for the exam, three topics to pay attention to are adherence to standards and policies, vulnerability scanning, and penetration testing.

All these policies are important, but those that relate to first responders and deal with data breaches are of elevated importance.

Active-Active Versus Active-Passive

When it comes to high availability, solutions fall into two types of approaches: active-active and active-passive. The difference between the two is pretty straightforward: if the devices to be used in the event of a failure are in use normally, that is *active-active* because they are both currently active. An example would be nodes in a cluster; they are all currently in use and can carry on in the event of a failure.

If, on the other hand, a device is not in use currently but becomes activated by a failure (a failover), that is *active-passive.* An example would be having multiple phone carriers or Internet service providers (ISPs) able to provide service to a facility and using only one unless their service goes out, in which case the others are activated.

> **ExamAlert**
>
> For the exam, make sure you know the differences between active-active and active-passive.

Redundancy and availability can be contracted with multiple ISPs to create diverse paths and make sure you can stay up in the event an ISP experiences a failure. You can also use redundancy with routers through the use of Virtual Router Redundancy Protocol (VRRP) and First Hop Redundancy Protocol (FHRP).

The *Virtual Router Redundancy Protocol (VRRP)* is used to automatically assign routers to hosts. It creates virtual routers (abstract representations of multiple routers) that act as a group. The default gateway on a host is configured to the virtual router rather than a physical router, so if the physical router fails, a redundant choice is already built in to the group.

There are several *First Hop Redundancy Protocols (FHRP)*; one of the more popular is the Hot Standby Router Protocol (HSRP), which is exclusive to Cisco. All of these protocols work by allowing a default router address to be configured to be used in the event that the primary router fails.

DR Testing

Disaster recovery (DR) testing is crucial for maintaining the resilience and continuity of a computer network in the face of unforeseen events or disruptions. It involves simulating various disaster scenarios, validating recovery procedures, and assessing the effectiveness of contingency plans to ensure that the organization can recover data, applications, and services in a timely manner. Two common testing methods are tabletop exercises and validation tests.

Tabletop exercises are scenario-based discussions or simulations conducted in a classroom or meeting room setting, where key stakeholders participate in guided discussions and decision-making exercises to evaluate their response to hypothetical disaster scenarios. Tabletop exercises focus on assessing preparedness; testing decision-making processes; and identifying gaps in communication, coordination, or resource allocation without actually executing recovery procedures.

Validation tests, also known as technical or operational tests, involve the actual execution of recovery procedures and the testing of DR infrastructure, systems, and processes in a controlled environment. Validation tests simulate real-world disaster scenarios and measure the effectiveness of recovery efforts, including data restoration, application failover, system recovery, and network reconfiguration. These tests validate the functionality and performance of DR solutions, identify technical issues or limitations, and validate recovery time objectives (RTOs) and recovery point objectives (RPOs).

ExamAlert

For the exam, know that tabletop exercises and validation tests serve different but complementary purposes in disaster recovery planning. Remember that in tabletop exercises the appropriate individuals and teams are simply brought together for a discussion.

Cram Quiz

1. Which backup methods clear the archive bit after the backup has been completed? (Choose two.)

   ○ **A.** Full

   ○ **B.** Differential

   ○ **C.** Incremental

   ○ **D.** GFS

2. You come to work on Thursday morning to find that the server has failed and you need to restore the data from backup. You finished a full backup on Sunday and incremental backups on Monday, Tuesday, and Wednesday. How many media sets are required to restore the backup?

   ○ **A.** Four

   ○ **B.** Two

   ○ **C.** Three

   ○ **D.** Five

3. Which of the following recovery sites might require the delivery of computer equipment and an update of all network data?

   ○ **A.** Cold site

   ○ **B.** Warm site

   ○ **C.** Hot site

   ○ **D.** None of the above

4. As part of your network administrative responsibilities, you have completed your monthly backups. As part of backup best practices, where should the media be stored?

   ○ **A.** In a secure location in the server room

   ○ **B.** In a secure location somewhere in the building

   ○ **C.** In an offsite location

   ○ **D.** In a secure offsite location

5. As network administrator, you have been tasked with designing a disaster recovery plan for your network. Which of the following might you include in a disaster recovery plan?

   ○ **A.** RAID 5

   ○ **B.** Offsite media storage

   ○ **C.** Mirrored hard disks

   ○ **D.** UPS

6. Which type of recovery site mirrors the organization's production network and can assume network operations on a moment's notice?

   ○ **A.** Warm site

   ○ **B.** Hot site

   ○ **C.** Cold site

   ○ **D.** Mirrored site

7. Which of the following are used to find weaknesses in your systems before others do? (Choose two.)

   ○ **A.** Data breaches

   ○ **B.** Vulnerability scanners

   ○ **C.** Penetration testers

   ○ **D.** First responders

8. Which of the following is a type of policy in which employees and other network users give consent to be monitored?

   ○ **A.** Consent to monitoring

   ○ **B.** Acceptable use

   ○ **C.** Memorandum of understanding

   ○ **D.** Service-level agreement

9. Which of the following is used to automatically assign routers to hosts and creates virtual routers that act as a group?

   ○ **A.** VRRP

   ○ **B.** RAID 5

   ○ **C.** RAID 10

   ○ **D.** NIC teaming

10. Which type of DR testing involves the actual execution of recovery procedures and the testing of DR infrastructure?

- ○ **A.** Tabletop exercises
- ○ **B.** Validation tests
- ○ **C.** Active-active
- ○ **D.** Active-passive

Cram Quiz Answers

1. **A and C.** The archive bit is reset after a full backup and an incremental backup. Answer B is incorrect because the differential backup does not reset the archive bit. Answer D is wrong because GFS is a rotation strategy, not a backup method.

2. **A.** Incremental backups save all files and directories that have changed since the last full or incremental backup. To restore, you need the latest full backup and all incremental media sets. In this case, you need four sets of media to complete the restore process.

3. **A.** A cold site provides an alternative location but typically not much more. A cold site often requires the delivery of computer equipment and other services. A hot site has all network equipment ready to go if a massive failure occurs. A warm site has most equipment ready but still needs days or weeks to have the network up and running.

4. **D.** Although not always done, it is a best practice to store backups in a secure offsite location in case of fire or theft. Answer A is incorrect because if the server room is damaged by fire or flood, the backups and the data on the server can be compromised by the same disaster. Similarly, answer B is incorrect because storing the backups onsite does not eliminate the threat of a single disaster destroying the data on the server and backups. Answer C is incorrect because of security reasons. The offsite media sets must be secured.

5. **B.** Offsite storage is part of a disaster recovery plan. The other answers are considered fault-tolerance measures because they are implemented to ensure data availability.

6. **B.** A hot site mirrors the organization's production network and can assume network operations at a moment's notice. Answer A is incorrect because warm sites have the equipment needed to bring the network to an operational state but require configuration and potential database updates. Answer C is incorrect because cold sites have the space available with basic services but typically require equipment delivery. Answer D is incorrect because a mirrored site is not a valid option.

7. **B and C.** Use vulnerability scanning and penetration testing to find the weaknesses in your systems before others do. Answer A is incorrect because data breaches are invalid. Answer D is incorrect because first responders are typically those who are first on the scene after an incident.

8. **A.** A consent to monitoring policy is one in which employees and other network users acknowledge that they know they're being monitored and consent to it. Answer B is incorrect because acceptable use policies describe how the employees in an organization can use company systems and resources. Answers C and D are incorrect because a memorandum of understanding and service-level agreements are standard business documents.

9. **A.** The Virtual Router Redundancy Protocol (VRRP) is used to automatically assign routers to hosts. It creates virtual routers (abstract representations of multiple routers) that act as a group. Popular fault-tolerant RAID implementations include RAID 1, RAID 5, and RAID 10. NIC teaming is the process of combining multiple network cards for performance and redundancy (fault-tolerance) reasons.

10. **B.** Validation tests, also known as technical or operational tests, involve the actual execution of recovery procedures and the testing of DR infrastructure, systems, and processes in a controlled environment. Validation tests simulate real-world disaster scenarios and measure the effectiveness of recovery efforts, including data restoration, application failover, system recovery, and network reconfiguration. Tabletop exercises are scenario-based discussions or simulations conducted in a classroom or meeting room setting, where key stakeholders participate in guided discussions and decision-making exercises to evaluate their response to hypothetical disaster scenarios. In tabletop exercises the appropriate individuals and teams are simply brought together for a discussion. Active-active and active-passive are high-availability approaches, not DR testing techniques. The difference between the two is pretty straightforward: if the devices to be used in the event of a failure are in use normally, that is active-active because they are both currently active. An example would be nodes in a cluster; they are all currently in use and can carry on in the event of a failure. If, on the other hand, a device is not in use currently but becomes activated by a failure (a failover), that is active-passive.

Network Access and Management Methods

▶ **3.5 Compare and contrast network access and management methods.**

CramSaver

If you can correctly answer these questions before going through this section, save time by skimming the ExamAlerts in this section and then completing the Cram Quiz at the end of the section.

1. What is the primary purpose of a site-to-site VPN?

2. What distinguishes a client-to-site VPN from a site-to-site VPN?

3. What role does a jump box or jump host serve in network security?

4. What distinguishes in-band management from out-of-band management?

Answers

1. Site-to-site VPNs are designed to create secure, encrypted connections between different physical locations or networks, allowing them to communicate securely over the Internet or other untrusted networks.

2. Client-to-site VPNs are typically used by individual users to securely connect to a corporate network from remote locations, while site-to-site VPNs connect entire networks or subnets across different physical locations.

3. A jump box or jump host is a designated, hardened server that serves as an intermediary for accessing internal network resources from external networks, providing an additional layer of security by controlling and monitoring access to sensitive systems.

4. In-band management refers to managing network devices using the same network infrastructure used for production traffic, while out-of-band management involves using a separate, dedicated management network or channel to administer devices, providing isolation from production traffic and enhanced security.

Efficient access and effective management are paramount for maintaining robust and secure network infrastructures. Network access and management methods play pivotal roles in shaping the performance, security, and scalability of networks across various industries and organizational settings. Understanding the nuances and implications of different approaches is essential for network administrators and IT professionals.

In the sections that follow, we compare and contrast network access and management methods, exploring their functionalities, strengths, limitations, and the impact they have on network operations and security.

Site-to-Site VPN

A *site-to-site VPN* establishes a secure and encrypted connection between two or more geographically separate networks or sites over the Internet or other untrusted networks. It enables secure communication and data exchange between the connected sites, allowing them to function as if they were part of the same local network. In relation to network access and management, here are the most important things to know about them:

▶ **Secure connectivity:** Site-to-site VPNs provide secure connectivity between remote sites or branch offices, allowing users and devices in different locations to access resources and services securely over the Internet.

▶ **Encryption:** Site-to-site VPNs encrypt data traffic between the connected sites, ensuring that sensitive information remains confidential and protected from interception or eavesdropping by unauthorized parties.

▶ **Authentication and authorization:** Site-to-site VPNs employ authentication and authorization mechanisms to verify the identities of devices and users accessing the network, preventing unauthorized access and ensuring that only authenticated entities can establish VPN connections.

▶ **Tunneling protocols:** Site-to-site VPNs use tunneling protocols such as Internet Protocol Security (IPsec) or Secure Sockets Layer/Transport Layer Security (SSL/TLS) to encapsulate and encrypt data traffic, providing a secure communication channel between the connected networks.

▶ **Scalability and flexibility:** Site-to-site VPNs are scalable and flexible, allowing organizations to easily add or remove sites, expand network connectivity, and accommodate changes in network infrastructure without significant reconfiguration or overhead.

▶ **Centralized management:** Site-to-site VPNs often feature centralized management capabilities, enabling administrators to configure, monitor, and manage VPN connections, policies, and settings from a central management console or interface.

▶ **Performance considerations:** While site-to-site VPNs provide secure connectivity, they may introduce latency and overhead due to encryption and decryption processes. It's essential to consider performance requirements and bandwidth constraints when deploying and configuring site-to-site VPNs to ensure optimal network performance.

▶ **Redundancy and failover:** Site-to-site VPNs can be configured with redundancy and failover mechanisms to ensure high availability and reliability. Redundant VPN tunnels, multiple VPN gateways, and dynamic routing protocols help maintain continuous connectivity and mitigate the impact of network failures or outages.

By leveraging site-to-site VPN technology effectively, organizations can establish seamless and secure communication channels between remote sites and enhance data privacy and integrity.

Client-to-Site VPN

A *client-to-site VPN*, also known as a remote-access VPN or a client VPN, enables individual users or remote devices to securely connect to a corporate network or resources from external locations, such as home offices, coffee shops, or airports. Unlike site-to-site VPNs that establish secure connections between entire networks or sites, client-to-site VPNs allow individual users to access network resources securely. The most important things to know about them include

▶ **Remote access:** Client-to-site VPNs provide remote-access capabilities, allowing authorized users to connect to the corporate network or resources from remote locations using their personal devices, such as laptops, smartphones, or tablets.

▶ **Secure connectivity:** Client-to-site VPNs establish secure and encrypted connections between remote devices and the corporate network, ensuring that data traffic remains protected from interception or eavesdropping while traversing untrusted networks.

▶ **Authentication and authorization:** Client-to-site VPNs utilize authentication and authorization mechanisms to verify the identities of users and devices accessing the network, ensuring that only authorized individuals with valid credentials can establish VPN connections and access network resources.

▶ **Client software:** Client-to-site VPNs typically require users to install VPN client software or applications on their devices to establish VPN connections. These client applications facilitate secure authentication, encryption, and tunneling of data traffic between the user's device and the VPN gateway.

▶ **Clientless VPN:** Some client-to-site VPN solutions offer *clientless* VPN access, allowing users to connect to the corporate network using a web browser without the need for installing dedicated VPN client software. Clientless VPNs leverage web-based portals or SSL/TLS protocols to provide secure access to network resources.

▶ **Split tunneling versus full tunneling:** In client-to-site VPNs, split tunneling and full tunneling are two tunneling modes that determine how network traffic is routed between the user's device and the corporate network. With *split tunneling*, only traffic destined for the corporate network is routed through the VPN tunnel, while Internet-bound traffic is sent directly to the Internet without passing through the VPN gateway. This can improve performance and reduce VPN overhead but may pose security risks if not properly configured. In contrast, *full tunneling* routes all traffic from the user's device, including both corporate and Internet-bound traffic, through the VPN tunnel to the corporate network. Full tunneling provides enhanced security by ensuring that all traffic is encrypted and inspected by corporate security policies but may result in increased VPN overhead and Internet latency.

> **ExamAlert**
>
> For the exam, know the differences between site-to-site and client-to-site VPNs.

By leveraging client-to-site VPN technology effectively, organizations can provide secure and seamless remote access to corporate resources while maintaining data privacy, integrity, and compliance with security policies.

Connection Methods

To manage a network, you need to connect to it, and there are many ways of doing so. The four most common are listed here; the differences between them lie in their respective interfaces, functionalities, and modes of interaction:

▶ **Secure Shell (SSH):** This connection method allows secure remote access and management of network devices using a command-line interface. Users interact with network devices by typing commands into a terminal window. SSH provides encryption for data transmission, authentication using passwords or public-key cryptography, and secure remote access over insecure networks. It is commonly used for remote configuration, monitoring, troubleshooting, and administration of network devices such as routers, switches, and servers. You use port 22 for SSH to make secure connections to remote devices and run commands.

▶ **Graphical user interface (GUI):** This connection method facilitates user-friendly interaction and management of network devices through a graphical interface by clicking, dragging, and navigating through menus and graphical elements. An advantage is that it offers intuitive navigation, visual feedback, and ease of use, particularly for users who are less familiar with command-line interfaces. It is used for configuration, monitoring, and management tasks that require visual representation and interactive control, such as configuring firewall rules, managing VLANs, and monitoring network performance. Microsoft Windows, for example, offers Remote Desktop (RDP) and Remote Desktop Connection (RDC) to allow users to access and control a computer from a remote location. RDP allows users to interact with the desktop environment of the remote computer as if they were physically present at that machine. RDC is the client software used to connect using the Remote Desktop Protocol. Users can initiate a Remote Desktop Connection session by entering the IP address or hostname of the remote computer and providing valid credentials (username and password) for authentication.

▶ **Application programming interface (API):** Developers and software applications interact with network devices programmatically by sending requests and receiving responses using API calls. This enables automation, integration, and programmable control of network devices and services and allows customization, integration with third-party systems, and development of custom applications for network management and automation. This approach is often used for automating repetitive tasks, orchestrating network workflows, integrating with management systems, and building custom network management applications.

▶ **Console:** Users interact with network devices using a dedicated console port, serial cable, or physical console interface. If nothing else, the console offers a fallback mechanism for accessing network devices locally, particularly in scenarios where network connectivity is disrupted or remote management interfaces are inaccessible. The console is often used for configuring the initial device, recovering passwords, troubleshooting network connectivity issues, and performing maintenance tasks directly on network devices.

> **ExamAlert**
>
> Know that SSH provides secure command-line access, a GUI offers user-friendly visual interaction, an API enables programmable automation, and a console provides direct local access to network devices, each catering to different user preferences and use cases.

Jump Box

A *jump box*, also known as a jump host or a bastion host, is a server deployed within a network infrastructure to provide secure access to other internal network resources, such as servers, routers, switches, or virtual machines. It acts as an intermediary or gateway for accessing internal resources from external or less trusted networks, offering an additional layer of security by controlling and monitoring access to sensitive systems.

> **ExamAlert**
>
> A jump box serves as a secure gateway or entry point for accessing internal network resources from external networks, such as the Internet or less trusted networks.

Jump boxes enforce access control policies and authentication mechanisms to verify the identities of users or administrators attempting to access internal resources. Users must authenticate themselves to the jump box before gaining access to other network resources. Jump boxes facilitate auditability and monitoring of access to internal resources by logging user activities, commands executed, and network traffic traversing through the jump box. This helps in tracking and reviewing user actions for security and compliance purposes.

By consolidating access through a single jump box, organizations can better control and manage access to internal resources, reducing the attack surface and minimizing the exposure of critical systems to potential threats. Jump boxes are typically configured with hardened security measures, such as restricted user privileges, firewall rules, intrusion detection/prevention systems, and security patches to mitigate potential security risks and vulnerabilities.

Jump boxes can be deployed within segmented network architectures to enforce network segmentation and isolation, limiting lateral movement of attackers within the network. To enhance security, jump boxes often support multifactor authentication (MFA) methods, requiring users to provide additional authentication factors, such as one-time passwords or biometric verification, in addition to passwords. In the event of security incidents, emergencies, or network outages, jump boxes provide a secure mechanism for authorized personnel to access and troubleshoot critical systems remotely.

In-Band Versus Out-of-Band Management

In-band and out-of-band management are two approaches used to monitor, configure, and troubleshoot network devices. *In-band management* involves accessing and managing network devices using the same data path or network infrastructure that carries regular user traffic. In other words, management traffic shares the same network path as the user traffic. In-band management uses the same network interfaces and connections that regular data traffic uses, typically over the organization's primary network infrastructure. It offers simplicity and ease of configuration, since it leverages existing network connectivity for management tasks. In-band management may be susceptible to network congestion or performance issues if there is heavy data traffic on the network. It may introduce security risks if the primary network is compromised, potentially allowing attackers to interfere with management traffic.

Out-of-band management involves accessing and managing network devices through a dedicated or separate management network or communication channel that is independent of the primary data network. Out-of-band management provides a dedicated and separate communication path for managing network devices, typically using dedicated management interfaces, serial connections, or separate network infrastructure. It offers increased security and reliability because management traffic is isolated from regular data traffic, reducing the risk of interference or disruption from user traffic or network congestion. Out-of-band management may require additional infrastructure and resources to maintain a separate management network or communication channels, increasing complexity and cost. It enables network administrators to access and manage network devices even when the primary data network is unavailable or experiencing issues, enhancing resilience and uptime.

ExamAlert

If devices are monitored remotely, this is known as out-of-band management; otherwise, it is known as in-band management.

Out-of-band management generally offers higher security than in-band management because it provides isolation from regular user traffic and potential security threats. The choice between in-band and out-of-band management depends on factors such as security requirements, network architecture, reliability considerations, and the organization's specific use cases and operational needs.

Cram Quiz

1. Which of the following best describes in-band management?

   ○ **A.** Management traffic is sent through a dedicated network separate from regular data traffic.

   ○ **B.** Management traffic shares the same network path as regular user traffic.

   ○ **C.** Management traffic is encrypted for secure communication.

   ○ **D.** Management traffic is accessible only through physical console connections.

2. What is the primary difference between split tunnel and full tunnel VPN configurations?

   ○ **A.** Full tunnel VPNs route only specific network traffic through the VPN.

   ○ **B.** Split tunnel VPNs route all network traffic through the VPN.

   ○ **C.** Full tunnel VPNs allow access to both public and private network resources.

   ○ **D.** Split tunnel VPNs provide faster connection speeds.

3. What role does a jump box play in network segmentation?

   ○ **A.** It connects different network segments.

   ○ **B.** It accelerates network traffic.

   ○ **C.** It duplicates network traffic for monitoring purposes.

   ○ **D.** It isolates internal resources from external networks.

4. What is a key advantage of out-of-band management compared to in-band management?

   ○ **A.** Lower complexity and cost

   ○ **B.** Higher network performance

   ○ **C.** Enhanced security and reliability

   ○ **D.** Simplified configuration and management

5. You would like to establish an authenticated and encrypted connection between a remote client and a host system. Which of the following will best suit your needs?

   ○ **A.** Telnet
   ○ **B.** SSH
   ○ **C.** A jump box
   ○ **D.** Authentication and authorization

Cram Quiz Answers

1. **B.** In in-band management, management traffic shares the same network path as regular user traffic, utilizing the same network infrastructure.

2. **A.** Full tunnel VPN configurations route all network traffic, including Internet-bound traffic, through the VPN connection.

3. **D.** Jump boxes help enforce network segmentation by acting as gateways that control access between different network segments, isolating internal resources from external networks.

4. **C.** Out-of-band management provides increased security and reliability by utilizing a separate communication path for managing network devices, isolated from regular user traffic.

5. **B.** The Secure Shell (SSH) utility establishes a session between the client and host computers using an authenticated and encrypted connection over port 22. SSH requires encryption of all data, including the login portion. Telnet is incorrect because SSH is the secure replacement for Telnet. Using Telnet is ill advised because a Telnet session is not encrypted. A jump box is a server deployed within a network infrastructure to provide secure access to other internal network resources, such as servers, routers, switches, or virtual machines. Site-to-site VPNs employ authentication and authorization mechanisms to verify the identities of devices and users accessing the network, preventing unauthorized access and ensuring that only authenticated entities can establish VPN connections. However, of the answer options provided, SSH is the best choice in this scenario.

What's Next?

The primary goals of today's network administrators are to design, implement, and maintain secure networks. These tasks are not always easy, so they are the topic of Chapter 9, "Network Security." No network can ever be labeled "secure." Security is an ongoing process involving a myriad of protocols, procedures, and practices.

CHAPTER 9

Network Security

This chapter covers the following official Network+ objectives:

▶ 4.1 Explain the importance of basic network security concepts.

▶ 4.2 Summarize various types of attacks and their impact to the network.

▶ 4.3 Given a scenario, apply network security features, defense techniques, and solutions.

This chapter covers CompTIA Network+ objectives 4.1, 4.2, and 4.3. For more information on the official Network+ exam topics, see the "About the Network+ Exam" section in the Introduction.

Network security is one of the toughest areas of IT to be responsible for. It seems as if a new threat surfaces on a regular basis and that you are constantly needing to learn new things just a half a step ahead of potential problems. This chapter focuses on some of the elements that administrators and technicians use to keep their networks as secure as possible.

Common Security Concepts

▶ **4.1 Explain the importance of basic network security concepts.**

CramSaver

If you can correctly answer these questions before going through this section, save time by skimming the ExamAlerts in this section and then completing the Cram Quiz at the end of the section.

1. True or false: Filtering network traffic using a system's MAC address typically is done using an access control list (ACL).

2. True or false: LDAP is a protocol that provides a mechanism to access and query directory services systems.

3. Which access control model uses an ACL to determine access?

4. True or false: A honeypot is a trap that allows the intruder in but does not allow access to sensitive data.

5. Which IEEE standard defines port-based security for wireless network access control?

6. What enables a user to log in to a system and access multiple systems or resources without the need to repeatedly reenter the username and password?

Answers

1. True. Filtering network traffic using a system's MAC address typically is done using an access control list (ACL).

2. True. Lightweight Directory Access Protocol (LDAP) is a protocol that provides a mechanism to access and query directory services systems.

3. Discretionary access control (DAC) uses an access control list to determine access. The ACL is a table that informs the operating system of the rights each user has to a particular system object, such as a file, a folder, or a printer.

4. True. A honeypot is a trap that allows the intruder in but does not allow access to sensitive data.

5. The IEEE 802.1X standard defines port-based security for wireless network access control.

6. Single sign-on (SSO) enables a user to log in to a system and access multiple systems or resources without the need to repeatedly reenter the username and password.

There are three concepts you will see throughout this chapter and probably implied in every security-related book you will ever read: *confidentiality, integrity, and availability*, often referred to as the *CIA* triad of security. All security measures should affect one or more of these areas. Confidentiality means preventing unauthorized users from accessing data: passwords, encryption, and access control all support confidentiality. Integrity means ensuring that data has not been altered: hashing and message authentication codes are the most common methods of accomplishing this task (as well as ensuring nonrepudiation with digital signatures). Simply making sure that the data and systems are available for authorized users is what availability is all about: data backups, redundant systems, and disaster recovery plans all support fault tolerance and increased availability.

Threats can come from anywhere. They can include *external* entities such as bored individuals wanting to inflict harm to any system they can find access to and yours just happened to let them in, or *internal* entities such as disgruntled employees unhappy they were passed over for a promotion and feeling as if they have a right to cause harm. Often, the internal threats are the hardest to prevent, since these users are already granted access to some resources and have an advantage over those outside the organization.

Vulnerabilities are discovered on a regular basis and, as an administrator, you must stay current on information related to discovered problems and be aware of common vulnerabilities and exposures (CVEs). When a hole is found in a web browser or other software, and miscreants begin exploiting it the very day it is discovered by the developer (bypassing the one- to two-day response time many software providers need to put out a patch after the hole has been found), it is known as an *exploit attack* or *zero-day attack*. Zero-day attacks are incredibly difficult to respond to. If attackers learn of the weakness the same day as the developer, they have the ability to exploit it until a patch is released. Often, the only thing you, as a security administrator, can do between the discovery of the exploit and the release of the patch is to turn off the service. Although this approach can be a costly undertaking in terms of productivity, it is the only way to keep the network safe.

> **Note**
>
> Vulnerability scanners heavily depend on lists of identified vulnerabilities. It's crucial that you, as a security analyst, are acquainted with two main catalogs: the common vulnerabilities and exposures (CVEs), which lists publicly known vulnerabilities with an ID, description, and reference; and the Common Vulnerability Scoring System (CVSS), which assigns a severity score from 0 to 10 to each vulnerability. This numerical score can then be translated into qualitative labels like low, medium, high, and critical, aiding organizations in effectively evaluating and prioritizing their vulnerability management procedures.
>
> Two recommended resources that can be helpful are https://cve.mitre.org/ and https://www.sans.org/blog/what-is-cvss/.

> **ExamAlert**
>
> Be ready to identify types of attacks and common vulnerabilities and exposures (CVEs) such as the one just described. You can expect questions about these types of attacks on the Network+ exam.

Encryption

Encryption is the process of converting data into a ciphertext (unreadable form) using cryptographic techniques, such that only authorized parties can decipher or decrypt the data back into its original form. Encryption plays a crucial role in protecting the confidentiality, integrity, and privacy of data. Data needs to be protected (often through encryption) when it is *in transit* (actively moving from one location to another over a network) and *at rest* (stored or saved in a nonvolatile storage medium). To protect data in transit, encryption and secure communication protocols such as Transport Layer Security (TLS), Secure Shell (SSH), or virtual private networks (VPNs) are commonly used. To protect data at rest, encryption, access controls, data masking, and data loss prevention (DLP) solutions are commonly used to safeguard the data from unauthorized access or disclosure.

Access Control

Access control describes the mechanisms used to filter network traffic to determine who is and who is not allowed to access the network and network resources. Firewalls, proxy servers, routers, and individual computers all can maintain access control to some degree by protecting the *edges* of the network. Because the security strategy limits who can and cannot access the network and its resources, it is easy to understand why access control plays a critical role. Several types of access control strategies exist, as discussed in the following sections.

> **ExamAlert**
>
> Be sure that you can identify the purpose and types of access control.

Mandatory Access Control

Mandatory access control (MAC) is the most secure form of access control. In systems configured to use mandatory access control, administrators dictate who can access and modify data, systems, and resources. MAC systems are commonly used in military installations, financial institutions, and, because of new privacy laws, medical institutions.

MAC secures information and resources by assigning sensitivity labels or attributes to objects and users. When users request access to an object, their sensitivity level is compared to the object's. A label is a feature applied to files, directories, and other resources in the system. It is similar to a confidentiality stamp. When a label is placed on a file, it describes the level of security for that specific file. It permits access by files, users, programs, and so on that have a similar or higher security setting.

Discretionary Access Control

Unlike mandatory access control, discretionary access control (DAC) is not enforced from the administrator or operating system. Instead, access is controlled by an object's owner. For example, the secretary who creates a folder decides who will have access to that folder. This access is configured using permissions and an access control list (ACL).

DAC uses an ACL to determine access. The ACL is a table that informs the operating system of the rights each user has to a particular system object, such as a file, a folder, or a printer. Each object has a security attribute that identifies its ACL. The list has an entry for each system user with access privileges. The most common privileges include the ability to read a file (or all the files in a folder), to write to the file or files, and to execute the file (if it is an executable file or program).

Microsoft Windows servers/clients, Linux, UNIX, and macOS are among the operating systems that use ACLs. The list is implemented differently by each operating system.

In Windows Server products, an ACL is associated with each system object. Each ACL has one or more *access control entries (ACEs)* consisting of the name of a user or group of users. The user can also be a role name, such as "secretary" or "research." For each of these users, groups, or roles, the access privileges are stated in a string of bits called an *access mask*. Generally, the system administrator or the object owner creates the ACL for an object.

> **Note**
>
> A server on a network that has the responsibility of being a repository for accounts (user/computer) is often referred to as a *network controller*. A good example is a domain controller on a Microsoft Active Directory–based network.

Rule-Based Access Control

Rule-based access control (RBAC) controls access to objects according to established rules. The configuration and security settings established on a router or firewall are a good example.

When a firewall is configured, rules are set up to control access to the network. Requests are reviewed to see if the requestor meets the criteria to be allowed access through the firewall. For instance, if a firewall is configured to reject all addresses in the 192.166.x.x IP address range, and the requestor's IP is in that range, the request would be denied.

In a practical application, RBAC is a variation on MAC. Administrators typically configure the firewall or other device to allow or deny access. The owner or another user does not specify the conditions of acceptance, and safeguards ensure that an average user cannot change settings on the devices.

Role-Based Access Control

> **Note**
>
> Both *rule-based* and *role-based* access control use the acronym *RBAC*.

In role-based access control (RBAC), access decisions are determined by the roles that individual users have within the organization. Role-based access requires the administrator to have a thorough understanding of how a particular organization operates, the number of users, and each user's exact function in that organization.

> **ExamAlert**
>
> Because the CompTIA objectives specifically call out role-based access control, be sure you know that with RBAC access decisions are determined by the roles that individual users have within the organization.

Because access rights are grouped by role name, the use of resources is restricted to individuals who are authorized to assume the associated role. For example, within a school system, the role of teacher can include access to certain data, including test banks, research material, and memos. School administrators might have access to employee records, financial data, planning projects, and more.

The use of roles to control access can be an effective means of developing and enforcing enterprise-specific security policies and for streamlining the security management process.

Roles should receive only the privilege level necessary to do the job associated with that role. This general security principle is known as the concept of *least privilege*. When people are hired in an organization, their roles are clearly defined. A network administrator creates a user account for a new employee and places that user account in a group with people who have the same role in the organization.

Least privilege is often too restrictive to be practical in business. For instance, using teachers as an example, some more experienced teachers might have more responsibility than others and might require increased access to a particular network object. Customizing access to each individual is a time-consuming process.

ExamAlert

Once just a Security+ objective and concept, this topic has now been added to the Network+ objectives as well. Because you might be asked about the concept of least privilege, know that it refers to assigning network users the privilege level necessary to do the job associated with their role—nothing more and nothing less.

Closely related to least privilege is the concept of a *zero trust* network. As the name implies, it simply means that you don't automatically trust anyone and instead always authenticate and authorize. Everything must be verified explicitly, data protection is held to utmost ideal, and all sessions are encrypted end to end.

Table 9.1 highlights the different methods of access control as well as their advantages and disadvantages.

TABLE 9.1 **Access Control Summary**

	Mandatory Access Control (MAC)	Discretionary Access Control (DAC)	Rule-Based Access Control (RBAC)	Role-Based Access Control (RBAC)
Description	Enforces access controls based on security labels assigned to subjects and objects. Access decisions are determined by a central authority, often the operating system or security policy.	Allows owners of resources to set access permissions and control access to their resources. Access decisions are based on the discretion of resource owners.	Uses predefined rules or conditions to determine access permissions. Access decisions are made based on matching rules or conditions with specific access requests.	Assigns access permissions to users based on their roles within an organization. Users inherit permissions associated with their roles, simplifying access management.
Implementation	Implemented at the operating system level	Implemented through access control lists (ACLs)	Configured through rule sets or policies	Implemented by mapping roles to permissions
Advantages	Provides strict control over access to sensitive resources, mitigates the risk of unauthorized data disclosure	Offers flexibility in granting permissions, allows resource owners to manage access based on their discretion	Provides granular control over access permissions, allows for customization of access policies based on specific criteria	Simplifies access management by assigning permissions based on organizational roles, reduces administrative overhead
Disadvantages	Complex to administer and maintain, requires ongoing management of security labels and policies	Prone to security risks if resource owners grant overly permissive access, lacks centralized control over access policies	May lead to rule explosion if not properly managed, can be complex to configure and maintain in large environments	Requires careful role definition and assignment, may lead to excessive permissions if roles are not properly defined or managed

Know that the attribute-based access control (ABAC) authorization process operates by assessing rules and policies against attributes linked to an entity, like the subject, object, operation, or environmental conditions. ABAC is particularly suitable for large and federated enterprises but tends to be more complex and costly to set up and maintain compared to simpler access control models.

Mandatory access control (MAC) is commonly employed in governmental systems, where resource access is determined by categorical assignments such as classified, secret, or top secret. Discretionary access control (DAC) allows individual resources to be selectively accessible or secured from access. Here, access rights are set at the discretion of account holders who have authority over each resource, including the ability to grant administrative rights through the same mechanism.

In a role-based access control scenario, access rights are initially allocated to roles. Subsequently, accounts are linked to these roles, rather than directly assigning access rights to resources.

Defense in Depth

Defense in depth is based on the premise that implementing security at different levels or layers to form a complete security strategy provides better protection and greater resiliency than implementing an individual security defense. This level of defense can be accomplished in a number of different ways; some of the most popular are through the use of network segmentation, screened subnets, separation of duties, network access control, and honeypots. Each variant of defense in depth is discussed in the sections that follow.

> **Note**
>
> Defense in depth is a well-established strategy that offers robust protection against various threats. Yet, it can be intricate and expensive to put into practice. Conversely, zero trust presents a more proactive method that minimizes the attack surface and restricts access to critical systems and data. For this reason, a zero trust strategy is usually suggested over defense in depth in recent implementations.

Network Segmentation

Dividing one network into smaller subnetworks enables you to optimize it in a number of ways. The segmentation is accomplished with switches and VLANs, and the separation can be done to isolate such things as heavy load systems or certain protocols. If you move them to their own subnetwork, the result should be performance increases for other parts of the network.

> **Note**
>
> One popular approach is using *virtual machines (VMs)* to segment or separate systems (software) from the main OS (host versus guest) and the rest of the network through virtual switches.

The *Internet of Things (IoT)* and the *Industrial Internet of Things (IIoT)* present unique challenges and considerations for network segmentation due to the large number of interconnected devices, diverse communication protocols, and varied security requirements associated with these technologies. Network segmentation enables organizations to isolate IoT and IIoT devices into separate network segments or VLANs based on factors such as device type, function, criticality, or security requirements. This isolation helps contain the impact of security breaches or compromised devices and prevents lateral movement of threats within the network.

IoT and IIoT devices generate a significant amount of network traffic, including data transmission, command and control messages, and device management communications. Network segmentation allows organizations to segregate IoT and IIoT traffic into dedicated network segments or subnets, enabling better traffic management, prioritization, and monitoring. This segregation helps optimize network performance, minimize congestion, and improve overall network security posture.

Effective access control is essential for securing IoT and IIoT environments, as unauthorized access to connected devices can pose significant risks to data confidentiality, integrity, and availability. Network segmentation facilitates the implementation of access control policies and firewall rules to regulate communication between IoT/IIoT devices and other network resources. By restricting access to specific network segments based on user identity, device type, or security posture, organizations can prevent unauthorized access and unauthorized activities.

Network segmentation enables organizations to establish distinct security zones or trust domains within their network architecture to enforce different security policies and controls based on the sensitivity and criticality of IoT/IIoT devices and applications. By segmenting IoT and IIoT environments into separate security zones with appropriate security measures such as intrusion detection/prevention systems (IDS/IPSs), network segmentation helps contain security incidents, limit the scope of potential breaches, and facilitate compliance with regulatory requirements.

IoT and IIoT devices have diverse communication requirements and resource dependencies, including bandwidth, latency, and reliability considerations.

Network segmentation allows organizations to allocate network resources more efficiently by dedicating separate network segments or Quality of Service (QoS) policies to IoT/IIoT traffic, ensuring adequate bandwidth, low latency, and reliable connectivity for critical applications and devices.

Many industries and sectors have specific regulatory requirements and compliance standards that govern the security and privacy of IoT and IIoT deployments. Network segmentation supports compliance efforts by enabling organizations to implement segmentation controls and boundary protections to meet regulatory requirements, demonstrate due diligence, and protect sensitive data from unauthorized access or disclosure.

ExamAlert

The best defenses for IoT and IIoT are firmware updates, isolation or segmentation, and IDS/IPSs.

Supervisory control and data acquisition (SCADA), *industrial control systems (ICSs)*, and *operational technology (OT)* are critical components of infrastructure in sectors such as manufacturing, energy, transportation, and utilities. These systems play a vital role in controlling and monitoring industrial processes, machinery, and infrastructure. Network segmentation is essential for securing SCADA, ICS, and OT environments due to their unique characteristics, requirements, and security considerations.

SCADA, ICS, and OT environments typically consist of interconnected control systems, sensors, actuators, and other devices responsible for managing industrial processes and operations. Network segmentation allows organizations to isolate control networks from enterprise IT networks and external-facing systems to minimize the risk of cyber threats, unauthorized access, and disruptions to critical operations. By segregating control networks into separate network segments or zones, organizations can contain security incidents, prevent lateral movement of threats, and maintain the integrity and availability of industrial processes.

SCADA, ICS, and OT environments often comprise multiple operational zones or areas with distinct functions, such as production lines, distribution networks, and remote facilities. Network segmentation enables organizations to segment operational zones into separate network segments or VLANs based on functional boundaries, security requirements, and operational dependencies. This segmentation helps optimize network performance, facilitate traffic management, and enforce access controls tailored to the specific needs of each operational zone.

All three environments generate a significant volume of real-time data traffic, including sensor data, control signals, and monitoring information. Network segmentation facilitates the isolation and prioritization of critical traffic within industrial networks by creating dedicated network segments or Quality of Service (QoS) policies for SCADA, ICS, and OT traffic. This ensures that essential data transmissions receive priority treatment, low latency, and reliable connectivity, thereby minimizing the risk of network congestion, delays, and disruptions to critical operations.

Access control is crucial for securing SCADA, ICS, and OT environments, as unauthorized access to control systems and industrial assets can have severe consequences for safety, reliability, and compliance. Network segmentation enables organizations to enforce access controls and boundary protections to regulate communication between control systems, devices, and external networks. By implementing firewall rules, intrusion detection/prevention systems (IDS/IPSs), and network access controls (NAC) at network boundaries, organizations can prevent unauthorized access, detect malicious activities, and enforce security policies tailored to industrial environments.

Note

An industrial control system (ICS) is overseen through a SCADA system, offering a Human-Machine Interface (HMI) for operators to track system status. Additional ICS systems encompass industrial automation and control systems (IACS), distributed control systems (DCS), programmable logic controllers (PLCs), and remote terminal units (RTUs).

A crucial defense against attacks on supervisory control and data acquisition (SCADA) systems and industrial control systems (ICS) is to establish physical segregation between internal and external networks. This segregation reduces the attack surface by isolating the SCADA network from the corporate LAN.

One last item of note related to segmentation: *bring your own device (BYOD) policies* allow employees to use their personal devices, such as smartphones, tablets, and laptops, for work purposes, accessing corporate networks, applications, and data. While BYOD offers flexibility and productivity benefits, it also introduces security challenges, because personal devices may not adhere to the same security standards as corporate-owned devices. Network segmentation plays a crucial role in mitigating the security risks associated with BYOD by isolating personal devices from sensitive corporate resources and controlling access based on device type, user identity, and security posture.

Mobile device management (MDM) and mobile application management (MAM) policies are crucial components of BYOD initiatives aimed at protecting and supporting the use of personal devices in the workplace. MDM policies apply to devices such as smartphones, tablets, and laptops that employees bring into the workplace, whereas MAM policies focus on the applications installed on those mobile devices. Unlike MDM, MAM allows IT to control only the corporate data and applications on the device, leaving personal data untouched.

> **Note**
>
> In this bowl of acronym soup, a few others that are often used with BYOD, MDM, and MAM are CYOD (choose your own device), COPE (company owned/personally enabled), and COBO (company owned/business only).

Screened Subnet

An important firewall-related concept is that of the *screened subnet*. This concept was formerly known as a *demilitarized zone (DMZ)* and sometimes called a *perimeter network*. A screened subnet is part of a network where you place servers that must be accessible by sources both outside and inside your network. However, the screened subnet is not connected directly to either network, and it must always be accessed through the firewall. The military term *DMZ* was used previously because it describes an area that has little or no enforcement or policing.

Using screened subnets gives your firewall configuration an extra level of flexibility, protection, and complexity. Figure 9.1 shows a screened subnet (aka DMZ) configuration.

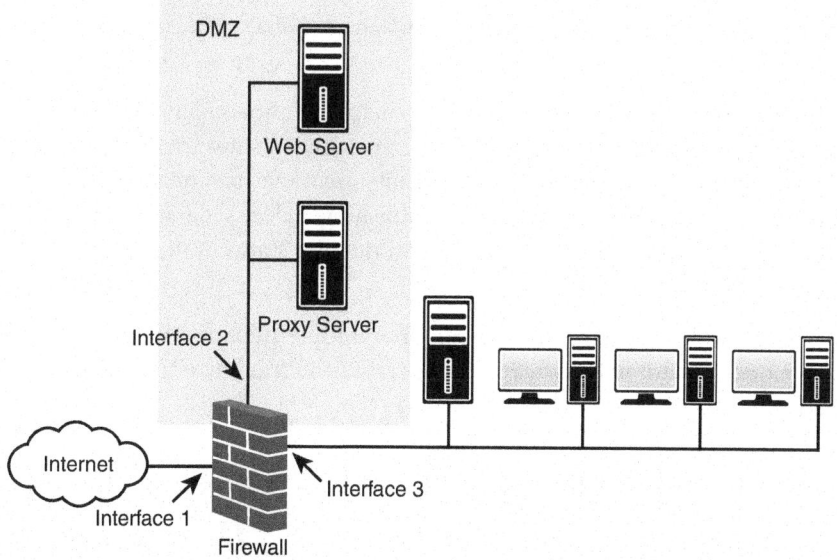

FIGURE 9.1   **A Screened Subnet (aka DMZ) Configuration**

By using a screened subnet, you can create an additional step that makes it more difficult for an intruder to gain access to the internal network. In Figure 9.1, for example, an intruder who tried to come in through Interface 1 would have to spoof a request from either the web server or proxy server into Interface 2 before it could be forwarded to the internal network. Although it is not impossible for an intruder to gain access to the internal network through a screened subnet, gaining access is difficult.

> **ExamAlert**
>
> Be prepared to identify the purpose of a screened subnet and that this concept was previously known as a DMZ.

Separation of Duties

Separation of duties policies are designed to reduce the risk of fraud and to prevent other losses in an organization. A good policy will require more than one person to accomplish key processes. This may mean that the person who processes an order from a customer isn't the same person who generates the invoice or deals with the billing.

Separation of duties helps prevent various problems, such as an individual embezzling money from a company. To embezzle funds successfully, an individual would need to recruit others to commit an act of collusion—that is, an agreement between two or more parties established for the purpose of committing deception or fraud. Collusion, when part of a crime, is also a criminal act in and of itself.

In addition, separation of duties policies can help prevent accidents from occurring in an organization. Suppose that you're managing a software development project. You want someone to perform a quality assurance test on a new piece of code before it's put into production. Establishing a clear separation of duties prevents development code from entering production status until quality testing is accomplished.

Many banks and financial institutions require multiple steps and approvals to transfer money. These policies help reduce errors and minimize the likelihood of fraud.

> **ExamAlert**
>
> Know that separation of duties policies help mitigate the risk of fraud, errors, and unauthorized activities by ensuring that critical tasks and responsibilities are divided among multiple individuals or roles within an organization.

Deception Technologies: Honeypots and Honeynets

Deception technologies involve the strategic deployment of decoy systems and resources to detect, divert, or deceive potential attackers. These technologies offer proactive defense capabilities that complement traditional security measures such as firewalls, antivirus software, and intrusion detection systems.

When we talk about network security and deception technologies, honeypots and honeynets are the most often mentioned. Honeypots are a rather clever approach to network security but perhaps a bit expensive. A *honeypot* is a system set up as a decoy to attract and deflect attacks from hackers. The server decoy appears to have everything a regular server does—OS, applications, and network services. Attackers think they are accessing a real network server, but they are in a network trap.

The honeypot has two key purposes. It can give administrators valuable information on the types of attacks being carried out. In turn, the honeypot can secure the real production servers according to what it learns. Also, the honeypot deflects attention from working servers, allowing them to function without being attacked.

A honeypot can do the following:

▶ Deflect the attention of attackers from production servers

▶ Deter attackers if they suspect their actions may be monitored with a honeypot

▶ Allow administrators to learn from the attacks to protect the real servers

▶ Identify the source of attacks, whether from inside the network or outside

ExamAlert

Think of a honeypot as a trap that allows the intruder in but does not allow access to sensitive data.

One step up from the honeypot is the honeynet. The *honeynet* is an entire network set up to monitor attacks from outsiders. All traffic into and out of the network is carefully tracked and documented. This information is shared with network professionals to help isolate the types of attacks launched against networks and to proactively manage those security risks. Honeynets function as a production network, using network services, applications, and more. Attackers don't know that they are actually accessing a monitored network.

RADIUS and TACACS+

Among the potential issues network administrators face when implementing remote access are utilization and the load on the remote-access server. As a network's remote-access implementation grows, reliance on a single remote-access server might be impossible, and additional servers might be required. *Remote Authentication Dial-In User Service (RADIUS)* can help in this scenario.

> **ExamAlert**
>
> RADIUS is a protocol that enables a single server to become responsible for all remote-access authentication, authorization, and auditing (or accounting) services. While it was originally used for dial-in connections, now the user authenticates either through a mobile VPN (with IPsec) or through a browser-based HTTPS connection (over port 4100).

RADIUS functions as a client/server system. The remote user accesses the remote-access server, which acts as a RADIUS client, or *network access server (NAS)*, and connects to a RADIUS server. The RADIUS server performs authentication, authorization, and auditing (or accounting) functions and returns the information to the RADIUS client (which is a remote-access server running RADIUS client software); the connection is either established or rejected based on the information received.

Terminal Access Controller Access Control System (TACACS) is a security protocol designed to provide centralized validation of users who are attempting to gain access to a router or NAS. Like RADIUS, TACACS is a set of security protocols designed to provide AAA of remote users. *Terminal Access Controller Access Control System Plus (TACACS+)* is a proprietary version of TACACS from Cisco and is the implementation commonly in use in networks today. TACACS+ uses TCP port 49 by default.

Although both RADIUS and TACACS+ offer AAA services for remote users, some noticeable differences exist:

▶ TACACS+ relies on TCP for connection-oriented delivery. RADIUS uses connectionless UDP for data delivery.

▶ RADIUS combines authentication and authorization, whereas TACACS+ can separate their functions.

> **ExamAlert**
>
> Both RADIUS and TACACS+ provide authentication, authorization, and auditing/accounting services. One notable difference between TACACS+ and RADIUS is that TACACS+ relies on the connection-oriented TCP, whereas RADIUS uses the connectionless UDP.

Kerberos Authentication

Kerberos is an *Internet Engineering Task Force (IETF)* standard for providing authentication. It is an integral part of network security. Networks, including the Internet, can connect people from all over the world. When data travels from one point to another across a network, it can be lost, stolen, corrupted, or misused. Much of the data sent over networks is sensitive, whether it is medical, financial, or otherwise. A key consideration for those responsible for the network is maintaining the confidentiality of the data. In the networking world, Kerberos plays a significant role in data confidentiality.

In a traditional authentication strategy, a username and password are used to access network resources. In a secure environment, it might be necessary to provide a username and password combination to access each network service or resource. For example, a user might be prompted to type in a username and password when accessing a database, and again for the printer, and again for Internet access. This is a time-consuming process, and it can also present a security risk. Each time the password is entered, there is a chance that someone will see it being entered. If the password is sent over the network without encryption, it might be viewed by malicious eavesdroppers.

Kerberos was designed to fix such problems by using a method requiring only a *single sign-on (SSO)*. This single sign-on enables a user to log in to a system and access multiple systems or resources without the need to repeatedly reenter the username and password. Additionally, Kerberos is designed to have entities authenticate themselves by demonstrating possession of secret information.

Kerberos is one part of a strategic security solution that provides secure authentication services to users, applications, and network devices by eliminating the insecurities caused by passwords stored or transmitted across the network. Kerberos is used primarily to eliminate the possibility of a network "eavesdropper" tapping into data over the network—particularly usernames and passwords. Kerberos ensures data integrity and blocks tampering on the network. It employs message privacy (encryption) to ensure that messages are not visible to eavesdroppers on the network.

For the network user, Kerberos eliminates the need to repeatedly demonstrate possession of private or secret information.

Kerberos is designed to provide strong authentication for client/server applications by using secret key cryptography. Cryptography is used to ensure that a client can prove its identity to a server (and vice versa) across an unsecure network connection. After a client and server have used Kerberos to prove their identity, they can also encrypt all their communications to ensure privacy and data integrity.

The key to understanding Kerberos is to understand its secret key cryptography. Kerberos uses *symmetric key cryptography*, in which both client and server use the same encryption key to cipher and decipher data.

In secret key cryptography, a plain-text message can be converted into cipher text (encrypted data) and then converted back into plain text using one key. Thus, two devices share a secret key to encrypt and decrypt their communications. Figure 9.2 shows the symmetric key process.

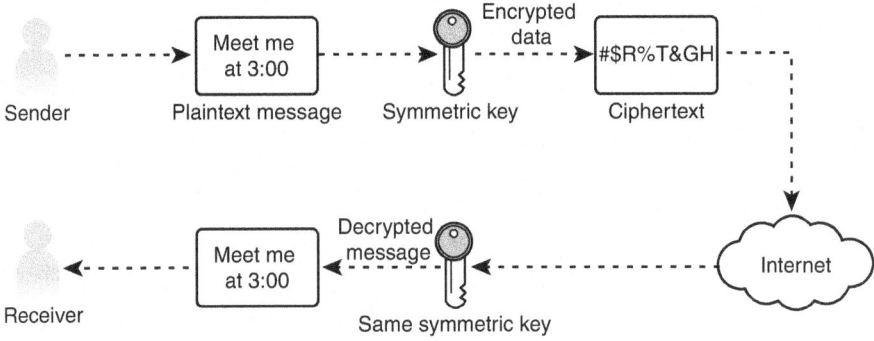

FIGURE 9.2 **The Symmetric Key Process**

> **ExamAlert**
>
> Another cryptography method in use is asymmetric key cryptography, or public key cryptography. In this method, a device has both a public and a private key. The private key is never shared. The public key is used to encrypt the communication, and the private key is used for decrypting.

Kerberos authentication works by assigning a unique key, called a *ticket*, to each client that successfully authenticates to a server. The ticket is encrypted and contains the user's password, which is used to verify the user's identity when a particular network service is requested. Each ticket is time stamped. It expires after a period of time, and a new one is issued. Kerberos works in the same way that you go to a movie. First, you go to the ticket counter, tell the person what movie you want to see, and get your ticket. After that, you go to a turnstile and hand the ticket to someone else, and then you're "in." In simplistic terms, that's Kerberos.

> **ExamAlert**
>
> You should know that the security tokens used in Kerberos are known as *tickets*.

Local Authentication

Most of the time, the goal is to authenticate the user using a centralized authentication server or service of some type. When that cannot be done—such as when there is no Internet connectivity available—authentication is done locally by the operating system using values stored within it. In Windows, for example, the Local Authentication Subsystem (LASS) performs this function and allows users access to the system after their stored username and password variables match.

Lightweight Directory Access Protocol

Lightweight Directory Access Protocol (LDAP) provides a mechanism to access and query directory services systems. In the context of the Network+ exam, these directory services systems are most likely to be UNIX based or Microsoft Active Directory based. Although LDAP supports command-line queries executed directly against the directory database, most LDAP interactions are via utilities such as an authentication program (network logon) or locating a resource in the directory through a search utility. By default, LDAP traffic is unsecured. LDAP over SSL (LDAPS) is a method to secure LDAP by enabling communication over SSL/TLS.

> **ExamAlert**
>
> Know that LDAP, by default, uses port 389.

Using Certificates

A *public key infrastructure (PKI)* is a collection of software, standards, and policies combined to enable users from the Internet or other unsecured public networks to securely exchange data. PKI uses a public and private cryptographic key pair obtained and shared through a trusted authority. Services and components work together to develop the PKI. Some of the key components of a PKI include the following:

▶ **Certificates:** Certificates are electronic credentials that validate users, computers, or devices on the network. They are digitally signed statements that associate the credentials of a public key to the identity of the person, device, or service that holds the corresponding private key.

▶ **Certificate authorities (CAs):** CAs issue and manage certificates. They validate the identity of a network device or user requesting data. CAs can be either independent third parties, known as *public CAs*, or they can be organizations running their own certificate-issuing server software, known as *private CAs*. Whether public or private, the CAs' job is to issue certificates, verify the holder of a digital certificate, and ensure that holders of certificates are who they claim to be. CAs are trusted entities and an integral part of public key infrastructure (PKI).

▶ **Certificate templates:** These templates are used to customize certificates issued by a certificate server. This customization includes a set of rules and settings created on the CA and used for incoming certificate requests.

▶ **Certificate revocation list (CRL):** This list identifies certificates that were revoked before they reached the certificate expiration date. Certificates are often revoked because of security concerns, such as a compromised certificate.

> **Note**
>
> *Self-signed* certificates are certificates that are signed by their own private key, rather than by a trusted third-party certificate authority. They are often used in scenarios where the entity generating the certificate needs to secure communication but does not have access to or does not want to use a CA-signed certificate.

Checking for common certificate issues is essential for ensuring the security and reliability of communication over HTTPS. Three of the most common problems include SSL Certificate Not Trusted or Signed (when the certificate authority that issued the certificate is not recognized by the client, or if the certificate has been tampered with), Name Mismatch Error (when the hostname in the URL does not match the Common Name or Subject Alternative Name specified in the SSL certificate), and Expired Certificate (when the certificate is not renewed in a timely manner and should be replaced with a new one issued by the CA).

Identity and Access Management (IAM)

Identity and access management (IAM) is a framework of policies, processes, and technologies used to manage digital identities and control access to resources within an organization's IT infrastructure. IAM solutions enable organizations to ensure that only authorized individuals or entities have access to specific systems, applications, data, and services, while also managing the identities and permissions of users, employees, contractors, partners, and customers.

Key components of IAM include

▶ **Identity lifecycle management:** IAM solutions manage the entire lifecycle of digital identities, including provisioning (creating new accounts), authentication (verifying user identities), authorization (defining access permissions), and deprovisioning (disabling or deleting accounts).

▶ **Single sign-on (SSO):** IAM systems often include SSO functionality, allowing users to access multiple applications and services with a single set of credentials (username and password). SSO improves user experience, simplifies authentication, and reduces the need for users to remember multiple passwords.

ExamAlert

Remember that with single sign-on, a user can log in to multiple applications during a session after authenticating only once. Access to cloud-based applications has ushered in widespread use of SSO technology in large enterprises.

▶ **Access control and authorization:** IAM solutions enforce access control policies to determine who can access what resources and under what conditions. This includes defining roles, groups, and permissions that govern access to systems, applications, data, and services based on user attributes, roles, and business requirements.

▶ **Multifactor authentication (MFA):** IAM systems support MFA mechanisms to enhance security by requiring users to provide additional authentication factors beyond passwords, such as one-time passwords (OTP), biometric authentication (fingerprint or facial recognition), or hardware tokens.

▶ **Identity federation:** IAM enables identity federation, allowing organizations to establish trust relationships with external identity providers (IdPs) or identity federation protocols (such as SAML or OAuth) to enable seamless and secure authentication and authorization across organizational boundaries.

▶ **Identity governance and compliance:** IAM solutions provide tools for identity governance and compliance management, helping organizations enforce regulatory compliance, audit access controls, and monitor user activities to detect and mitigate security risks and violations.

▶ **Privileged access management (PAM):** IAM systems include PAM capabilities to manage and secure privileged accounts (such as administrator or root accounts) with elevated access privileges. PAM solutions enforce strict controls, monitoring, and auditing of privileged access to prevent misuse or abuse.

▶ **Identity analytics and risk assessment:** IAM platforms leverage identity analytics and risk assessment capabilities to analyze user behavior, detect anomalies, assess access risks, and identify potential security threats or insider risks in real time.

Security Assertion Markup Language (SAML)

Security Assertion Markup Language (SAML) is an XML-based open standard for exchanging authentication and authorization data between identity providers (IdPs) and service providers (SPs). SAML enables single sign-on (SSO) functionality, allowing users to authenticate once with an identity provider and then access multiple service providers without having to reenter credentials. When a user attempts to access a service provider, the SP redirects the user to the identity provider's authentication endpoint. The identity provider then authenticates the user through various methods (e.g., username/password, multifactor authentication) and generates a SAML assertion containing information about the user's identity and authentication status. The identity provider sends the SAML assertion back to the service provider, typically through the user's web browser, using a POST request. The service provider validates the SAML assertion, extracts the user's identity information, and grants access to the requested resource or application.

> **ExamAlert**
>
> Understand that identity and access management (IAM) provides a centralized framework for managing and controlling access to resources within an organization's IT environment. By implementing IAM solutions, organizations can enforce security policies, ensure compliance with regulatory requirements, and protect sensitive data from unauthorized access or breaches.

Multifactor Authentication Factors

Multifactor authentication (MFA) is defined as having two or more access methods included as part of the authentication process; for example, using both smartcards and passwords. An ATM card and PIN are another common example representing multifactor authentication (in this case, two-factor authentication), as opposed to just one (which constitutes single-factor authentication). The ATM card is something you have, and the PIN is something you know.

The factors used in authentication systems or methods are based on one or more of these five factors:

▶ **Something you know**, such as a password or PIN

▶ **Something you have**, such as a smartcard, token, or identification device

▶ **Something you are**, such as your fingerprints or retinal pattern (often called *biometrics*)

▶ **Somewhere you are** (based on geolocation)

▶ **Something you do**, such as an action you must take to complete authentication

> **ExamAlert**
>
> Be able to identify the five types of authentication factors used in multifactor authentication.

Often associated with MFA, *time-based authentication* is a method of authentication that relies on the current time as a factor for validating the identity of a user or entity attempting to access a system or resource. In time-based authentication systems, users are typically required to provide a time-based one-time password (TOTP) or similar credential that changes periodically based on a predefined algorithm and synchronized clock (achieved through time synchronization protocols such as the Network Time Protocol [NTP] or by using a trusted time source).

Auditing and Regulatory Compliance

Auditing is the process of monitoring occurrences and keeping a log of what has occurred on a system. A system administrator determines which events should be audited. Tracking events and attempts to access the system helps prevent unauthorized access and provides a record that administrators can analyze to make security changes as necessary. This record, or log, also provides administrators with solid evidence if they need to look into improper user conduct.

> **Caution**
>
> Be sure that you can identify the purpose of authentication, authorization, and accounting.

The first step in auditing is to identify what system events to monitor. After the system events are identified, in a Windows environment, the administrator can choose to monitor the success or failure of a system event. For instance, if "logon" is the event being audited, the administrator might choose to log all unsuccessful logon attempts, which might indicate that someone is attempting to gain unauthorized access. Conversely, the administrator can choose to audit all successful attempts to monitor when a particular user or user group is logging on. Some administrators prefer to log both events. However, overly ambitious audit policies can reduce overall system performance.

Regulatory compliance requirements often dictate the standards and practices that organizations must adhere to in order to protect their networks and sensitive data from unauthorized access, breaches, and misuse. Compliance with regulatory mandates is essential for maintaining the confidentiality, integrity, and availability of data, as well as for mitigating legal and financial risks associated with noncompliance.

Regulatory frameworks such as the General Data Protection Regulation (GDPR), *Health Insurance Portability and Accountability Act (HIPAA)*, Payment Card Industry Data Security Standard (PCI DSS), and others impose requirements for safeguarding sensitive data. These regulations mandate measures such as encryption, access controls, data masking, and secure transmission protocols to protect data both in transit and at rest within the network. One important component that must factor in is data locality—the requirement or practice of storing and processing data within specific geographic jurisdictions or regions to comply with data protection regulations, privacy laws, and contractual obligations.

> **Note**
>
> The General Data Protection Regulation (GDPR) strengthens and unifies data protection for individuals and companies within the European Union (EU). The Payment Card Industry Data Security Standard (PCI DSS) is designed to reduce fraud and protect customer credit card information. The Health Insurance Portability and Accountability Act (HIPAA) of 1996 sets national standards for protecting health information. The Gramm-Leach-Bliley Act (GLBA) establishes privacy rules for the financial industry. Sarbanes-Oxley (SOX) governs financial and accounting disclosure information.

Many countries and regions have laws and regulations governing the collection, storage, processing, and transfer of personal data. These regulations often require organizations to store and process data within the boundaries of specific jurisdictions, known as *data sovereignty* or *data residency requirements*. During network audits, organizations may be required to demonstrate compliance with data locality requirements by providing evidence that sensitive data is stored and processed within authorized geographic regions. This may involve documenting data storage locations, access controls, encryption measures, and data transfer mechanisms to ensure compliance with regulatory and contractual obligations. Failure to comply with data locality requirements can result in legal and financial consequences, including regulatory fines, legal sanctions, reputational damage, and loss of customer trust. Therefore, organizations conduct network audits to assess and mitigate risks associated with data residency and ensure compliance with applicable laws and regulations.

Compliance regulations often require organizations to implement security monitoring and incident response capabilities to detect and respond to security incidents in a timely manner. This includes continuous monitoring of network traffic, log data, and security events to identify anomalies, as well as the development of incident response plans and procedures for containing and mitigating security breaches. Organizations are often required to assess and manage the security risks posed by third-party vendors, suppliers, and service providers that have access to their networks or handle sensitive data. Compliance regulations may mandate due diligence assessments, contractually binding agreements, and security requirements for third parties to ensure the security of shared networks and data.

Additional Access Control Methods

Today, there are many ways to establish access into networks. You could fill an entire tome with a discussion of the possibilities. What follows are some of the more important ones to know for this exam (others appear on other CompTIA exams, such as Security+).

802.1X

The IEEE standard 802.1X defines port-based security for wireless network access control. As such, it offers a means of authentication and defines the Extensible Authentication Protocol (EAP) over IEEE 802, which is often known as EAP over LAN (EAPOL). The biggest benefit of using 802.1X is that the access points and the switches do not need to do the authentication but instead rely on the authentication server to do the work.

> **ExamAlert**
>
> Remember that IEEE 802.1X authentication allows only authorized devices to connect to the network. The most secure form of IEEE 802.1X authentication is certificate-based authentication.

Extensible Authentication Protocol (EAP)

Choosing the correct authentication protocol for remote clients is an important part of designing a secure remote-access strategy. After they are authenticated, users have access to the network and servers. *Extensible Authentication Protocol (EAP)* provides a framework for authentication that is often used with wireless networks. Among the EAP types adopted by the WPA/WPA2 standard are PEAP, EAP-FAST, and EAP-TLS. EAP was developed in response to an increasing demand for authentication methods that use other types of security devices, such as token cards, smartcards, and digital certificates.

To simplify network setup, a number of small office/home office (SOHO) routers use a series of EAP messages to allow new hosts to join the network and use WPA/WPA2. Known as Wi-Fi Protected Setup (WPS), this setup often requires the user to do something to complete the enrollment process: press a button on the router within a short time period, enter a PIN, or bring the new device close by (so that near-field communication can take place).

> **Note**
>
> WPA3 uses the Simultaneous Authentication of Equals (SAE) to replace WPA2's preshared key (PSK) exchange protocol.

Cisco, RSA, and Microsoft worked together to create Protected Extensible Authentication Protocol (PEAP). There is now native support for it in Windows (which previously favored EAP-TLS). Although many consider PEAP and EAP-TLS to be similar, PEAP is more secure because it establishes an encrypted channel between the server and the client.

EAP-FAST (Flexible Authentication via Secure Tunneling) was designed by Cisco to allow for the use of certificates to establish a TLS tunnel in which client credentials are verified.

To put EAP implementations in a chronological order, think EAP-TLS, EAP-FAST, and then PEAP. In between came EAP-TTLS, a form of EAP-TLS that adds tunneling (Extensible Authentication Protocol—Tunneled Transport Layer Security). EAP-TLS is an ideal solution, since it uses digital certificates as opposed to credentials, but it may not be supported with legacy systems or organizations that don't want to manage PKI. Of all the choices, PEAP is the one with more vendors than just Cisco and thus is currently favored for use today. EAP-FAST, while similar to PEAP, uses a Protected Access Credential (PAC) to set up a TLS tunnel.

> **Note**
>
> The number one reason a network administrator may use EAP-TLS, EAP-FAST, or PEAP is for secure authentication. These protocols provide robust authentication mechanisms that help ensure the confidentiality and integrity of data transmitted over the network, safeguarding against unauthorized access and potential security threats.

Network Access Control (NAC)

Network access control (NAC) is a method to restrict access to the network based on identity or posture (discussed later in this chapter). This method was created by Cisco to enforce privileges and make decisions on a client device based on information gathered from it (such as the vendor and version of the antivirus software running). If the wanted information is not found (such as that the antivirus definitions are a year old), the client can be placed in a *quarantine network* area to keep it from infecting the rest of the network. It can also be placed in a *guest network* and/or allowed to run *nonpersistent* (versus persistent) *agents*.

A *posture assessment* is any evaluation of a system's security based on settings and applications found. In addition to looking at such values as settings in the Registry or dates of files, NACs can also check *802.1X* values—the group of networking protocols associated with authentication of devices attempting to connect to the network. 802.1X works with EAP.

> **ExamAlert**
>
> As you prepare for the exam, be sure that you can identify posture assessment as any evaluation of a system's security based on settings and applications found.

MAC Filtering

Another name for a network card or network adapter is a *network controller*. Every controller has a unique MAC address associated with it. Filtering network traffic using a system's MAC address typically is done using an ACL. This list keeps track of all MAC addresses and is configured to allow or deny access to certain systems based on the list. As an example, look at the MAC ACL from a router. Figure 9.3 shows the MAC ACL screen. Specific MAC addresses can be either denied or accepted, depending on the configuration. It would be possible, for example, to configure it so that only the system with the MAC address of 02-00-54-55-4E-01 can authenticate to the router.

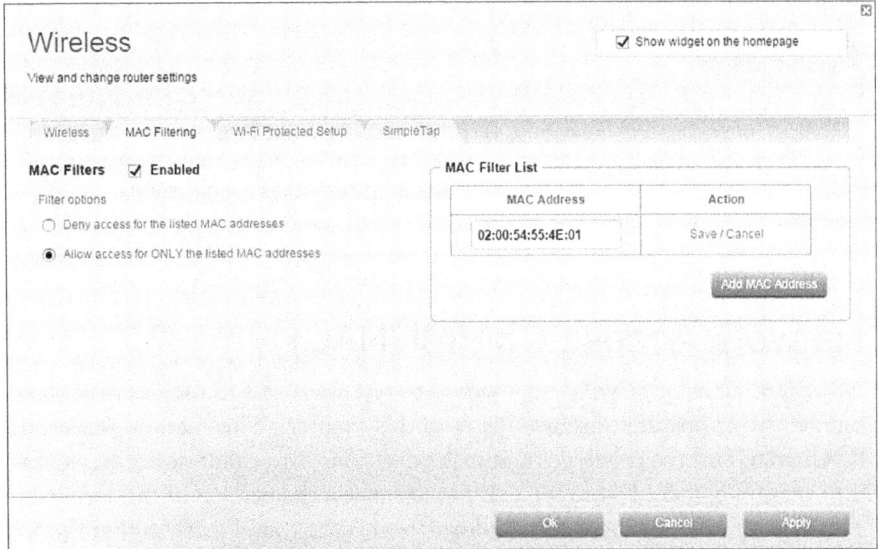

FIGURE 9.3 **A MAC ACL**

> **Note**
>
> When administrators are configuring security for wireless networks, filtering by MAC address is a common practice. Typically, in MAC filtering security, MAC addresses can be added to an "allow" ACL or "deny" ACL.

The MAC address can be used with a number of other specifications as well. For example, in DHCP, you can configure a reservation to ensure a device with a specific MAC address is always assigned the same IP address. Similarly, port security can be configured on a switch interface to limit and identify the MAC addresses of workstations that are allowed to connect to that specific port, and if a violation is detected, the switch can take a number of different actions, such as shutting down the port, sending an SNMP trap, or sending an email alert.

Geofencing

Geofencing can be used in network authentication to enhance security by restricting access to resources based on the geographic location of users or devices. It allows organizations to define virtual boundaries or geographical regions and enforce access control policies that permit or deny access to network resources based on the location of users or devices relative to these boundaries. Geofencing relies on geolocation data collected from users' devices, such as smartphones, tablets, or laptops, to determine their current geographic location. This data may be obtained through various sources, including GPS, Wi-Fi triangulation, cellular network signals, or IP address geolocation.

Organizations define access control policies based on geographic locations or regions using geofencing rules configured in the network authentication system. These policies specify which resources are accessible or restricted based on the geographic coordinates or boundaries defined for each resource. When users attempt to access network resources, they are required to authenticate using their credentials, such as usernames and passwords, biometric data, or time-based tokens (e.g., TOTP). Additionally, the authentication system may collect geolocation data from users' devices to verify their location.

The network authentication system compares a user's geolocation data with the predefined geofencing rules to determine whether access should be granted or denied. If the user's location falls within the allowed geographic boundaries specified in the policy, access is granted. Otherwise, access is denied, and the user may be prompted to provide additional authentication factors or receive a notification informing of the restriction.

Geofencing can be combined with other authentication factors, such as device posture assessment, user behavior analytics, or time-based restrictions, to implement adaptive access controls. These controls dynamically adjust access privileges based on contextual factors, including the user's location, device characteristics, and activity patterns.

Access attempts and geolocation data are logged and audited to track user activities, enforce compliance with security policies, and investigate security incidents or policy violations. Detailed logs provide visibility into access patterns, geographic locations, and potential anomalies for security analysis and monitoring.

Physical Security

Having great network security does little good if everything can be compromised by someone walking in your office, picking up your server, and walking out the front door with it. Physical security of the premises is equally important to an overall security implementation.

Ideally, your systems should have a minimum of three physical barriers:

▶ The external entrance to the building, referred to as a *perimeter*, which is protected by one or more *detection methods*. Detection methods usually involve a *camera* and *motion detection*. The protection of the perimeter can be accomplished with burglar alarms, external walls, fencing, surveillance, and so on. This type of protection should be used with an access list, which should exist to specifically identify who can enter a facility and can be verified by a security guard or someone in authority. An *access control vestibule* (previously known as a *mantrap*) can be used to limit access to only one or two people going into the facility at a time. A properly developed vestibule includes bulletproof glass, high-strength doors, and locks. In high-security and military environments, an armed guard, as well as video surveillance (IP cameras and CCTVs), should be used at the access control vestibule. After a person is inside the facility, additional security and authentication may be required for further entrance.

▶ A locked door with door access controls protecting the computer center and network closets. You should also rely on such *access control hardware* as radio frequency identification (RFID) badge readers (also known as *ID badge readers, proximity readers*), key fobs, or keys to gain access. *Biometrics*, such as fingerprint or retinal scans, can be used for authentication.

▶ The entrance to the computer room itself. This entrance should be another locked door that is carefully monitored and protected by keypads and cipher locks. Although you try to keep as many intruders out with the other two barriers, many who enter the building could be posing as people they are not—heating and air-conditioning technicians, representatives of the landlord, and so on. Although these pretenses can get them past the first two barriers, they should still be stopped by the locked computer room door. If they do manage to gain entry, *locking racks*, *locking cabinets*, and even *smart lockers* (secure storage and distribution systems with computer and sensors built in) should be used to protect hardware and keep it from being absconded.

▶ Assets should have *asset tags* (also known as *tracking tags*) or *tamper tags* attached to them that have unique identifiers for each client device in your environment (usually just incrementing numbers corresponding to values in a database) to help you identify and manage your IT assets. Additionally, tamper detection devices should be installed to protect against unauthorized chassis cover and component removal.

The objective of any physical barrier is to prevent access to computers and network systems. The most effective physical barrier implementations require that more than one physical barrier be crossed to gain access. This type of approach is called a *multiple barrier system*.

One step to never overlook in the security equation is that of *employee training*. All employees should be trained to look out for social engineering techniques that could be tried on them, the importance of the assets (data and hardware), and that security—also associated with job security—is everyone's responsibility.

Risk Management

Risk management involves recognizing and acknowledging that risks exist and then determining what to do about them. Sometimes, the best solution is to do nothing. If, for example, there is a risk that some data could be lost in the event of a fire but the value placed on that data is below the cost of protecting it from that harm, the economically feasible solution could be to accept the risk and do nothing further. In most cases, acceptance alone is not enough and two likely approaches are mitigation (trying to minimize the risk) and/or transference (shifting a part of the risk to another party such as an insurance provider).

> **Note**
>
> Purchasing cybersecurity insurance serves as a compensating security control by mitigating financial losses associated with security incidents. While it doesn't directly prevent or mitigate cyber threats, it provides financial coverage for damages, helping organizations recover from incidents such as data breaches or cyberattacks.

To determine how to manage risk, assessments are usually employed. An assessment can be done on threats, vulnerabilities, or business operations (process assessment and vendor assessment).

Penetration Testing

It is becoming more common for companies to hire penetration testers to test their system's defenses. Essentially, a penetration tester will use the same techniques that a hacker would use to find any flaws in a system's security. These flaws may be discovered by means other than directly accessing the system, such as collecting information from public databases, talking to employees/ partners, dumpster diving, and social engineering. This approach is known as *passive reconnaissance*. In contrast, *active reconnaissance* directly focuses on the system (port scans, traceroute information, network mapping, and so forth) to identify weaknesses that could be used to launch an attack.

When you're doing penetration testing, it is important to have a scope document outlining the extent of the testing that is to be done. It is equally important to have permission from an administrator who can authorize such testing—in writing—enabling the testing to be conducted.

One weakness a good penetration test looks for is escalation of privilege—that is, a hole created when code is executed with higher privileges than those of the user running it. By breaking out of the executing code, the users are left with higher privileges than they should have.

Three types of penetration testing are black box/unknown environment (the tester has absolutely no knowledge of the system and is functioning in the same manner as an outside attacker), white box/known environment (the tester has significant knowledge of the system, which simulates an attack from an insider—a rogue employee), and gray box/partially known environment (a middle ground between the first two types of testing. In gray box testing, the tester has some limited knowledge of the target system). While the "box" analogy is used to describe the environment, "hats" have often been used to describe the person doing the testing, but that terminology is now being replaced by level of authorization. Therefore, a white hat tester is equivalent to an authorized tester, a black hat tester to an unauthorized tester, and a gray hat tester to a semi-authorized tester.

> **Note**
>
> With so many security-related topics now appearing on the Network+ exam, you have a good head start on Security+ certification study after you successfully finish taking this exam.

Security Information and Event Management

Security information and event management (SIEM) products provide notifications and real-time analysis of security alerts and can help you head off problems quickly. Chapter 8, "Network Operations," discussed event management and looking at log files, including the security log.

SIEM tools collect, correlate, and display data feeds that support response activities. A SIEM can take individual benign events and tie them together to reveal more details about what occurred and why it's of concern.

SIEM dashboards are an important element in SIEM solutions, presenting data (and analysis) in actionable format. Most SIEM packages allow organizations to customize their dashboard(s) based on their needs and requirements. One need of any SIEM is to be able to interpret (and visualize) data coming from different sources and, thus, different formats. Normalizing logs involves extracting and processing entries to put them into a readable and structured format that can be displayed and acted upon.

Cram Quiz

1. Which of the following ports is used by TACACS+ by default?

   ○ **A.** 49

   ○ **B.** 51

   ○ **C.** 53

   ○ **D.** 59

2. Which of the following is the main authentication protocol used with Windows servers?

   ○ **A.** LDAP

   ○ **B.** Kerberos

○ **C.** L2TP

○ **D.** TFTP

3. What are the security tokens used with Kerberos known as?

○ **A.** Coins

○ **B.** Vouchers

○ **C.** Tickets

○ **D.** Gestures

4. Which of the following is NOT one of the factors associated with multifactor authentication?

○ **A.** Something you like

○ **B.** Something you are

○ **C.** Something you do

○ **D.** Something you have

5. Which of the following is one step up from the honeypot?

○ **A.** Geofence

○ **B.** VLAN

○ **C.** DMZ

○ **D.** Honeynet

6. Which of the following is based on the premise that implementing security at different levels or layers to form a complete security strategy provides better protection and greater resiliency than implementing an individual security defense?

○ **A.** Screened subnet

○ **B.** Separation of duties

○ **C.** EAP

○ **D.** Defense in depth

7. What is a list of publicly available security flaws that you should be familiar with for security otherwise known as?

○ **A.** CVEs

○ **B.** MAC filter

○ **C.** SIEM

○ **D.** CRL

8. What type of control is present if access decisions are determined by the roles that individual users have within the organization?

   ○ **A.** Passive reconnaissance

   ○ **B.** RBAC

   ○ **C.** Active reconnaissance

   ○ **D.** SSO

9. What is it called when a hole is found in a web browser or other software, and miscreants begin exploiting it the very day it is discovered by the developer?

   ○ **A.** Zero trust

   ○ **B.** Perimeter network

   ○ **C.** Screened subnet

   ○ **D.** Zero-day attack

10. You go to the store and put your bank card into an ATM and enter your PIN. What examples of multifactor authentication (MFA) are you using? (Choose two.)

    ○ **A.** Something you are

    ○ **B.** Something you have

    ○ **C.** Something you know

    ○ **D.** Somewhere you are

11. Which of the following strengthens and unifies data protection for individuals and companies within the European Union (EU)?

    ○ **A.** PCI DSS

    ○ **B.** HIPAA

    ○ **C.** GLBA

    ○ **D.** GDPR

Cram Quiz Answers

1. **A.** TACACS+ uses TCP port 49 by default.

2. **B.** Kerberos is the main authentication protocol used with Windows servers.

3. **C.** The tokens used for security in Kerberos are known as tickets.

4. **A.** Something you like is not one of the five factors used in multifactor authentication. The five legitimate possibilities are the following: something you know, something you have, something you are, somewhere you are, and something you do.

5. **D.** One step up from the honeypot is the honeynet.

6. **D.** Defense in depth is based on the premise that implementing security at different levels or layers to form a complete security strategy provides better protection and greater resiliency than implementing an individual security defense.

7. **A.** Common vulnerabilities and exposures (CVEs) are a list of publicly available security flaws that you should be familiar with for security.

8. **B.** In role-based access control (RBAC), access decisions are determined by the roles that individual users have within the organization. Role-based access requires the administrator to have a thorough understanding of how a particular organization operates, the number of users, and each user's exact function in that organization.

9. **D.** When a hole is found in a web browser or other software, and miscreants begin exploiting it the very day it is discovered by the developer (bypassing the one- to two-day response time many software providers need to put out a patch after the hole has been found), it is known as an exploit attack *or* zero-day attack. Zero trust is considered a more proactive approach than defense in depth because it reduces the attack surface and limits access to sensitive data. An important firewall-related concept is that of the screened subnet. This concept was formerly known as a demilitarized zone (DMZ) and sometimes called a perimeter network.

10. **B, C.** Automated teller machines (ATMs) use a common example of a multifactor authentication (MFA) system, requiring both a "something you have" physical key (your ATM card) and a "something you know" personal identification number (PIN).

11. **D.** The General Data Protection Regulation (GDPR) strengthens and unifies data protection for individuals and companies within the European Union (EU). The Payment Card Industry Data Security Standard (PCI DSS) is designed to reduce fraud and protect customer credit card information. The Health Insurance Portability and Accountability Act (HIPAA) of 1996 sets national standards for protecting health information. The Gramm-Leach-Bliley Act (GLBA) establishes privacy rules for the financial industry.

Common Networking Attacks

▶ **4.2 Summarize various types of attacks and their impact to the network.**

Note

Insider threat actors can be malicious, as in the case of a disgruntled employee, or simply careless. In many cases, this includes employees who have the right intentions but either are unaware of an organization's security policy or simply ignore it.

Malware encompasses many types of malicious software and all were intended by their developer to have an adverse effect on the network. The following sections look at some of the more malevolent types.

Denial-of-Service and Distributed Denial-of-Service Attacks

Denial-of-service (DoS) attacks are designed to tie up network bandwidth and resources and eventually bring the entire network to a halt. This type of attack is done simply by flooding a network with more traffic than it can handle. When more than one computer is used in the attack, it is technically known as a *distributed DoS (DDoS)* attack. These attacks typically use a *botnet* to launch a *command and control attack* through a traffic spike. Almost every attack of this type today is DDoS, and we will use DoS to mean both.

A DoS attack is not designed to steal data but rather to cripple a network and, in doing so, cost a company huge amounts of money. It is possible, in fact, for the attack to be an unintentional, or *friendly*, DoS attack coming from the inside. Just as with friendly fire in combat, it matters not that the attack was unintentional; it does just as much damage.

The effects of DoS attacks include the following:

▶ Saturating network resources, which then renders those services unusable

▶ Flooding the network media, preventing communication between computers on the network

▶ Causing user downtime because of an inability to access required services

▶ Causing potentially huge financial losses for an organization because of network and service downtime

Types of DoS Attacks

Several types of DoS attacks exist, and each seems to target a different area. For instance, they might target bandwidth, network service, memory, CPU, or hard drive space. When a server or other system is overrun by malicious requests, one or more of these core resources break down, causing the system to crash or stop responding. A *permanent DoS attack* continues for more than a short period of time and requires you to change your routing, IP addresses, or other configurations to get around it.

Fraggle

In a *Fraggle attack*, spoofed UDP packets are sent to a network's broadcast address. These packets are directed to specific ports, such as port 7 or port 19, and, after they are connected, can flood the system.

Smurfing

The *Smurf attack* is similar to a Fraggle attack. However, a ping request is sent to a broadcast network address, with the sending address spoofed so that many ping replies overload the victim and prevent it from processing the replies.

Ping of Death

In a *ping of death attack*, an oversized *Internet Control Message Protocol (ICMP)* datagram is used to crash IP devices that were manufactured before 1996.

SYN Flood

In a typical TCP session, communication between two computers is initially established by a three-way handshake, referred to as a SYN, SYN/ACK, ACK. At the start of a session, the client sends a SYN message to the server. The server acknowledges the request by sending a SYN/ACK message back to the client. The connection is established when the client responds with an ACK message.

In a *SYN attack*, the victim is overwhelmed with a flood of SYN packets. Every SYN packet forces the targeted server to produce a SYN/ACK response and then wait for the ACK acknowledgment. However, the attacker doesn't respond with an ACK, or spoofs its destination IP address with a nonexistent address so that no ACK response occurs. The result is that the server begins filling up with half-open connections. When all the server's available resources are tied up on half-open connections, it stops acknowledging new incoming SYN requests, including legitimate ones.

Buffer Overflow

A *buffer overflow* is a type of DoS attack that occurs when more data is put into a buffer (typically a memory buffer) than it can hold, thereby overflowing it (as the name implies). Buffer overflows can also be a result of improper input validation. The most common result of improper input validation is buffer overflow exploitation.

Distributed Reflective DoS

A *distributed reflective DoS (DRDoS)* attack is also called an *amplification attack*, and it targets public UDP servers. Two of the most common protocols/servers that a DRDoS attack usually goes after are Domain Name Service (DNS) and Network Time Protocol (NTP) servers, but Simple Network Management Protocol (SNMP), NetBIOS, and other User Datagram Protocols (UDPs) are also susceptible.

> **ExamAlert**
>
> As you study for the exam, be sure to focus on the difference between DoS and DDoS and know that a botnet is formed by many bots (infected systems) and controlled by a command and control server to carry out a DDoS attack.

ICMP Flood

An *ICMP flood*, also known as a *ping flood*, is a DoS attack in which large numbers of ICMP messages are sent to a computer system to overwhelm it. The result is a failure of the TCP/IP protocol stack, which cannot tend to other TCP/IP requests.

Other Common Attacks

This section details some of the more common attacks used today.

> **ExamAlert**
>
> Know all the common types of attacks detailed throughout this section. The Network+ exam is extremely likely to ensure your knowledge here is what an administrator should know.

Social Engineering

Social engineering is a common form of cracking. It can be used by both outsiders and people within an organization. *Social engineering* is a hacker term for tricking people into revealing their password or some form of security information. It might include trying to get users to send passwords or other information over email, following someone closely into a secured area (known as *tailgating*), walking in with them (known as *piggybacking*), looking over someone's shoulder at their screen (known as *shoulder surfing*), finding sensitive data—such as passwords—in the trash (known as *dumpster diving*), or any other method that tricks users into divulging information. Social engineering is an attack that attempts to take advantage of human behavior.

Table 9.2 highlights the different methods of social engineering you should know for the exam and mitigation strategies for each.

TABLE 9.2 **Social Engineering Tactics**

	Phishing	Dumpster Diving	Shoulder Surfing	Tailgating
Description	Involves the use of deceptive emails, messages, or websites to trick individuals into divulging sensitive information or performing certain actions	Involves physically rummaging through trash or recycling bins to find discarded documents or materials containing sensitive information	Involves covertly observing individuals as they enter passwords, PINs, or other sensitive information on electronic devices or physical keypads	Involves gaining unauthorized physical access to a restricted area by closely following an authorized individual through a secure entry point
Characteristics	Typically involves impersonation of legitimate entities, urgent or enticing messages, and malicious links or attachments	Relies on lax disposal practices, such as not shredding documents or failing to secure dumpsters or bins	Often requires close proximity to the target, such as standing behind or beside them in crowded areas	Exploits the natural tendency to be polite or helpful, often relying on the goodwill of authorized individuals
Risks	Risk of credential theft, data breaches, financial fraud, or malware infection	Risk of identity theft, unauthorized access to confidential information, or compromise of corporate secrets	Risk of unauthorized access to personal or confidential information, compromise of security credentials	Risk of unauthorized access to physical facilities, theft of sensitive equipment or information, or compromise of security measure
Mitigation Strategies	Employee training on recognizing phishing attempts, use of email filtering and spam detection tools, implementation of multifactor authentication (MFA)	Implementation of secure disposal procedures, including shredding sensitive documents and acquiring a certificate of destruction from a licensed disposal/shredding company and implementing secure waste disposal bins	Encouraging awareness of surroundings, using privacy screens/filters or shields on electronic devices, and employing techniques like "covering" or "shielding" when entering sensitive information	Implementing access control measures such as access control vestibules (formerly called mantraps), badge authentication, plus enforcing strict visitor policies, and promoting a culture of security awareness and vigilance

MAC Flooding

Media Access Control (MAC) flooding is an attack that exploits the limitations of switches in Ethernet networks. In Ethernet networks, each device connected to the network has a unique MAC address assigned to its NIC, and switches use the MAC addresses to forward data frames to the appropriate destination devices. During a MAC flooding attack, an attacker floods the switch's MAC address table with a large number of fake or random MAC addresses. This flood of MAC addresses overwhelms the switch's capacity to store MAC address mappings, and it enters a fail-open state, where it treats all incoming frames as broadcast frames and forwards them out to all ports, rather than selectively forwarding them based on MAC address mappings. This behavior effectively turns the switch into a hub, causing network congestion, broadcast storms, and potential DoS conditions.

> **ExamAlert**
>
> Remember that MAC flooding is an attack that targets switches on a local-area network (LAN). It involves sending multiple packets with fake MAC addresses to overflow the switch's address table. This causes the buffer to overflow, making the switch unable to process legitimate traffic.

Logic Bomb

Software programs or code snippets that execute when a certain predefined event occurs are known as *logic bombs*. Such a bomb may send a note to an attacker when a user is logged on to the Internet and is accessing a certain application, for example. This message could inform the attacker that the user is ready for an attack and then open a backdoor. Similarly, a programmer could create a program that always makes sure their name appears on the payroll roster; if it doesn't, then key files begin to be erased. Any code that is hidden within an application and causes something unexpected to happen constitutes a logic bomb.

Rogue DHCP

A *rogue DHCP* server added to a network has the potential to issue an address to a client, isolating it on an unauthorized network where its data can be captured. Although it sounds like a bad thing, *DHCP snooping* is just the opposite. It is the capability for a switch to look at packets and drop DHCP traffic that it determines to be unacceptable based on the defined rules. The purpose is to prevent rogue DHCP servers from offering IP addresses to DHCP clients.

Rogue Access Points and Evil Twins

A *rogue access point* is a wireless access point that has been placed on a network without the administrator's knowledge. The result is that it is possible to remotely access the rogue access point because it likely does not adhere to company security policies. So, all security can be compromised by a cheap wireless router placed on the corporate network. An *evil twin attack* is one in which a rogue wireless access point poses as a legitimate wireless service provider to intercept information that users transmit.

> **Note**
>
> Rogue access points can also serve as a type of on-path (formerly known as man-in-the-middle) attack. Regular wireless site surveys should be performed to seek out or reveal and remove rogue access points or evil twins.

Advertising Wireless Weaknesses

Attacks that advertise wireless weaknesses start with *war driving*—driving around with a mobile device looking for open wireless access points with which to communicate and looking for weak implementations that can be cracked (called *WEP cracking* or *WPA cracking*). They then lead to *war chalking*—those who discover a way in to the network leave signals (often written in chalk) on, or outside, the premise to notify others that the vulnerability is there. The marks can be on the sidewalk, the side of the building, a nearby signpost, and so on.

Phishing

Often users receive a variety of emails offering products, services, information, or opportunities. Unsolicited email of this type is called *phishing* (pronounced "fishing"). This technique involves a bogus offer sent to hundreds of thousands or even millions of email addresses. The strategy plays the odds. For every 1,000 emails sent, perhaps one person replies. Phishing can be dangerous because users can be tricked into divulging personal information such as credit card numbers or bank account information. Today, phishing is performed in several ways. Phishing websites and phone calls are also designed to steal money or personal information.

> **Note**
>
> The best way to prevent phishing, and other malicious email attacks, is through user education, training, and adherence to company policy.

Ransomware

With *ransomware*, software—often delivered through a Trojan horse—takes control of a system and demands that a third party be paid. The "control" can be accomplished by encrypting the hard drive, by changing user password information, or via any of a number of other creative ways. Users are usually assured that by paying the extortion amount (the ransom), they will be given the code needed to revert their systems to normal operations.

> **Note**
>
> Trojan horses appear as helpful or harmless programs but, when installed, carry and deliver a malicious payload.

Spoofing

Spoofing is a technique in which the real source of a transmission, file, or email is concealed or replaced with a fake source. This technique enables an attacker, for example, to misrepresent the original source of a file available for download. Then the attacker can trick users into accepting a file from an untrusted source, believing it is coming from a trusted source. *MAC spoofing* is the act of faking the MAC address of a machine—faking its physical identity. While the real MAC address cannot be changed, it is possible for drivers to give values provided to them and fool a system into believing that the NIC it is talking to has the MAC address of a recognized/authorized host. *IP spoofing* does a similar action with the IP address rather than the MAC address.

ARP Spoofing

With *ARP spoofing*, the Media Access Control (MAC) address of the data is faked. When this value is faked, it is possible to make it look as if the data came from a network that it did not. This technique can be used to gain access to the network, to fool the router into sending data here that was intended for another host, or to launch a DoS attack. In all cases, the address being faked is an address of a legitimate user, and that makes it possible to get around such measures as allow/deny lists.

ARP Poisoning

Address Resolution Protocol (ARP) poisoning is a subset of ARP spoofing that tries to convince the network that the attacker's MAC address is the one associated with an IP address so that traffic sent to that IP address is wrongly sent to the attacker's machine.

ExamAlert

Know that ARP spoofing is a technique where an attacker sends falsified Address Resolution Protocol (ARP) messages over a network that contain incorrect or spoofed MAC address-to-IP address mappings, tricking other devices on the network into associating the attacker's MAC address with the IP address of a legitimate device. ARP poisoning is a specific form of ARP spoofing where the attacker continuously sends falsified ARP messages to update the ARP cache (also known as the ARP table) of targeted devices on the network with the goal of corrupting the ARP cache of one or more devices on the network, causing them to associate the attacker's MAC address with the IP address of a legitimate device.

DNS Poisoning and Spoofing

With *DNS poisoning*, the DNS server is given information about a name server that it thinks is legitimate when it isn't. This type of attack can send users to a website other than the one to which they wanted to go, reroute mail, or do any other type of redirection wherein data from a DNS server is used to determine a destination. This is a subset of *DNS spoofing*: a broader term that encompasses various techniques used to forge or manipulate DNS responses to redirect network traffic. DNS spoofing attacks can be conducted using techniques such as on-path attacks (formerly known as man-in-the-middle [MITM] attacks), DNS cache poisoning, DNS hijacking, DNS tunneling, or DNS rebinding.

ExamAlert

DNS poisoning specifically refers to the injection of false information into the DNS cache to corrupt DNS resolution. *DNS spoofing* is a broader term that encompasses various techniques used to manipulate DNS responses and redirect network traffic. Both techniques exploit vulnerabilities in the DNS infrastructure to deceive users, redirect traffic to malicious destinations, and facilitate unauthorized access to sensitive information.

Deauthentication

Deauthentication is also known as a *disassociation attack*. With this type of attack, the intruder sends a frame to the AP with a spoofed address to make it look as if it came from the victim but disconnects the user from the network. Because the victim is unable to keep a connection with the AP, this attack increases the chances of the victim choosing to use another AP—a rogue one or one in a hotel or other venue that requires extra pay to use. The Federal Trade Commission has filed suits against a number of hotels for launching attacks of this type and generating revenue by requiring their guests to pay for "premium" services rather than being able to use the free Wi-Fi.

Brute Force

There are a number of ways to ascertain a password, but one of the most common is a *brute-force attack* in which one value after another is guessed until the right value is found. Although that could take forever if done manually, software programs can take lots of values in a remarkably short period of time. Since they systematically attempt to guess passwords or encryption keys by trying every possible combination until the correct one is found, they can be highly effective with weak or commonly used passwords. When attackers gain access, they can then compromise sensitive data, systems, or accounts, leading to unauthorized access, data breaches, financial losses, and reputational damage.

On-Path Attack

In an *on-path attack* (previously known as a *man-in-the-middle attack*), intruders place themselves between the sending and receiving devices and capture the communication as it passes by. The interception of the data is invisible to those sending and receiving the data. The intruders can capture the network data and manipulate it, change it, examine it, and then send it on. Wireless communications are particularly susceptible to this type of attack. A rogue access point, a wireless AP that has been installed without permission, is an example of an on-path attack. If the attack is done with FTP (using the **port** command), it is known as an *FTP bounce* attack.

VLAN Hopping

VLAN hopping, as the name implies, is an exploit of resources on a virtual LAN that is made possible because the resources exist on said virtual LAN. There is more than one method by which this occurs, but in all of them, the result is the same: an attacking host on a VLAN gains access to resources on other VLANs that are not supposed to be accessible to them (becoming a compromised system). Again, regardless of the method employed, the solution is to properly configure the switches to keep this from happening.

> **ExamAlert**
>
> A compromised system is any that has been adversely impacted (intentionally or unintentionally) by an untrusted source. The compromise can relate to confidentiality, integrity, or availability (CIA).

Vulnerabilities and Prevention

The threat from malicious code is a real concern. You need to take precautions to protect your systems. Although you might not eliminate the threat, you can significantly reduce it.

One of the primary tools used in the fight against malicious software is antivirus/antimalware software. Antimalware software (which includes antivirus software and more) is available from a number of companies, and each offers similar features and capabilities. You can find solutions that are host based, cloud/server based, or network based. Common features and characteristics of antivirus software are as follows:

- ▶ **Real-time protection:** An installed antivirus program should continuously monitor the system looking for viruses. If a program is downloaded, an application opened, or a suspicious email received, the real-time virus monitor detects and removes the threat. The virus application sits in the background, largely unnoticed by the user.

- ▶ **Virus scanning:** An antivirus program must scan selected drives and disks, either locally or remotely. You can manually run scanning or schedule it to run at a particular time.

- ▶ **Scheduling:** It is a best practice to schedule virus scanning to occur automatically at a predetermined time. In a network environment, this time typically is off hours, when the overhead of the scanning process won't impact users.

- ▶ **Live updates:** New viruses and malicious software are released with alarming frequency. It is recommended that the antivirus software be configured to regularly receive virus updates.

- ▶ **Email vetting:** Emails represent one of the primary sources of virus delivery. It is essential to use antivirus software that provides email scanning for both inbound and outbound email.

▶ **Centralized management:** If antivirus software or antimalware is used in a network environment, it is a good idea to use software that supports managing the virus program from the server. Virus updates and configurations need to be made only on the server, not on each client station.

Managing the threat from viruses is considered a proactive measure, with antivirus software only part of the solution. A complete security protection strategy requires many aspects to help limit the risk of malware, viruses, and other threats:

▶ **Develop in-house policies and rules:** In a corporate environment or even a small office, you need to establish what information can be placed on a system. For example, should users download programs from the Internet? Can users bring in their own storage media, such as USB flash drives? Is there a corporate BYOD policy restricting the use of personal smartphones and other mobile devices?

▶ **Monitor virus threats:** With new viruses coming out all the time, you need to check whether new viruses have been released and what they are designed to do. Network administrators should keep up with CVEs (see https://cve.mitre.org/), and developers should keep up with secure coding practices and threats.

▶ **Educate users:** One of the keys to a complete antivirus solution is to train users in virus prevention and recognition techniques. If users know what to look for, they can prevent a virus from entering the system or network. Back up copies of important documents. Keep in mind that no solution is absolute, so care should be taken to ensure that the data is backed up. In the event of a malicious attack, redundant information is available in a secure location. Educate users not to open email attachments from unknown senders.

▶ **Automate virus scanning and updates:** You can configure today's antivirus software to automatically scan and update itself. Because such tasks can be forgotten and overlooked, it is recommended that you have these processes scheduled to run at predetermined times.

▶ **Don't run unnecessary services:** Know every service that is running on your network and the reason for it. If you can avoid running the service, equate it with putting another lock on the door and do so. For example, to turn off services in Windows, open the Services app (click the Start menu and type **services.msc** into the search field), find the service you want to disable, double-click, and click **Stop**.

▶ **Keep track of open ports:** Just as you don't want unnecessary services running, you don't want unnecessary ports left open. Every one of them is an unguarded door through which a miscreant can enter. Both the **netstat** and **nmap** commands can be used to scan for open ports, and both are discussed in Chapter 10, "Network Troubleshooting."

> **Note**
>
> Run the command **netstat -ab** in an elevated command prompt, PowerShell, or Terminal window to see a list of applications and their associated ports.

▶ **Avoid unencrypted channels and cleartext credentials:** The days when these were acceptable have passed. Continuing to use them is tantamount to inviting an attack.

▶ **Shun unsecure protocols:** Once upon a time, cleartext passwords were okay to use because risks of anyone getting on your network who should not were minimal. During those days, it was okay to use unsecure protocols as well because the priority was on ease of use as opposed to data protection. These practices went out of acceptance decades ago, so you must—in the interest of security—be careful with the following unsecure protocols: Telnet, HTTP, SLIP, FTP, TFTP, and SNMP (v1 and v2). Secure alternatives—offering the same functionality but adding acceptable levels of security—are available for each. Where possible, opt instead for SSH, SNMPv3, TLS/SSL, SFTP, HTTPS, and IPsec.

▶ **Install patches and updates:** All applications—including productivity software, virus checkers, and especially the operating system—release patches and updates often, designed to address potential security weaknesses. Administrators must keep an eye out for these patches and install them when they are released. Pay particular attention to unpatched/legacy systems and keep them as secure as possible. And keep device firmware updated as well.

One of the best tools to use when dealing with problems is knowledge. In several locations, CompTIA stresses that user education is important, but even more important is that administrators know what is going on and keep learning from what is going on now and what has gone on in the past. As an example, all administrators should be familiar with *TEMPEST*, which is the name of a project that the U.S. government commenced in the late 1950s. TEMPEST was concerned with reducing electronic noise from devices that would divulge intelligence about systems and information. This program has become a standard for computer systems certification. *TEMPEST shielding protection* means that a computer system doesn't emit any significant amounts of electromagnetic interference (EMI) or radio frequency interference (RFI), known as

RF emanation. For a device to be approved as a TEMPEST, it must undergo extensive testing, done to exacting standards that the U.S. government dictates. Today, control zones and white noise are used to accomplish the shielding. TEMPEST-certified equipment often costs twice as much as non-TEMPEST equipment.

> **ExamAlert**
>
> Know some of the ways to prevent networking attacks and mitigate vulnerabilities.

Cram Quiz

1. Which of the following is an attack in which a rogue wireless access point poses as a legitimate wireless service provider to intercept information that users transmit?

   - ○ **A.** Zero day
   - ○ **B.** Phishing
   - ○ **C.** Evil twin
   - ○ **D.** Social engineering

2. Which of the following is a type of DoS attack that occurs when more data is put into a buffer than it can hold?

   - ○ **A.** Dictionary attack
   - ○ **B.** Buffer overflow
   - ○ **C.** Worm
   - ○ **D.** Trojan horse

3. Which of the following is an attack in which something that appears as a helpful or harmless program carries and delivers a malicious payload?

   - ○ **A.** Worm
   - ○ **B.** Phish
   - ○ **C.** Evil twin
   - ○ **D.** Trojan horse

4. Which of the following is an attack in which users are tricked into revealing their passwords or some form of security information?

   - ○ **A.** Bluesnarfing
   - ○ **B.** Phishing
   - ○ **C.** Evil twin
   - ○ **D.** Social engineering

5. Which of the following protocols are considered insecure? (Choose all that apply.)

    ○ **A.** Telnet

    ○ **B.** HTTP

    ○ **C.** FTP

    ○ **D.** SNMPv1

6. Which of the following attacks has occurred when malicious intruders place themselves between the sending and receiving devices and capture the communication as it passes by?

    ○ **A.** Brute-force attack

    ○ **B.** Phishing

    ○ **C.** VLAN hopping

    ○ **D.** On-path attack

Cram Quiz Answers

1. **C.** An evil twin attack is one in which a rogue wireless access point poses as a legitimate wireless service provider to intercept information users transmit.

2. **B.** A buffer overflow is a type of DoS attack that occurs when more data is put into a buffer than it can hold.

3. **D.** Trojan horses appear as helpful or harmless programs but, when installed, carry and deliver a malicious payload.

4. **D.** Social engineering is a way of tricking people (users) into revealing their passwords or some form of security information.

5. **A, B, C, and D.** Telnet, HTTP, FTP, and SNMPv1 are considered insecure and should not be used. Secure alternatives—offering the same functionality but adding acceptable levels of security—are available for each. Where possible, opt instead for SSH, HTTPS, SFTP, and SNMPv3.

6. **D.** In an on-path attack, intruders place themselves between the sending and receiving devices and capture the communication as it passes by. The interception of the data is invisible to those sending and receiving the data. The intruders can capture the network data and manipulate it, change it, examine it, and then send it on. There are a number of ways to ascertain a password, but one of the most common is a brute-force attack in which one value after another is guessed until the right value is found. Often users receive a variety of emails offering products, services, information, or opportunities. Unsolicited email of this type is called *phishing*. With VLAN, an attacking host on a VLAN gains access to resources on other VLANs that are not supposed to be accessible to them (becoming a compromised system). Again, regardless of the method employed, the solution is to properly configure the switches to keep this from happening.

Applying Network Security

▶ **4.3 Given a scenario, apply network security features, defense techniques, and solutions.**

CramSaver

If you can correctly answer these questions before going through this section, save time by skimming the ExamAlerts in this section and then completing the Cram Quiz at the end of the section.

1. True or false: All unnecessary services on a server should be disabled.

2. A system that uses any two items, such as smartcards and passwords, for authentication is referred to as a _____ system.

3. True or false: Common passwords should be used on similar system devices within the same geographic confines.

Answers

1. True. Unnecessary services serve no purpose and take up overhead. They also represent extra possibilities (expand the attack surface) for an attacker to exploit and use to gain access to your system(s).

2. A system that uses any two items, such as smartcards and passwords, for authentication is referred to as a two-factor, or multifactor, authentication system.

3. False. Common passwords should be avoided at all cost because they serve to weaken the security of the system.

ExamAlert

Remember that this objective begins with "Given a scenario." This means that you may receive a drag-and-drop, matching, or "live OS" scenario where you have to click through to complete a specific objective-based task.

Earlier in this chapter, we discussed several network security features such as 802.1X, MAC filtering, and access control lists (ACLs). Rather than repeating the same information here, the focus is on what has not been discussed before.

Disposing of Assets

Almost every asset reaches the end of its life at some point in time. Whether that asset is a workstation, a hard drive, a piece of backup media, or something else altogether, care needs to be taken when disposing of it to reduce the risk of sensitive data falling into the wrong hands. As an example, consider that the capacity of a flash drive (also known as a *thumb drive, USB drive, jump drive,* and so on) is now greater than that of many hard drives only a few years ago, and many users now store all of their files on a small drive that they transport with them.

A policy should be created, and implemented, governing the disposal of all assets that hold data. Options of what can be done include performing factory resets (for laptops, tablets, and similar devices), wiping the data, and sanitizing the device before disposal. Sanitizing is simply removing the data and the traces of it; this is usually done with storage devices, such as hard drives, and is often referred to as *purging*. Wiping goes further than purging and is also known as *overwriting* (or *shredding*). With wiping, the data that was there is first replaced with something else and then removed. That way, if the data is somehow recovered, what comes back is the overwritten data rather than the original data. The simplest overwrite technique writes a pattern of zeros over the original data.

> **Note**
>
> Today, many companies employ professional shredding and disposal companies for the disposal of personal information (PI) or restricted financial information. Company policy will often dictate that you are provided with a certificate of destruction (COD) to guarantee the information has been properly disposed of.

Secured Versus Unsecured Protocols

Every network needs a number of protocols to function. This includes both LAN and WAN protocols. Not all protocols are created the same, however. Some are designed for secure transfer, whereas others are not. Table 9.3 lists several protocols and describes their use.

> **Note**
>
> NIST provides a white paper on securing servers and reducing threats by removing insecure and unnecessary protocols that is highly recommended. It can be found at https://nvlpubs.nist.gov/nistpubs/legacy/sp/nistspecialpublication800-123.pdf.

TABLE 9.3 **Protocol Summary**

Protocol	Name	Description
FTP	File Transfer Protocol	An insecure protocol for uploading and downloading files to and from a remote host. It also accommodates basic file management tasks. FTP uses ports 20 and 21.
SFTP	Secure File Transfer Protocol	A protocol for securely uploading and downloading files to and from a remote host. It is based on SSH security. SFTP uses port 22.
HTTP	Hypertext Transfer Protocol	A protocol for retrieving files from a web server. Data is sent in clear text. HTTP uses port 80.
HTTPS	Hypertext Transfer Protocol Secure	A secure protocol for retrieving files from a web server. HTTPS uses SSL/TLS to encrypt data between the client and host. HTTPS uses port 443.
Telnet	Telnet	A protocol that enables sessions to be opened on a remote host. Telnet is not considered secure. Telnet uses port 23.
SSH	Secure Shell	A secure alternative to Telnet that enables secure sessions to be opened on a remote host. SSH uses port 22.
TLS	Transport Layer Security	A cryptographic protocol whose purpose is to verify that secure communications between a server and a client remain secure. TLS is an enhancement/replacement for SSL.
IPsec	IP security	IP security that encrypts data during communication between two computers.

ExamAlert

You will most certainly be asked questions on secure protocols and when they might be used. Review Table 9.3 before taking the Network+ exam.

Key Management

Key management refers to the processes and procedures involved in generating, distributing, storing, and revoking cryptographic keys used to protect sensitive data, communications, and network resources. It encompasses various tasks and functions aimed at managing cryptographic keys throughout their lifecycle to mitigate the risk of unauthorized access, data breaches, and cryptographic attacks.

Several components of management include

▶ **Key generation:** The process of creating cryptographic keys using random or pseudo-random algorithms. Secure key generation techniques are essential to ensure that keys have sufficient entropy and unpredictability to resist cryptographic attacks.

▶ **Key distribution:** The secure transmission of cryptographic keys from the key generator or key management system to authorized users or devices. Key distribution mechanisms should ensure confidentiality and integrity to prevent interception, tampering, or unauthorized access to keys during transmission.

▶ **Key storage:** The secure storage of cryptographic keys to prevent unauthorized access, theft, or loss. Key storage mechanisms may include hardware security modules (HSMs), secure key vaults, encrypted databases, or cryptographic key management systems.

▶ **Key usage:** The proper use of cryptographic keys in encryption, decryption, digital signing, and verification operations according to established security policies and protocols. Key usage should adhere to cryptographic best practices and avoid common pitfalls such as key reuse, weak key management, or insecure key storage.

▶ **Key rotation:** The periodic changing of cryptographic keys to limit the exposure window in case of key compromise or cryptographic attacks. Key rotation helps mitigate the risk of long-term key compromise and enhances overall security posture by ensuring that cryptographic keys remain fresh and unpredictable.

▶ **Key revocation:** The process of invalidating or revoking cryptographic keys that are compromised, lost, or no longer needed. Key revocation mechanisms should promptly update key repositories, certificate revocation lists (CRLs), or key distribution centers to prevent unauthorized use of revoked keys.

▶ **Key destruction:** The secure deletion or destruction of cryptographic keys at the end of their lifecycle or when no longer needed. Key destruction ensures that keys cannot be recovered or exploited by adversaries and helps maintain compliance with data protection regulations.

Effective key management practices are essential for maintaining the confidentiality, integrity, and availability of sensitive data and network resources protected by cryptographic keys. Organizations should implement robust key management policies, procedures, and controls tailored to their specific security requirements and regulatory compliance obligations to mitigate the risk of cryptographic attacks and data breaches.

> **Note**
>
> An informative Key Management Cheat Sheet can be found at https:// cheatsheetseries.owasp.org/cheatsheets/Key_Management_Cheat_Sheet.html.

> **Note**
>
> While touched upon with Network+, key management is an important topic for Security+ certification study, which you may consider pursuing after you successfully finish taking this exam.

Hardening Best Practices

In addition to physically securing network devices and opting to run secure protocols in favor of unsecured ones, an administrator should take the following best practices steps to further network hardening:

▶ **URL filtering:** *Uniform Resource Locator (URL) filtering,* also known as *web filtering,* is a technique used to control access to websites and web content based on predefined policies or criteria. It involves monitoring and blocking or allowing access to specific URLs or website categories based on factors such as content, reputation, security risks, and compliance requirements. URL filtering is commonly implemented as part of a broader web security strategy to protect organizations from web-based threats, enforce acceptable use policies (AUPs), and ensure regulatory compliance.

▶ **Content filtering:** Similar to web filtering, *content filtering* is a security measure used to monitor, control, and manage access to digital content based on predefined criteria, policies, or rules. It involves inspecting data packets, files, emails, or web content to identify and filter out undesirable or unauthorized content, such as malicious software, spam, inappropriate material, or sensitive information. Content filtering is commonly employed to enforce acceptable use policies, protect against web-based threats, prevent data breaches, and ensure compliance with regulatory requirements. Network administrators can configure firewalls, mail servers, routers, and Domain Name System (DNS) servers to filter unwanted or malicious content.

> **Note**
>
> Cloud-based and mobile device content filtering is a growing area. Cloud-based proxy servers can intercept and filter Internet traffic, allowing organizations to enforce content policies and block access to inappropriate or malicious websites.

▶ **Utilize Secure SNMP:** The Simple Network Management Protocol (SNMP) collects a lot of data that could be of value to someone looking to compromise a network. Secure SNMP, utilizing SNMPv3, protects the data more by moving from ports 161 and 162 to 10161 and 10162 and enables secure authentication and communication between the SNMP manager and agent.

▶ **Configure Router Advertisement guard:** With IPv6, routers send out multicast messages (router advertisements) to announce their availability and associated information. This information is used by Neighbor Discovery Protocol (NDP) to detect what is available and configure accordingly. A problem is that the messages are unsecured, making them susceptible to spoofing. To increase security, you can configure IPv6 *Router Advertisement (RA) guard* to protect the network against RA messages generated by unauthorized routers (rogues) trying to join the network.

▶ **Employ port security and Dynamic ARP:** *Port security* works at Layer 2 of the OSI model and allows an administrator to configure switch ports so that only certain MAC addresses can use the port. This is a common feature on both Cisco's Catalyst as well as Juniper's EX Series switches and essentially differentiates so-called dumb switches from managed (or intelligent) switches. Similarly, *Dynamic ARP Inspection (DAI)* works with these and other smart switches to protect ports from ARP spoofing.

▶ **Implement Control Plane Policing:** Many routers and switches include a Control Plane Policing feature that enables you to configure a Quality of Service (QoS) filter to protect against DoS attacks. When enabled, the control plane (CP) can continue to forward packets despite an attack or abnormally heavy traffic load on the router/switch.

▶ **Use private VLANs:** Also known as *port isolation*, creating a *private VLAN* is a method of restricting switch ports (now called *private ports*) so that they can communicate only with a particular uplink. The private VLAN usually has numerous private ports and only one uplink, which is usually connected to a router, or firewall.

▶ **Disable unneeded ports:** Disabling unnecessary services (mentioned next) increases security by removing doors that someone could use to enter the server. Similarly, IP ports that are not needed for devices also represent doors that could be used to sneak in. It is highly recommended that unused ports be disabled to increase security along with device ports (both physical and virtual ports).

▶ **Disable unneeded services:** Every unnecessary service that is running on a server is akin to another door on a warehouse that some unauthorized person may choose to sneak in. Just as an effective way to secure a warehouse is to reduce the number of doors to only those needed, so too is it recommended that a server be secured by removing (disabling) services not in use.

▶ **Change default passwords:** The easiest way for any unauthorized individual to access a device is to use the default credentials. Many routers, for example, come configured with an "admin" account and a simple value for the password ("admin," "password," and so on). Anyone owning one of those routers knows those values and could use them to access any other of the same make if the values have not been changed. To make it more difficult for unauthorized users to access your devices, change those default usernames and passwords as soon as you start using them.

▶ **Avoid common passwords and increase complexity:** It is a good thing to preach password security to users, but often administrators are guilty of using too-simplistic passwords on network devices such as routers, switches, and the like. Given the large number of devices in question, sometimes the same passwords are also used on multiple devices. Common sense tells every administrator that this approach is wrong, but often it is done anyway with the hope that no miscreant will try to gain unauthorized access. Don't be that administrator: use complex passwords, with notable length, and use a different password for each device, increasing the overall security of your network.

▶ **Enable DHCP snooping:** As was mentioned when discussing rogue DHCP servers earlier in this chapter, DHCP snooping is a good thing. It provides a way for a switch to look at packets and drop DHCP traffic that it determines to be unacceptable based on a set of defined rules to prevent rogue DHCP servers from offering IP addresses to clients.

▶ **Change the default VLAN:** On switches, the native VLAN is the only VLAN that is not tagged in a trunk. This means that native VLAN frames are transmitted unchanged. By default, the native VLAN is port 1, and that default represents a weakness in that it is something known about your network that an attacker could use. To strengthen security, albeit a small amount, you can change the native VLAN to another port. The command or commands used to do so are dependent on the vendor and model of your switch but can be easily found online.

▶ **Utilize patch and firmware management:** There is a reason why each firmware update is written. Sometimes, it is to optimize the device or make it more compatible with other devices. Other times, it is to fix security issues and/or head off identified problems. Keep firmware on your production machines current after first testing the upgrades on lab machines and verifying that you're not introducing any unwanted problems by installing. Just as firmware upgrades are intended to strengthen or solve problems, patches and updates do the same with software (including operating systems). Test each release on a lab machine(s) to make sure you are not adding to network woes, and then keep your software current to harden it.

▶ **Put an access control list in place:** Discussed at length in this chapter, access control is necessary to limit who can access resources to only those who should have access and thus protect those resources. *Access control lists (ACLs)* enable devices in your network to ignore requests from specified users or systems or to grant them access to certain network capabilities. You may find that a certain IP address is constantly scanning your network. You can block this IP address at the router, and the IP address will automatically be rejected any time it attempts to utilize your network.

▶ **Apply role-based access:** It is recommended that role-based access be applied where possible. The difference between this approach and others was discussed earlier in this chapter.

▶ **Enforce firewall rules:** Firewall rules are used to dictate what traffic can pass between the firewall and the internal network. Three possible actions can be taken based on the rule's criteria: block the connection (*explicit deny*), accept the connection, or allow the connection if conditions are met (such as it being secured). It is this last condition that is the most difficult to configure, and conditions usually end with an implicit deny clause. An *implicit deny* clause means that if the proviso in question has not been explicitly granted, access is denied.

Note

Where it exists, an explicit deny takes precedence over all other settings.

▶ **Verify file hashes:** File hashing is used to verify that the contents of files are unaltered. A hash is often created on a file before it is downloaded, and that value remains constant once it is generated, serving as a unique identifier for the file that can be compared after the download to make sure the contents are the same. When downloading files—particularly upgrades, patches, and updates—check hash values and use this one test to keep from installing those entities that have had Trojan horses attached to them. Common checksum algorithms include MD5, SHA-1, SHA-256, and SHA-512

▶ **Generate new keys:** Keys are used as a part of the encryption process, particularly with public key infrastructure (PKI), to encrypt and decrypt messages. The longer you use the same key, the longer the opportunity becomes for someone to crack that key. To increase security, generate new keys on a regular basis. The commands to do so will differ based on the utility that you are creating the keys for.

Wireless Security

Wireless systems transmit data through the air and can be much more difficult to secure than those dependent on wires that can be physically protected. The growth of wireless systems in every workplace and home creates many opportunities for attackers looking for signals that can be easily intercepted. This section discusses various types of wireless systems that you'll likely encounter and some of the security issues associated with this technology.

Antenna Placement and Power Levels

Antenna placement can be crucial in allowing clients to reach the access point. There isn't a universal solution to this issue, and it depends on the environment in which the access point is placed. As a general rule, the greater the distance the signal must travel, the more it will attenuate, but you can lose a signal quickly over a short distance as well if the building materials reflect or absorb the signal. You should try to avoid placing access points near metal (which includes appliances) or near the ground. Placing them in the center of the area to be served and high enough to get around most obstacles is recommended. On the chance that the signal is actually traveling too far, some access points include *power level* controls, which allow you to reduce the amount of output provided. Both a heat map and a wireless site survey can be used to optimize wireless antenna placement.

> **Note**
>
> Some common problem areas with antennas can include the wireless signal being refracted by windows, the antenna's power level being set too high and overlapping coverage areas of adjacent access points (APs) or wireless routers, an omnidirectional antenna being used instead of a unidirectional antenna, and wireless access points using channels from the wrong spectrum. A great resource on wireless antenna best practices can be found at https://www.shure.com/en-US/performance-production/louder/all-about-wireless-antenna-positioning-best-practices.

Isolation

As the name implies, *isolation* is the process of keeping something from communicating with others. This can be done on a wireless network with a single client (*wireless client isolation*) or a guest network (*guest network isolation*). With the latter, guest users cannot see any other connected devices even though they are logged in to the network. Most wireless routers allow for creating a private network in each frequency and thus enabling temporary users (guests) to access the Internet and their own files without jeopardizing the rest of the network's resources.

Preshared Keys

During the authentication process, keyed security measures are applied before communication can take place. On many APs, authentication can be set to either *shared* key authentication or *open* authentication. The problem with open authentication is that if you do not have other protection or authentication mechanisms in place, your wireless network is totally open to intruders. When set to shared key mode, the client must meet security requirements before communication with the AP can occur.

Working with Zones

The first part of this chapter discussed creating zones and screened subnets. Similarly when it comes to zones, *trusted* and *untrusted* refer to different segments or areas of a network based on their security posture. These zones are defined by access controls, security policies, and trust boundaries to enforce appropriate security measures and protect against unauthorized access, data breaches, and cyber threats. The trusted zone encompasses network segments, devices, or resources considered secure, trustworthy, and under the organization's direct control, and access is restricted to authorized users, devices, and services based on identity, authentication, and access control mechanisms.

The untrusted zone comprises network segments, devices, or resources considered less secure, potentially compromised, or outside the organization's direct control, such as the Internet, public Wi-Fi networks, partner networks, guest networks, and external-facing services accessible from the Internet. Access to the untrusted zone is subject to stricter security controls, such as firewall rules, access lists, and network segmentation, to minimize the risk of unauthorized access, data leakage, and cyberattacks.

> **Note**
>
> A common Wi-Fi guest network today used in coffee shops, hospitals, and hotels is the captive portal. A *captive portal* is a web page/portal that is first launched when a user is connecting through a public guest network that usually requires some type of interaction before the user is allowed access to other network resources or Internet sites. Many businesses that offer a captive portal first require the user to view and acknowledge an acceptable use policy (AUP).

Security measures such as firewalls, intrusion detection/prevention systems (IDS/IPSs), encryption, and endpoint security solutions are deployed to protect assets in the trusted zone from external and internal threats. Security measures such as intrusion prevention systems (IPSs), web application firewalls (WAFs), email filtering, and malware protection are often deployed to protect assets in the untrusted zone from external threats and vulnerabilities.

Cram Quiz

1. Which of the following is used to verify that the contents of files are unaltered?

   - ○ **A.** Key generation
   - ○ **B.** File hashing
   - ○ **C.** Biometrics
   - ○ **D.** Asset tracking

2. Which of the following is a secure alternative to Telnet that enables secure sessions to be opened on a remote host?

   - ○ **A.** SSH
   - ○ **B.** RSH
   - ○ **C.** IPsec
   - ○ **D.** RDP

3. Which of the following can be used to limit access to only one or two people going into the facility at a time?

   ○ **A.** Access control vestibule

   ○ **B.** Min cage

   ○ **C.** Tholian web

   ○ **D.** Cloud minder

4. Which of the following systems use a credit card–sized plastic card that is read by a reader on the outside of the door?

   ○ **A.** Contiguity reader

   ○ **B.** Key fob

   ○ **C.** Swipe card

   ○ **D.** Cipher lock

5. Which of the following are considered hardening best practices? (Choose three.)

   ○ **A.** Utilize SNMPv2.

   ○ **B.** Disable unneeded ports.

   ○ **C.** Disable unneeded services.

   ○ **D.** Change default passwords.

6. Which of the following provides a score from 0 to 10 that indicates the severity of a vulnerability?

   ○ **A.** CVSS

   ○ **B.** CVE

   ○ **C.** COD

   ○ **D.** Cryptographic checksum

7. Which of the following is a Wi-Fi guest network that often requires the user to first acknowledge an acceptable use policy before gaining access?

   ○ **A.** Trusted zone

   ○ **B.** Isolation

   ○ **C.** Private VLAN

   ○ **D.** Captive portal

Cram Quiz Answers

1. **B.** File hashing is used to verify that the contents of files are unaltered.

2. **A.** SSH is a secure alternative to Telnet that enables secure sessions to be opened on a remote host. SSH uses port 22.

3. **A.** An access control vestibule can be used to limit access to only one or two people going into the facility at a time.

4. **C.** Swipe card systems use a credit card–sized plastic card read by a reader on the outside of the door. To enter the server room, you must swipe the card (run it through the reader), at which point it is read by the reader, which validates it.

5. **B, C, and D.** It is highly recommended that unused ports be disabled to increase security along with device ports (both physical and virtual ports). It is also recommended that a server be secured by removing (disabling) services not in use. To make it more difficult for unauthorized users to access your devices, change default usernames and passwords as soon as you start using them. SNMPv2 is not secure. SNMPv3 should be used in its place.

6. **A.** The Common Vulnerability Scoring System (CVSS) provides a score from 0 to 10 that indicates the severity of a vulnerability. The numerical score can then be translated into a qualitative representation (such as low, medium, high, and critical) to help organizations properly assess and prioritize their vulnerability management processes. File hashing is used to verify that the contents of files are unaltered. The common vulnerabilities and exposures (CVE) is a list of publicly known vulnerabilities containing an ID number, description, and reference. A certificate of destruction (COD) is a guarantee from a professional shredding/destruction company that your information has been destroyed. A cryptographic checksum is assigned to a file and is used to verify that the data in that file has not been tampered with or manipulated, possibly by a malicious entity.

7. **D.** A captive portal is a web page/portal that is first launched when a user is connecting through a public guest network that usually requires some type of interaction before the user is allowed access to other network resources or Internet sites. Many businesses that offer a captive portal first require the user to view and acknowledge an acceptable use policy (AUP). The trusted zone encompasses network segments, devices, or resources considered secure, trustworthy, and under the organization's direct control, and access is restricted to authorized users, devices, and services based on identity, authentication, and access control mechanisms. As the name implies, isolation is the process of keeping something from communicating with others. This can be done on a wireless network with a single client (wireless client isolation) or a guest network (guest network isolation). Also known as port isolation, creating a private VLAN is a method of restricting switch ports (now called private ports) so that they can communicate only with a particular uplink. The private VLAN usually has numerous private ports and only one uplink, which is usually connected to a router, or firewall.

What's Next?

The final chapter of this book, "Network Troubleshooting," focuses on all areas of network troubleshooting, including troubleshooting best practices and some of the tools and utilities you use to assist in the troubleshooting process.

No matter how well a network is designed and how many preventive maintenance schedules are in place, troubleshooting is always necessary. Consequently, network administrators must develop those troubleshooting skills.

CHAPTER 10

Network Troubleshooting

This chapter covers the following official Network+ objectives:

▶ 5.1 Explain the troubleshooting methodology.

▶ 5.3 Given a scenario, troubleshoot common issues with network services.

▶ 5.4 Given a scenario, troubleshoot common performance issues.

▶ 5.5 Given a scenario, use the appropriate tool or protocol to solve networking issues.

This chapter covers CompTIA Network+ objectives 5.1, 5.3, 5.4, and 5.5. For more information on the official Network+ exam topics, see the "About the Network+ Exam" section in the Introduction.

Many duties and responsibilities fall under the umbrella of network administration. Of these, one of the most practiced is that of troubleshooting. No matter how well a network is designed and how many preventive maintenance schedules are in place, troubleshooting is always necessary. Therefore, network administrators and technicians must develop those troubleshooting skills.

This chapter focuses on all areas of troubleshooting, including troubleshooting best practices and some of the tools and utilities you can use to assist in the troubleshooting process.

Troubleshooting Steps and Procedures

▶ **5.1 Explain the troubleshooting methodology.**

CramSaver

If you can correctly answer these questions before going through this section, save time by skimming the ExamAlerts in this section and then completing the Cram Quiz at the end of the section.

1. What are the key sources from which you can gain information about a computer problem?

2. What is the final step in the network troubleshooting methodology CompTIA expects test takers to follow?

Answers

1. It is important to get as much information as possible about the problem. You can glean information from three key sources: the computer (in the form of logs and error messages), the computer user experiencing the problem, and your own observation.

2. Document the findings, actions, outcomes, and lessons learned throughout the process.

Regardless of the problem, effective network troubleshooting follows some specific steps. These steps provide a framework in which to perform the troubleshooting process. When you follow them, they can reduce the time it takes to isolate and fix a problem. The following sections discuss the common troubleshooting methodology steps and procedures as identified by the CompTIA Network+ objectives:

1. Identify the problem.

   ▶ Gather information.

   ▶ Question users.

   ▶ Identify symptoms.

   ▶ Determine if anything has changed.

   ▶ Duplicate the problem, if possible.

   ▶ Approach multiple problems individually.

2. Establish a theory of probable cause.

   ▶ Question the obvious.

   ▶ Consider multiple approaches:

      ▶ Top-to-bottom/bottom-to-top OSI model.

      ▶ Divide and conquer.

3. Test the theory to determine the cause:

   ▶ If the theory is confirmed, determine the next steps to resolve the problem.

   ▶ If the theory is not confirmed, reestablish a new theory or escalate.

4. Establish a plan of action to resolve the problem and identify potential effects.

5. Implement the solution or escalate as necessary.

6. Verify full system functionality and implement preventive measures if applicable.

7. Document findings, actions, outcomes, and lessons learned throughout the process.

ExamAlert

You should expect questions asking you to identify the troubleshooting methodology steps in exact order.

Identify the Problem

The first step in the troubleshooting process is to establish exactly what the problem is. This stage of the troubleshooting process is all about gathering information, identifying symptoms, questioning users, and determining whether anything has changed. To get this information, you need knowledge of the operating system used, good communication skills, and a little patience. You need to get as much information as possible about the problem. You can glean information from three key sources: the computer (in the form of logs and error messages), the computer user experiencing the problem, and your own observation.

After you have listed the symptoms, you can begin to identify some of the potential causes of those symptoms.

> **ExamAlert**
>
> You do not need to know where error messages are stored on an operating system. You need to know only that the troubleshooting process requires you to read system-generated log errors.

Identify Symptoms

Some computer problems are isolated to a single user in a single location; others affect several thousand users spanning multiple locations. Establishing the affected area is an important part of the troubleshooting process, and it often dictates the strategies you use to resolve the problem.

> **ExamAlert**
>
> You might be provided with either a description of a scenario or a description augmented by a network diagram. In either case, you should carefully read the description of the problem, step by step. In most cases, the correct answer is fairly logical, and the wrong answers can be easily identified.

Problems that affect many users are often connectivity issues that disable access for many users. Such problems often can be isolated to wiring closets, network devices, and server rooms. The troubleshooting process for problems that are isolated to a single user often begins and ends at that user's workstation. The trail might indeed lead you to the wiring closet or server, but that is probably not where the troubleshooting process began. Understanding who is affected by a problem can give you the first clues about where the problem exists. For example, a change in Dynamic Host Configuration Protocol (DHCP) scope (or an exhausted scope) by a new administrator might affect several users, whereas a user playing with the TCP/IP settings of a single computer can affect only that person.

Determine If Anything Has Changed

Whether there is a problem with a workstation's access to a database or an entire network, they were working at some point. Although many people claim that their computer "just stopped working," that is unlikely. Far more likely is that changes to the system or network have caused the problem. Look for newly installed applications, applied patches or updates, new hardware, a physical move of the computer, or a new username and password. Establishing any recent changes to a system can often lead you in the right direction to isolate and troubleshoot a problem.

Duplicate the Problem if Possible

Every problem has a cause, and determining the cause of a problem is key to preventing it from happening again with either this machine or another. One way to know that you have identified the cause is to be able to duplicate the problem. The sole reason for duplicating the problem should be to verify that you have, indeed, found the cause so that you can then take steps to make certain that cause never becomes an issue again. As an example, if the cause of the problem turns out to be a new patch for an application not working properly with a particular operating system, you will want to make sure that the patch/OS combination is not implemented again (steps to consider would include updating the OS, installing only some of the patch, and so on).

Approach Multiple Problems Individually

One of the toughest situations to tackle is having to deal with more than one problem (a particular workstation won't load Application A, can't print from Application B, and so on). When this is the case, the only way to be able to solve the problems is to address each one individually. Problems should be ranked in order of importance (based on criteria such as business necessity) and solved in that order.

Establish a Theory of Probable Cause

> **ExamAlert**
>
> When approaching a problem, start by questioning the obvious (such as looking to see if the power light is on). If that fails, consider ways to tackle the issue from multiple approaches. Consider using a top-to-bottom or bottom-to-top model approach (such as working through the OSI model stack, pinging the default gateway and DNS server on a workstation that cannot connect to the Internet, and so on) and assigning any coworkers you have to divide and conquer the problem.

A single problem on a network can have many causes, but with appropriate information gathering, you can eliminate many of them. When you look for a probable cause, it is often best to look at the easiest solution first and then work from there. Even in the most complex of network designs, the easiest solution is often the right one. For instance, if a single user cannot log on to a network, it is best to confirm network settings before replacing the network interface card (NIC). Remember, though, that at this point you need to determine only the most probable cause, and your first guess might be incorrect. Determining the correct cause of the problem might take a few tries.

> **ExamAlert**
>
> Avoid discounting a possible answer because it seems too easy. Many of the troubleshooting questions are based on possible real-world scenarios, some of which do have easy or obvious solutions.

Test the Theory to Determine the Cause

After questioning the obvious, you need to establish a theory. After you formulate a theory, you should attempt to confirm it. An example might be a theory that users can no longer print because they downloaded new software that changed the print drivers, or that they can no longer run the legacy application they used to run after the latest service pack was installed.

If the theory can be confirmed, you must plot a course of action—a list of the next steps to take to resolve the problem. If the theory cannot be confirmed (in the example given, no new software was downloaded and no service pack was applied), you must establish a new theory or consider escalating the problem.

Establish a Plan of Action

After identifying a cause but before implementing a solution, you should establish a plan for the solution. This is particularly a concern for server systems in which taking the server offline is a difficult and undesirable prospect. After you identify the cause of a problem on the server, it is absolutely necessary to plan for the solution. The plan must include the details of when the server or network should be taken offline and for how long, what support services are in place, and who will be involved in correcting the problem.

Planning is an important part of the whole troubleshooting process and can involve formal or informal written procedures. Those who do not have experience troubleshooting servers might wonder about all the formality, but this attention to detail ensures the least amount of network or server downtime and the maximum data availability.

> **Tip**
>
> If part of an action plan includes shutting down a server or another similar event that can impact many users, it is a best practice to let users know when they will be shut out of the network. Having this information allows them to properly shut off any affected applications and not be frustrated by not being able to access the network or other services.

With the plan in place, you should be ready to implement a solution—that is, apply the patch, replace the hardware, plug in a cable, or implement some other solution. In an ideal world, your first solution would fix the problem; however, unfortunately, this is not always the case. If your first solution does not fix the problem, you need to retrace your steps and start again.

You must attempt only one solution at a time. Trying several solutions at once can make it unclear which one corrected the problem.

Implement the Solution or Escalate

After the corrective change has been made to the server, network, or workstation, you must test the results—never assume. This is when you find out if you were right and the remedy you applied worked. Don't forget that first impressions can deceive, and a fix that seems to work on first inspection might not have corrected the problem.

The testing process is not always as easy as it sounds. If you are testing a connectivity problem, it is not difficult to ascertain whether your solution was successful. However, changes made to an application or to databases you are unfamiliar with are much more difficult to test. It might be necessary to have people who are familiar with the database or application run the tests with you in attendance.

Sometimes the problems you encounter fall outside the scope of your knowledge. Few organizations expect their administrators to know everything, but organizations do expect administrators to fix any problem. To do this, you often need additional help.

> **Note**
>
> System administration is often as much about knowing whom and what to refer to in order to get information about a problem as it is about actually fixing the problem.

Technical escalation procedures do not follow a specific set of rules; rather, the procedures to follow vary from organization to organization and situation to situation. Your organization might have an informal arrangement or a formal one requiring documented steps and procedures to be carried out. Whatever the approach, general practices should be followed for appropriate escalation.

Unless otherwise specified by the organization, the general rule is to start with the closest help and work out from there. If you work in an organization that has an IT team, talk with others on your team; every IT professional has had different experiences, and someone else may know about the issue at hand. If you are still struggling with the problem, it is common practice to notify a supervisor or head administrator, especially if the problem is a threat to the server's data or can bring down the server.

Suppose that, as a server administrator, you notice a problem with a hard disk in a RAID 1 array on a Linux server. You know how to replace drives in a failed RAID 1 configuration, but you have no experience working with software RAID on a Linux server. This situation would most certainly require an escalation of the problem. The job of server administrator in this situation is to notice the failed RAID 1 drive and to recruit the appropriate help to repair the RAID failure within Linux.

> **Note**
>
> When you are confronted with a problem, it is yours until it has been solved or passed to someone else. Of course, the passing on of an issue requires that both parties know that it has been passed on.

Verify Full System Functionality

At times, you might apply a fix that corrects one problem but creates another. Many such circumstances are hard to predict—but not always. For instance, you might add a new network application, but the application requires more bandwidth than your current network infrastructure can support. The result would be that overall network performance would be compromised.

Everything done to a network can have a ripple effect and negatively affect another area of the network. Actions such as adding clients, replacing hubs or switches, and adding applications can all have unforeseen results. It is difficult to always know how the changes you make to a network might affect the network's functioning. The safest thing to do is assume that the changes you make will affect the network in some way and realize that you have to figure out how. At times like this, you might need to think outside the box and try to predict possible outcomes.

It is imperative that you verify full system functionality before you are satisfied with the solution. After you obtain that level of satisfaction, you should look at the problem and ascertain if any preventive measures should be implemented to keep the same problem from occurring again.

Document Findings, Actions, Outcomes, and Lessons Learned Throughout the Process

Although it is often neglected in the troubleshooting process, documentation is as important as any of the other troubleshooting procedures. Documenting a solution involves keeping a record of all the steps taken during the fix—not necessarily just the solution. The lessons learned can be just as important as the solution.

For the documentation to be of use to other network administrators in the future, it must include several key pieces of information. When documenting a procedure, you should include the following information:

▶ **When:** When was the solution implemented? You must know the date, because if problems occur after your changes, knowing the date of your fix makes it easier to determine whether your changes caused the problems.

▶ **Why:** Although it is obvious when a problem is being fixed why it is being done, a few weeks later, the reason why that solution was needed might become less clear. Documenting why the fix was made is important because if the same problem appears on another system, you can use this information to reduce the time needed to find the solution.

▶ **What:** The successful fix should be detailed, along with information about any changes to the configuration of the system or network that were made to achieve the fix. Additional information should include version numbers for software patches or firmware, as appropriate.

▶ **Results:** Many administrators choose to include information on both successes and failures. The documentation of failures might prevent you from going down the same road twice, and the documentation of successful solutions can reduce the time it takes to get a system or network up and running.

▶ **Who:** It might be that information is left out of the documentation or someone simply wants to ask a few questions about a solution. In both cases, if the name of the person who made a fix is in the documentation, that person can easily be tracked down. Of course, documenting who is more of a concern in environments that have a large IT staff or if system repairs are performed by contractors instead of company employees.

Cram Quiz

1. A user reports that they can no longer access a legacy database. What should be one of the first questions you ask?

   - ○ **A.** What has changed since the last time you accessed that database?
   - ○ **B.** How many help calls have you placed in the past few months?
   - ○ **C.** Who originally installed or created that database?
   - ○ **D.** How long have you worked here?

2. You've spent two hours trying to fix a problem and then realize that it falls outside of your area of expertise and ability to fix. What should you do in most organizations?

   - ○ **A.** Let the user immediately know that they need to call someone else; then exit the scene so another person can help.
   - ○ **B.** Formulate a workaround; then document the problem and bring it up at the next meeting.
   - ○ **C.** Escalate the issue with a supervisor or manager.
   - ○ **D.** Continue working on the problem, trying as many solutions as you can find, until you solve the problem.

3. You get numerous calls from users who cannot access an application. Upon investigation, you find that the application crashed. You restart the application, and it appears to run okay. What is the next step in the troubleshooting process?

   - ○ **A.** Email the users to let them know that they can use the application again.
   - ○ **B.** Test the application to ensure that it operates correctly.
   - ○ **C.** Document the problem and the solution.
   - ○ **D.** Reload the application executables and restart it.

4. A user tells you that they are having a problem accessing their email. What is the first step in the troubleshooting process?

   - ○ **A.** Document the problem.
   - ○ **B.** Make sure that the user's email address is valid.
   - ○ **C.** Discuss the problem with the user.
   - ○ **D.** Visit the user's desk to reload the email client software.

5. You have successfully fixed a problem with a server and have tested the application and let the users back on to the system. What is the next step in the troubleshooting process?

   ○ **A.** Document the problem.

   ○ **B.** Restart the server.

   ○ **C.** Document the problem and the solution.

   ○ **D.** Clear the error logs of any reference to the problem.

Cram Quiz Answers

1. **A.** Establishing any recent changes to a system can often lead you in the right direction to isolate and troubleshoot a problem.

2. **C.** When a problem is outside of your ability to fix, you must escalate the issue. Unless otherwise specified by the organization, the general rule is to start with the closest help and work out from there. None of the other options are acceptable choices.

3. **B.** After you fix a problem, you should test it fully to ensure that the network operates correctly before you allow users to log back on. Emailing the users and documenting the problem are valid but only after the application has been tested. Reloading the application is incorrect because you would reload the executable only as part of a systematic troubleshooting process. Because the application loads, it is unlikely that the executable has become corrupted.

4. **C.** Not enough information is provided for you to come up with a solution. In this case, the next troubleshooting step would be to talk to the user and gather more information about exactly what the problem is. All the other answers are valid troubleshooting steps but only after the information gathering has been completed.

5. **C.** After you have fixed a problem, tested the fix, and let users back on to the system, you should create detailed documentation that describes the problem and the solution. Documenting the problem only is incorrect because you must document both the problem and the solution. You do not need to restart the server, so that answer is incorrect. You would clear the logs only after the system's documentation has been created.

Troubleshooting Common Networking Issues

▶ **5.3 Given a scenario, troubleshoot common issues with network services.**

▶ **5.4 Given a scenario, troubleshoot common performance issues.**

CramSaver

If you can correctly answer these questions before going through this section, save time by skimming the ExamAlerts in this section and then completing the Cram Quiz at the end of the section.

1. What one, hard-coded address must be unique on a network for networking to function properly?

2. What can you try to do to handle DHCP exhaustion if you cannot increase the scope?

3. A client has an incorrect gateway configured. What is the most likely manifestation of this error?

4. You have noticed that connections between nodes on one network are inconsistent and suspect there may be another network using the same channel. What should you try first?

5. True or false: Weather conditions should not have a noticeable impact on wireless signal integrity.

6. True or false: When a client connects to an AP, it is said to associate with that AP, and disassociation is the process of it no longer associating with that AP.

Answers

1. The MAC address must be unique for each network interface card, and there can be no duplicates.

2. You can shorten the lease period for each client and, hopefully, recover addresses sooner for issue to other clients.

3. With an incorrect gateway, the client will not be able to access networking services beyond the local network.

4. If connections are inconsistent, try changing the frequency channel to another, nonoverlapping channel.

5. False. Weather conditions can have a huge impact on wireless signal integrity.

6. True. When a client connects to an AP, it is said to associate with that AP, and disassociation is the process of it no longer associating with that AP. Disassociation can happen any time the AP thinks it no longer needs to communicate with the client.

> **ExamAlert**
>
> Remember that both of the objectives covered here begin with "Given a scenario." This means that you may receive a drag-and-drop, matching, or "live OS" scenario where you have to click through to complete a specific objective-based task.

You will no doubt find yourself troubleshooting networking problems much more often than you would like to. When you troubleshoot these problems, a methodical approach is likely to pay off.

> **ExamAlert**
>
> Wiring problems are related to the cable used in a network. For the purposes of the exam, infrastructure problems are classified as those related to network devices, such as hubs, switches, and routers.

Common Considerations

There are a number of things to take into consideration when trying to troubleshoot a problem—mainly, whether you can isolate what you are trying to find and the extent of the issue. Five broad considerations are outlined in Table 10.1.

TABLE 10.1 Common Considerations

Option	Description
Device configuration review	Look at the device configuration and make certain that you are not creating, or experiencing, a bottleneck due to a configuration error.
Routing tables	Check routing tables to make certain that the routes within are the most cost effective in terms of hops and routes taken.
Interface status	Focus on optimization, redundancy, and performance as much as possible.
VLAN assignment	A virtual local-area network (VLAN) allows you to create groups of users and systems and segment them on the network. This segmentation lets you hide segments of the network from other segments and thereby control access. You can also set up VLANs to control the paths that data takes to get from one point to another. A VLAN is a good way to contain network traffic to a certain area in a network, but be sure that the resources you are using can support what you create.
Network performance baselines	Creating baselines is only a part of the equation; you also have to analyze them. However, too many administrators collect information that they never do anything with. Be sure to look at the baseline information regularly and track current conditions to see what needs to be improved upon.

Common Problems to Be Aware Of

In the eyes of CompTIA and the Network+ exam, you should be aware of some problems more than others. Although other sections have looked at problems in particular areas, pay special attention to those that fall within this section as you study for the exam.

Congestion/Contention

Congestion and *contention* are two related concepts that refer to situations where network resources become overloaded or shared among multiple users, leading to performance degradation and potential service disruptions. Generally, as more systems are added to a network, the possibility for more collisions to occur increases, and the network becomes slower. The type of collision detection used (discussed in Chapter 3, "Network Addressing, Routing, and Switching") can impact performance.

> **Note**
>
> *Runts* are frames that fail to meet the minimum size requirement of 64 bytes, often due to collisions. When an interface receives a frame smaller than 64 bytes, it's classified as a runt, which can occur from collisions or faulty connections. *Giants*, on the other hand, are frames larger than 1,518 bytes.

Congestion occurs when there is an excessive amount of traffic traversing a network segment, link, or device, leading to delays, packet loss, and reduced throughput. Contention occurs when multiple devices or users compete for access to the same shared network resource, such as a transmission medium or network interface. Congestion can be caused by factors such as high network utilization, insufficient bandwidth, bursty traffic patterns, network failures, or inefficient routing. Contention can arise in scenarios where network devices share a common communication channel or when multiple users attempt to access a shared resource simultaneously. It often occurs in Ethernet networks where devices contend for access to the network using Carrier Sense Multiple Access with Collision Detection (CSMA/CD) or Carrier Sense Multiple Access with Collision Avoidance (CSMA/CA) protocols.

Effective network management, traffic shaping, congestion control mechanisms, and Quality of Service (QoS) policies are essential for mitigating congestion and contention issues and ensuring optimal performance in computer networks.

Bottlenecking

Similar to congestion, *bottlenecking* is a situation where the performance of a network is limited or constrained by a specific component, resource, or point of congestion, causing a reduction in overall network throughput and efficiency. A bottleneck occurs when the capacity of one element in the network becomes insufficient to handle the volume of traffic passing through it, leading to delays, packet loss, and degraded performance for network services and applications.

Bottlenecks can be caused by a variety of factors, including insufficient bandwidth, network congestion, hardware limitations, software inefficiencies, or misconfigurations. Common sources of bottlenecks include overloaded network links, inadequate router or switch capacity, underpowered network devices, and poorly optimized network protocols.

Detecting and mitigating bottlenecks in computer networks require proactive monitoring, analysis, and optimization of network infrastructure. Network administrators use performance monitoring tools, traffic analysis software, and network management systems to identify bottlenecks, analyze traffic patterns, and troubleshoot performance issues. Strategies for mitigating bottlenecks may include upgrading network equipment, optimizing network configurations, implementing Quality of Service (QoS) policies, load balancing traffic, and deploying caching or content delivery solutions.

Bandwidth/Throughput Capacity

Throughput capacity is the maximum rate at which data can be transmitted over a network link or channel, typically measured in bits per second (bps) or its multiples (e.g., Gbps). Throughput capacity is a key metric that indicates the effective bandwidth available for data transfer and determines the maximum achievable data transfer rate between two network endpoints.

While the bandwidth is the capacity of a network link to carry data, network utilization is the actual throughput capacity achieved on a network link and it is heavily based on the presence of competing traffic. Network congestion, packet loss, collisions, and contention for shared resources can reduce the effective throughput capacity of a link, leading to performance degradation and slower data transfer rates.

Latency

Network *latency*, or the delay incurred during data transmission, can impact the achievable throughput capacity. Higher latency can reduce the efficiency of data transfer and limit the effective throughput capacity, particularly for small packet sizes or interactive applications.

Packet Loss

Packet loss is the failure of one or more data packets to reach their destination within a network. Packet loss can be caused by several factors, including network congestion, hardware failures, transmission errors, buffer overflows, and routing issues. Network congestion occurs when the volume of traffic exceeds the capacity of the network link, leading to dropped packets as routers and switches are unable to forward all incoming data. Hardware failures, such as faulty network interfaces or cables, can also result in packet loss. Transmission errors, such as bit errors or signal interference, can corrupt packets during transmission, causing them to be discarded. Buffer overflows can occur when network devices are overwhelmed with incoming traffic, leading to the dropping of packets that cannot be buffered. Routing issues, such as misconfigurations or link failures, can cause packets to be routed incorrectly or dropped along the path to their destination.

Packet loss can have detrimental effects on network performance and application quality. When packets are lost, the data they contain must be retransmitted, leading to increased latency and reduced throughput. For real-time applications such as voice over Internet Protocol (VoIP) and video streaming, packet loss can result in choppy audio/video playback, dropped calls, and poor call quality. In data-intensive applications such as file transfers and cloud computing, packet loss can lead to slow transfer speeds and degraded performance.

To mitigate packet loss, network administrators can implement several strategies, including increasing network capacity to reduce congestion, replacing faulty hardware components, optimizing network configurations to improve routing efficiency, implementing error correction techniques such as forward error correction (FEC), and deploying redundancy and failover mechanisms to provide alternate paths for traffic in case of link failures.

Jitter

Jitter refers to the variation in the delay of packet delivery within a network—the difference in the time it takes for packets to reach their destination, with variations occurring due to factors such as network congestion, packet buffering, routing changes, and transmission errors. Jitter is typically measured in milliseconds (ms) and represents the deviation from the expected or average delay. Jitter can have significant implications for network performance and the quality of real-time applications such as VoIP, video conferencing, and online gaming.

> **Note**
>
> Often, the best way to determine the cause of jitter and high latency is to configure SNMP traps on the network. With VoIP, hearing echoes of a receiver's voice can be caused from jitter or QoS misconfiguration. Metrics such as jitter, speed, and bandwidth are important, but they are not as directly impacted by sustained link saturation as latency.

Broadcast Storm

When an abnormally high number of broadcast packets is sent across the network within a short period of time, this is known as a *broadcast storm*, and it can degrade network performance as switches become overwhelmed with trying to keep up with the flood of packets. A practical method to prevent broadcast storms is to divide the broadcast domain into several subdomains. This effect can be achieved by partitioning the network into isolated segments that communicate exclusively through routing (which can be done with VLANs).

Multicast Flooding

Similar to broadcast storms, *multicast flooding* tends to be more prevalent on VLANs and occurs when a switch receives a multicast packet that has an IP address for a group it has not learned. Since the switch does not know what to do with it, the switch floods that packet out of all ports on the VLAN. Many switches have an option to disallow this behavior, and you configure them to only forward unregistered packets to ports on a VLAN that are connected to specific ports.

Asymmetrical Routing

When a packet travels from a source to a destination in one path and takes a different path when it returns to the source, it is known as *asymmetrical routing*. The biggest weakness to this is the risk that packets might not arrive in the right order.

> **Note**
>
> The opposite of asymmetrical routing is symmetrical routing: the network uses a single route for incoming and outgoing packets.

Switching Issues

Spanning Tree Protocol (STP) is a switching protocol used in computer networks to prevent *network loops* in Ethernet networks. STP operates at the data link layer (Layer 2) of the OSI model and is designed to ensure that there is only one active path (the *root bridge*) between any two network devices to prevent broadcast storms and network instability caused by redundant paths. When multiple paths exist between switches in a network, STP dynamically calculates and selects the most efficient path while blocking redundant paths. If the active path fails, STP automatically recalculates and activates an alternative path to maintain network connectivity. STP is standardized by the IEEE 802.1D specification, with several variants and improvements such as Rapid Spanning Tree Protocol (RSTP) and Multiple Spanning Tree Protocol (MSTP) enhancing its capabilities.

A switching loop can occur any time there are multiple paths between two endpoints: more than one connection between two switches, for example. When there is a loop, it creates broadcast storms as broadcasts, and multicasts are forwarded by switches out every port (the switch[es] will repeatedly rebroadcast the messages, thus flooding the network). Layer 2 headers do not support a time-to-live (TTL) value, so a frame sent into a looped topology can loop forever.

As mentioned, the root bridge is the central switch in the STP topology. All other switches in the network calculate their paths to the root bridge based on the least-cost path, which is determined by the cumulative cost of the network links between each switch and the root bridge. The root bridge selection process involves switches exchanging bridge protocol data unit (BPDU) messages to determine which switch has the lowest bridge ID (a combination of bridge priority and MAC address). The switch with the lowest bridge ID becomes the root bridge.

With STP, each port on a switch is assigned a specific *port role* and *port state*, which determines its behavior in the spanning tree topology. These port roles and port states play a crucial role in preventing loops and ensuring a loop-free network topology. The root port is the port on a nonroot switch that has the lowest cost path to the root bridge. Each nonroot switch selects one root port per VLAN. The designated port is the port on a segment that has been chosen as the designated forwarder for that segment. Each network segment (LAN) has one designated port. A blocking port is a port that is in a nonforwarding state, effectively blocking traffic. Blocking ports are present on redundant paths to prevent loops. An alternate port is a backup port that is not currently part of the active spanning tree topology but can quickly transition to a forwarding state if the root port fails. A backup port is similar to an alternate port but is designated as a backup to a designated port. Backup ports provide redundancy for designated ports.

Port states may be blocking, listening, learning, or forwarding. Ports in the blocking state do not forward traffic and are effectively disabled: they listen to BPDUs to participate in the spanning tree election process. Ports in the listening state listen to BPDUs to determine the topology of the network and are preparing to transition to the learning state. Ports in the learning state receive and process BPDUs, update their MAC address tables, but do not yet forward traffic. Ports in the forwarding state are fully operational and forward data packets: they actively participate in the data forwarding process.

Routing Selection Issues

Similar to switching issues, routing issues can wreak havoc: a *routing loop*, for example, can go on forever. This typically is a problem when the routing tables contain cyclical entries. For example, suppose router A needs to send data to router C and believes the best way to get there is to forward it first to router B. Router B gets the data, sees it is for C, and has in its own table that the best way to get there is to go through router A. In this scenario, a loop is created that can go on forever, preventing the data from reaching its destination.

Problems with route selection, especially concerning the *routing table* and *default route*, can lead to various network issues and disruptions. Some of the common problems with these are explored in the sections that follow.

Missing Route

If the routing table is incomplete, certain destinations may become unreachable. This can occur due to misconfigurations, routing protocol failures, or network topology changes not being properly reflected in the routing table. Inaccurate routing entries can result in traffic being forwarded through suboptimal paths or to incorrect destinations. This may lead to increased latency, packet loss, or even security vulnerabilities if traffic is inadvertently routed to unauthorized destinations.

Default Route

A missing default route (gateway of last resort) can cause traffic destined for unknown networks to be dropped, resulting in connectivity issues. Without a default route, the router has no way to forward packets to destinations outside of its directly connected network. Misconfigured default routes can lead to traffic being forwarded to the wrong next-hop router or gateway. This can result in packets being dropped or sent to unintended destinations, causing network inefficiencies and communication failures.

Low Optical Link Budget

While it may sound like a problem with the company's comptroller, in reality, *low optical link budget* refers to the optical power budget in a fiber-optic communication link. This is the allocation of available optical power considering such factors as attenuation, splice losses, and connector losses.

Incorrect VLAN Assignment

While VLANs offer a plethora of positives, problems can occur when a user moves or gets connected to the wrong one. On a regular basis, an administrator should ensure that the user system is plugged into the correct VLAN port and that there are no problems with users connecting to an interface that is not assigned to them.

DNS Issues

When the wrong Domain Name Service (DNS) values (typically primary and secondary) are entered during router configuration, users cannot take advantage of the DNS. Depending on where the wrong values are given, name resolution may not occur (if all values are incorrect), or resolution could take a long time (if only the primary value is incorrect), thus giving the appearance that the Web is taking a long time to load.

Make sure the correct values appear for DNS entries in the router configuration to avoid name resolution problems.

Incorrect Gateway

The default gateway configured on the router is the place where the data goes after it leaves the local network. Although many routes can be built dynamically, it is often necessary to add the first routes when installing or replacing a router. You can use the **ip route** command on most Cisco routers to do this from the command line, or most routers include a graphical interface for simplifying the process.

When you have the gateways configured, use the **ping** and **tracert/traceroute** utilities to verify connectivity and proper configuration.

Incorrect Subnet Mask

When the subnet mask is incorrect, the router thinks the network is divided into segments other than how it is actually configured. Because the purpose of the router is to route traffic, a wrong value here can cause it to try to route traffic to subnets that do not exist. The value of the subnet mask on the router must match the true configuration of the network.

Duplicate or Incorrect IP Address

Every IP address on a network must be unique. This is true not only for every host, but for the router as well, and every network card in general. The scope of the network depends on the size of the network that the card is connected to; if it is connected to the LAN, the IP address must be unique on that LAN, whereas if it is connected to the Internet, that address must be unique on it.

If there is a duplicate address, in the best scenario you will receive messages indicating duplicate IP addresses, and in the worst scenario, network traffic will become unreliable. In all cases, you must correct the problem and make certain duplicate addresses exist nowhere on your network, including the routers.

Duplicate MAC Addresses

The MAC address is hard-coded into the NIC and cannot be changed. It consists of two components: one identifies the vendor, and the other identifies a serial number so that it will be unique. Of all things on the network, this is the one value that must stay constant for ARP, RARP, and other protocols to be able to translate IP addresses to machines and have communication across the network.

Given that, the only way for a MAC address to not be unique is for someone to be trying to add a rogue device impersonating another (typically a server). If this is the case, there will be serious problems on the network, and you must find—and disable—the unauthorized device immediately.

Expired IP Address

DHCP leases IP addresses to clients and—when functioning properly—continues to renew those leases as long at the client needs them. An expired address can mean that the DHCP server is down or unavailable, and the client will typically lose its address, rendering it unable to continue communicating on the network.

Each system must be assigned a unique IP address so that it can communicate on the network. Clients on a LAN have a private IP address and matching subnet mask. Table 10.2 shows the private IP ranges. If a system has the wrong IP or subnet mask, it cannot communicate on the network. If the client system has *misconfigured DHCP settings*, such as an IP address in the 169.254.0.0 APIPA range, the system is not connected to a DHCP server and is not able to communicate beyond the network.

TABLE 10.2 **Private Address Ranges**

Class	Address Range	Default Subnet Mask
A	10.0.0.0 to 10.255.255.255	255.0.0.0
B	172.16.0.0 to 172.31.255.255	255.255.0.0
C	192.168.0.0 to 192.168.255.255	255.255.255.0

ExamAlert

You need to know the private address ranges in Table 10.2.

Note

To release an IP address on a Windows PC, type **ipconfig /release** in a command window (reached by clicking Start, Run and typing **cmd** in the Open field; then select Run as administrator) and press Enter. To renew the address, use the command **ipconfig /renew** instead.

Rogue DHCP Server

A *rogue DHCP server* is any DHCP server on the network that was added by an unauthorized party and is not under the administrative control of the network administrators. It can be used to give false values or to set up clients for network attacks, such as on-path (formerly called man-in-the-middle) attacks. To prevent rogue DHCP servers, look for IP address conflicts, keep a properly documented network, use Active Directory to authorize DHCP servers, and use DHCP snooping and trusted ports on your switches.

Certificate Issues

Certificate authorities (CAs) issue certificates, verify the holder of a digital certificate, and ensure that holders of certificates are who they claim to be. An *untrusted SSL certificate* is usually one that is not signed or that has expired. Sometimes, this issue can be caused by a client using an older browser or one that is not widely supported. As a general rule, though, users should be instructed to stop attempting to visit a site if they see this error. While certificate security was previously covered in Chapter 9, "Network Security," Table 10.3 lists the most common certificate issues and fixes.

TABLE 10.3 **Common Certificate Issues and Fixes**

Issue	Fix
Certificate Expired	Renew the certificate by obtaining a new one from the certificate authority (CA) and replace the expired certificate.
Certificate Revoked	Check the certificate revocation list (CRL) or Online Certificate Status Protocol (OCSP) for revocation information. If revoked, obtain a new certificate from the CA.
Certificate Not Trusted	Ensure that the root certificate of the issuing CA is trusted by the client's system or application.
Name Mismatch	Update the certificate with the correct Common Name (CN) or Subject Alternative Name (SAN) entries that match the hostname or URL.
Weak Encryption Algorithm	Reissue the certificate using a stronger encryption algorithm (e.g., SHA-256 or SHA-384) to enhance security.
Incomplete Certificate Chain	Ensure that the certificate chain is complete by including all necessary intermediate and root certificates.
Incorrect Certificate Usage	Validate that the certificate is being used for its intended purpose (e.g., server authentication, client authentication) and reissue if necessary.
Self-Signed Certificate	Obtain a certificate signed by a trusted CA instead of using a self-signed certificate for enhanced trust and security.
Misconfigured Certificate Store	Check the certificate store configuration to ensure that certificates are installed correctly and accessible by the intended applications.

NTP Issues/Incorrect Time

Incorrect time on a network can be more than just an annoyance because time-stamps are important if you're trying to document an attack. Most network devices use Network Time Protocol (NTP) to keep the system time as defined by a designated server. You should make sure that server has the correct time on it and is updated, patched, and secured just as you would any other network critical server.

> **Note**
>
> The Network Time Protocol (NTP) must be functioning properly in order for a network administrator to create an accurate timeline during a troubleshooting process.

Address Pool Exhaustion

The DHCP scope is the pool of possible IP addresses a DHCP server can issue. If that pool becomes exhausted and not enough addresses are available for the devices needing to connect, devices will not be given the values they need (many will then resort to using APIPA addresses in the 169.254 range, as discussed in Table 10.2).

The only solution is to increase the scope and/or decrease the lease time. If you reduce the lease time from days to hours, more addresses should become available as hosts leave the network at the end of their shifts, and those values become available for use by others.

Blocked Ports, Services, or Addresses

As a security rule, only needed ports should be enabled and allowed on a network. Unfortunately, you don't always have a perfect idea of which ports you need, and it is possible to inadvertently have some blocked TCP/UDP ports that you need to use.

If you find your firewall is blocking a needed port, you should open that port (make an exception) and allow it to be used.

Incorrect Firewall Settings

Incorrect firewall settings typically fall under the category of blocking ports that you need open (previously addressed) or allowing ports that you don't need. From a security perspective, the latter situation is worse because every open port represents a door that an intruder could use to access the system or at least a vulnerability. Be sure to know which ports are open, and close any that are not needed.

Incorrect ACL Settings

The purpose of an access control list (ACL) is to define who or what can access your system. Incorrect ACL settings could keep too many off, but typically the error is allowing too many on. Used properly, an ACL can enable devices in your network to ignore requests from specified users or systems or to grant them certain network privileges. You may find that a certain IP address is constantly scanning your network, and you can block this IP address. If you block it at the router, the IP address will automatically be rejected anytime it attempts to use your network.

Unresponsive Service

When a service does not respond, the reason could be that it is overloaded, is down, or has bad configuration. The first order of business is to ascertain which of these three the situation is and then decide what you need to do to fix it. If the server/service is overloaded, you can look for a way to increase the capacity or balance the load. If the server/service is down, you can investigate why and what needs to be done to bring it back up again. If the server/service is misconfigured, you can make the necessary changes to configure it properly.

BYOD Challenges

Bring-your-own-device (BYOD) challenges occur when employees are allowed to bring personally owned mobile devices (laptops, tablets, and smartphones) to their workplace and use them on the network. Good onboarding and off-boarding procedures (discussed in Chapter 8, "Network Operations") as well as mobile device management (MDM) policies should be used to help protect network resources and still allow these devices to be used in the workplace.

Licensed Feature Issues

Many devices, such as switches, have features that are available with them only if licensed. Many times, you can enable the feature and use it on a trial basis before purchasing (a grace period, if you will), but you must purchase and install the number of licenses required for that feature before the grace period ends; otherwise, the feature will disable itself. To keep legal, you should always be cognizant of licensing issues and careful to not run afoul of them.

Hardware Failure

If you are looking for a challenge, troubleshooting hardware infrastructure problems is for you. It is often not an easy task and usually involves many processes, including baselining and performance monitoring. One of the keys to identifying the hardware failure is to know what devices are used on a particular network and what each device is designed to do. Table 10.4 lists some of the common hardware components used in a network infrastructure, as well as some common problem symptoms and troubleshooting methods.

TABLE 10.4 **Common Network Hardware Components, Their Functions, and Troubleshooting Strategies**

Networking Device	Function	Troubleshooting and Failure Signs
Firewall	A firewall monitors and controls incoming and outgoing network traffic based on predetermined security rules.	Review firewall rule configurations for accuracy. Monitor firewall logs for unusual traffic patterns or denied connections.
Hub	Hubs are used with a star/hub-and-spoke network topology and UTP cable to connect multiple nodes.	Because hubs connect multiple network devices, if many devices are unable to access the network, the hub may have failed. When a hub fails, all devices connected to it cannot access the network. In addition, hubs use broadcasts and forward data to all the connected ports, increasing network traffic. When network traffic is high, and the network is operating slowly, it may be necessary to replace slow hubs with switches.
Load Balancer	A load balancer distributes incoming network traffic across multiple servers to optimize resource utilization and ensure high availability.	Monitor server health and performance metrics. Review load balancer configuration for balancing algorithms and health checks.
Network cable	A network cable connects devices within a network.	Inspect cable connectors for damage or wear. Use cable testing tools to check for continuity and signal strength. Replace damaged cables or connectors as needed.
Network interface card (NIC)	A NIC enables devices to connect to a network.	Ensure NIC drivers are installed and up-to-date. Check network cable connections. Verify NIC settings and TCP/IP configuration.

Networking Device	Function	Troubleshooting and Failure Signs
Router	Routers are used to separate broadcast domains and to connect different networks.	If a router fails, network clients are unable to access remote networks connected by the router. For example, if clients access a remote office through a network router, and the router fails, the remote office is unavailable. You can test router connectivity using utilities such as **ping** and **tracert**.
Switch	Like hubs, switches are used with a star/hub-and-spoke topology to create a central connectivity device.	The inability of several network devices to access the network may indicate a failed switch. If the switch fails, all devices connected to the switch cannot access the network. Switches forward data only to the intended recipient, allowing them to manage data better than hubs.
Wireless access point	Wireless access points provide the bridge between the wired and wireless network.	If wireless clients cannot access the wired network, the AP may have failed. However, you should check many configuration settings first.
Network-attached storage (NAS)	NAS provides centralized storage accessible over a network.	Verify NAS connectivity and accessibility. Check NAS storage status and available disk space. Review NAS access permissions and user accounts.

ExamAlert

Be familiar with the devices listed in Table 10.4 and their failure signs.

Note

With network device issues, consider using network monitoring tools such as Syslog and SNMP: the logs they create are very helpful when troubleshooting.

For more information on network hardware devices and their functions, refer to Chapter 4, "Network Implementations."

Network Performance Issues

Chapter 8 looked at many performance issues, and most of the issues discussed in this section already have been addressed as they relate to networking. Be aware of those issues and that it is always a balancing act trying to get optimum performance from a network when working with so many disparate devices.

Wireless Issues

Poor communication between wireless devices has many different potential causes. Some of these problems, such as *latency* and *jitter*, are similar to those that exist with wired connections and were discussed earlier in this chapter. Others are characteristic only of wireless connectivity and are discussed in the following sections.

To put a lot of information into a format that is coherent, this discussion starts with a review checklist of wireless troubleshooting and then moves into some individual topics:

▶ **Signal loss:** The cause of signal loss, known as *attenuation*, can be any-thing from distance to obstacles to interference. The *signal-to-noise ratio* should be examined to measure the desired signal against the background noise interfering with it. Look for signs of saturation with either the device or the bandwidth.

> **ExamAlert**
>
> Signal-to-noise ratio can be used to measure that which the name implies.

▶ **Wireless enabled:** Some laptops make it incredibly easy to turn wireless on and off. A user may accidentally press a button that they are not aware of and then suddenly not be able to access the network. Although this is a simple problem to fix, it is one that you need to identify as quickly as possible. Figure 10.1 shows the wireless light on an HP laptop. This light is also a button that toggles wireless on and off. When the light is blue, wireless is enabled, and when it is not blue (orange), it is disabled.

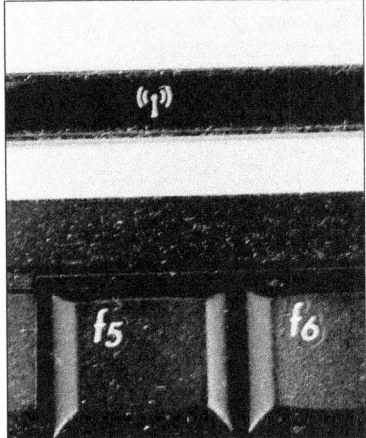

FIGURE 10.1 **A Light also Serves As a Button, Enabling Wireless to Be Quickly Turned On and Off**

▶ **Untested updates:** Never apply untested updates to the network. This is especially true with AP updates, which should always be tested in nonproduction environments before being applied to live machines.

▶ **Wrong wireless standard:** Make sure that the standard you are using supports the rates and attributes you are striving for. This is particularly important in terms of throughput, frequency, distance, and channels.

▶ **Auto transfer rate:** By default, wireless devices are configured to use the strongest, fastest signal. If you experience connectivity problems between wireless devices, try using the lower transfer rate in a fixed mode to achieve a more stable connection. For example, you can manually choose the wireless transfer rate. Also, instead of using the highest transfer rate available, try a lesser speed. The higher the transfer rate, the shorter the connection distance.

▶ **AP placement and configuration:** If signal strength is low, try moving the AP to a new location. Moving it just a few feet can make a difference. You can also try to *bounce* a signal, as needed, off reflective surfaces. The configuration of the AP should take into account the use of *Lightweight Access Point Protocol (LWAPP)*, which can allow you to monitor the network and reduce the amount of time needed to configure and troubleshoot it—and whether the authentication/configuration will be done at the AP (known as *thick*) or it will be passed on up (known as *thin*).

Within the 802.11 standard, signal strength is measured in terms of *Received Signal Strength Indication (RSSI)*. This value is an indicator of the power level being received by the receiving host after any antenna or cable loss. The greater the RSSI value, the stronger the signal.

> **Note**
>
> Anytime an AP is doing key functions—authentication, filtering, QoS enforcement, and so on—it is said to be *thick*. If it is not doing these key functions—even though it might be doing others—it is usually said to be *thin*. Although there is no 100 percent sure method of distinguishing what a vendor will label thick or thin, one good rule is to question whether the AP is dependent on another device (thin) or not (thick).

▶ **Antenna:** The default antenna shipped with wireless devices may not be powerful enough for a particular client system. Better-quality antennas can be purchased for some APs, which can boost the distance the signal can go. Make sure you do not use the wrong antenna type or have other incompatibilities.

Effective Isotropic Radiated Power (EIRP) is used to measure the combination of the power emitted by the transmitter and the ability of the antenna to direct that power in a given direction. It is the total power—expressed in watts—that would need to be radiated by a half-wave dipole antenna to give the same signal strength as the actual source antenna at a distant receiver located in the direction of the antenna's strongest beam.

▶ **Environmental obstructions:** Wireless RF communications are weakened if they have to travel through obstructions such as metal studs, window film, and concrete walls. Wireless site surveys can be performed to troubleshoot RF signal loss issues as well as assist in planning optimal locations for new wireless networks.

▶ **Conflicting devices:** Any device that uses the same frequency range as the wireless device can cause interference. For example, 2.4 GHz phones, appliances, or Bluetooth devices can cause interference with devices using the 802.11g or single-band 802.11n wireless standards.

▶ **Wireless channels:** If connections are inconsistent, try changing the channel to another, nonoverlapping channel. Make certain you do not have mismatched channels between devices.

▶ **Protocol issues:** If an IP address is not assigned to the wireless client, a wrong SSID or incorrect WEP/WPA/WPA2/WPA3 settings can prevent a system from obtaining IP information.

▶ **SSID:** The SSID number used on the client system must match the one used on the AP. You might need to change it if you are switching a laptop or other wireless device between different WLANs.

▶ **Encryption type:** If encryption is enabled on the connecting system, the encryption type must match what is set in the AP. For example, if the AP uses WPA2/WPA3-AES, the connecting system must also use WPA2/WPA3-AES.

▶ **Captive portal issues:** Most public networks, including Wi-Fi hotspots, use a *captive portal*, which is a web page that requires users to agree to some condition before they use the network or Internet. The condition could be to agree to the acceptable use policy (AUP), payment charges for the time they are using the network, and so forth. Security vulnerabilities have been reported with captive portals, so administrators should be on the alert for any new problems that are reported.

▶ **Client disassociation issues:** When a client connects to an AP, it is said to associate with that AP, and disassociation is the process of it no longer associating with that AP. Disassociation can happen any time the AP thinks it no longer needs to communicate with the client—due to going into hibernation mode, powering down, leaving the building, and so on. Most unintentional disassociations can be traced to weak signals, but relocating the AP (or boosting the signal) can often help. The goal of *seamless roaming* is for a mobile device to be able to maintain a continuous and uninterrupted connection to a wireless network while moving between different access points (APs) or coverage areas within the same network. This can be accomplished with preauthentication (the device authenticates with nearby access points in advance, allowing for quicker transitions when roaming occurs), Layer 2 roaming (roaming occurs at the data link layer (Layer 2) of the OSI model, enabling the client device to maintain its IP address and session state during the transition), and intelligent steering (access points or controllers dynamically steer client devices to the most suitable AP based on factors like signal strength, traffic load, and RF interference).

Note

Disassociation also factors into a common wireless attack: the attacker disassociates the user from the authenticating wireless access point and then carries out another attack to obtain the user's valid credentials. A main purpose of this attack is to get the user to connect to an evil twin access point where the attacker can steal the user's credentials or data.

ExamAlert

Captive portals are common in public places such as airports and coffee shops. The user simply clicks Accept, views an advertisement, provides an email address, or performs some other required action. The network then grants access to the user and no longer holds the user captive to that portal.

Most router configuration interfaces allow you to run basic diagnostics through them, as illustrated in Figure 10.2. You can also usually change the security settings and configure the firewall, as shown in Figure 10.3.

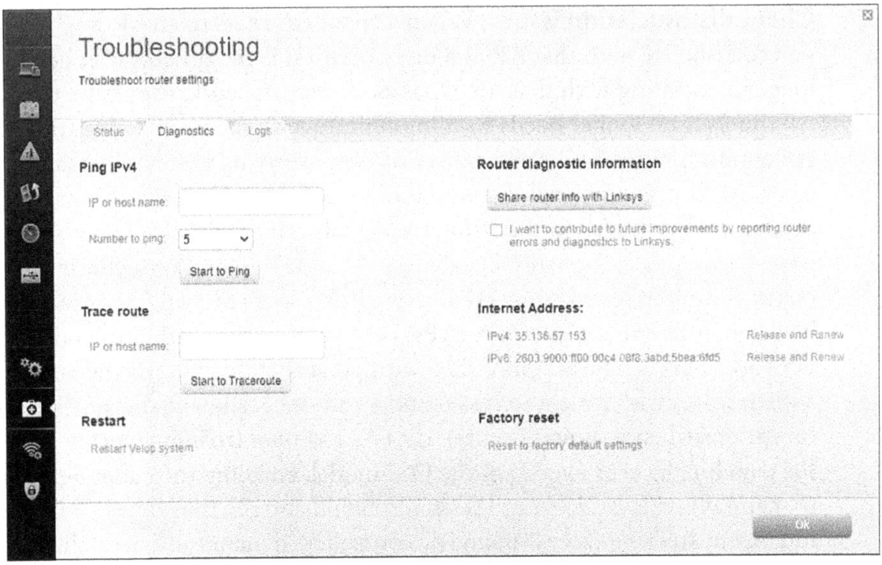

FIGURE 10.2 **Wireless Router Diagnostic Options**

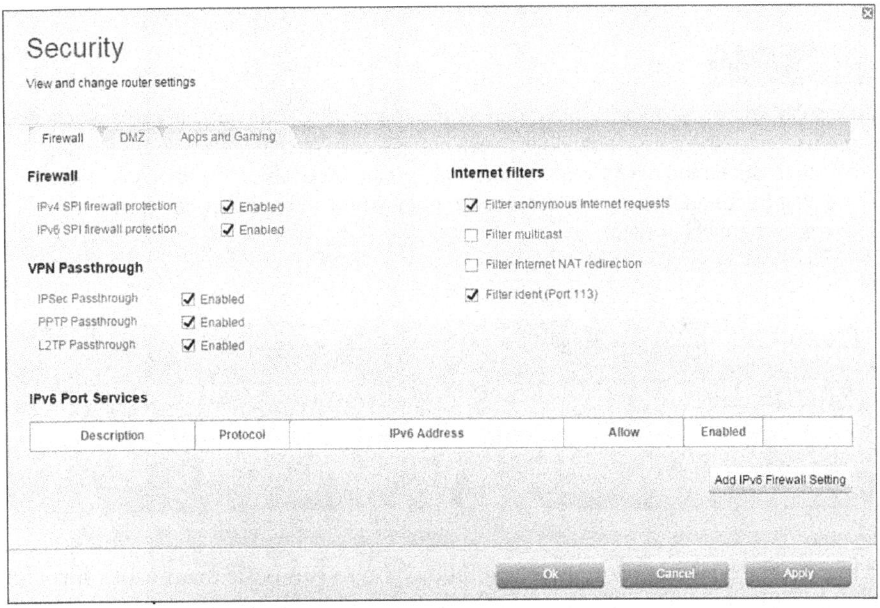

FIGURE 10.3 **Configuring Security Settings**

Site Surveys

As more networks go wireless, you need to pay special attention to issues associated with them. Wireless survey tools can be used to create heat maps showing the quantity and quality of wireless network coverage in areas. They can also allow you to see access points (including rogues) and security settings. These tools can be used to help you design and deploy an efficient network, and they can also be used (by you or others) to find weaknesses in your existing network (often marketed for this purpose as wireless analyzers).

Insufficient wireless coverage refers to areas within a wireless network where the signal strength or quality is inadequate for reliable connectivity. It means that wireless devices such as laptops, smartphones, or IoT devices may experience poor or no wireless connectivity in those areas, leading to communication issues, slow performance, or dropped connections.

Factors Affecting Wireless Signals

Because wireless signals travel through the atmosphere, they are susceptible to different types of interference than are standard wired networks. Interference weakens wireless signals and therefore is an important consideration when working with wireless networking.

Interference

Wireless interference is an important consideration when you plan a wireless network. Interference is, unfortunately, inevitable, but the trick is to minimize the levels of interference. Wireless LAN communications typically are based on radio frequency signals that require a clear and unobstructed transmission path.

The following factors can cause interference:

▶ **Physical objects:** Trees, masonry, buildings, and other physical structures are some of the most common sources of interference. The *density* of the materials used in a building's construction determines the number of walls the RF signal can pass through and still maintain adequate coverage. Concrete and steel walls are particularly difficult for a signal to pass through. These structures weaken or at times completely prevent wireless signals.

ExamAlert

Be sure that you understand that physical objects are a common source of interference. A wireless site survey can be used to test for interference.

▶ **Radio frequency interference:** Wireless technologies such as 802.11n can use an RF range of 2.4 GHz, and so do many other devices, such as cordless phones, microwaves, Bluetooth devices, and so on. Devices that share the channel can cause noise and weaken the signals.

▶ **Electrical interference:** Electrical interference comes from devices such as computers, refrigerators, fans, lighting fixtures, or any other motorized devices. The impact that electrical interference has on the signal depends on the proximity of the electrical device to the wireless AP. Advances in wireless technologies and in electrical devices have reduced the impact that these types of devices have on wireless transmissions.

▶ **Environmental factors:** Weather conditions can have a huge impact on wireless signal integrity. Lightning, for example, can cause electrical interference, and fog can weaken signals as they pass through.

Channel overlap can occur where adjacent or overlapping wireless channels in the same frequency band interfere with each other, leading to degraded performance (*signal degradation*) and reduced throughput in the network. When wireless channels are adjacent or closely spaced, the signals transmitted on one channel can spill over into neighboring channels: this interference occurs due to imperfect filtering and signal attenuation, causing devices operating on adjacent channels to experience interference from each other. Co-channel interference occurs when multiple devices share the same channel and attempt to transmit data simultaneously: this can lead to collisions and packet loss as devices contend for access to the shared channel.

Reflection, Refraction, and Absorption

The line differentiating between interference and reflection can be blurry when it comes to wireless networking. The key difference between them is that *interference* is a conflict with something else (usually another signal), whereas *reflection* is a problem caused by a bouncing of the same signal off an object. A subset of this is refraction, which involves a change in direction of the wave as a result of its traveling at different speeds at different points. Put in simple terms, reflection happens when the signal hits a piece of metal and cannot pass through, and refraction happens when the signal goes through a body of water.

If the wave is completely swallowed by the object it hits (not reflected, or refracted), then it is said to be absorbed. Where security is concerned, items known to absorb wireless signals can be used to prevent the signal from traveling beyond an established perimeter. Shielding paint (sometimes called RF paint) can be used for this purpose, as can copper plates and aluminum sheets.

Many wireless implementations are found in the office or at home. Even when outside interference such as weather is not a problem, every office has plenty of wireless obstacles. Table 10.5 highlights a few examples to be aware of when implementing a wireless network indoors.

TABLE 10.5 **Wireless Obstacles Found Indoors**

Obstruction	Obstacle Severity	Sample Use
Wood/wood paneling	Low	Inside a wall or hollow door
Drywall	Low	Inside walls
Furniture	Low	Couches or office partitions
Clear glass	Low	Windows
Tinted glass	Medium	Windows
People	Medium	High-volume traffic areas that have considerable pedestrian traffic
Ceramic tile	Medium	Walls
Concrete blocks	Medium/high	Outer wall construction
Mirrors	High	Mirror or reflective glass
Metals	High	Metal office partitions, doors, metal office furniture
Water	High	Aquariums, rain, fountains

ExamAlert

Be sure that you understand the severity of obstructions given in Table 10.5.

Troubleshooting AP Coverage

Like any other network medium, APs have a limited transmission distance. This limitation is an important consideration when you decide where an AP should be placed on the network. When troubleshooting a wireless network, pay close attention to how far the client systems are from the AP.

ExamAlert

Distance limitations from the AP are among the first things to check when troubleshooting AP coverage.

When faced with a problem in which client systems cannot consistently access the AP, you could try moving the AP to better cover the area, but then you may disrupt access for users in other areas. So what can be done to troubleshoot AP coverage?

Depending on the network environment, the quick solution may be to throw money at the problem and purchase another access point, cabling, and other hardware to expand the transmission area. However, you can try a few things before installing another wireless AP. The following list starts with the least expensive solution and progresses to the most expensive:

▶ **Increase transmission power:** Some APs have a setting to adjust the transmission power output (power levels). By default, most of these settings are set to the maximum output; however, this is worth verifying just in case. You can decrease the transmission power if you are trying to reduce the dispersion of radio waves beyond the immediate network. Increasing the power gives clients stronger data signals and greater transmission distances.

▶ **Relocate the AP:** When wireless client systems suffer from connectivity problems, the solution may be as simple as relocating the AP. You could relocate it across the room, a few feet away, or across the hall. Finding the right location will likely take a little trial and error.

▶ **Adjust or replace antennas:** If the AP distance is insufficient for some network clients, you might need to replace the default antenna used with both the AP and the client with higher-end antennas. Upgrading an antenna can make a big difference in terms of transmission range. Unfortunately, not all APs have replaceable antennas.

▶ **Tweak the signal amplification:** *Radio frequency (RF)* amplifiers add significant distance to wireless signals. An RF amplifier increases the strength and readability of the data transmission. The amplifier improves both the received and transmitted signals, resulting in an increase in wireless network performance.

▶ **Use a repeater:** Before installing a new AP, you might want to think about a wireless repeater. When set to the same channel as the AP, the repeater takes the transmission and repeats it. So, the AP transmission gets to the repeater, and then the repeater duplicates the signal and passes it on. This is an effective strategy to increase wireless transmission distances.

ExamAlert

Be prepared to answer questions on AP coverage and possible reasons to relocate or replace APs.

Lastly, know that *roaming misconfiguration* (issues or errors in the configuration of roaming parameters that affect the seamless transition of wireless clients) can lead to problems such as connectivity issues, dropped connections, or degraded performance for roaming clients. Roaming thresholds dictate the signal strength at which a client should switch to a different AP. If the roaming thresholds are set too high, clients may stay connected to an AP with a weak signal for too long, leading to poor performance. Conversely, if the thresholds are set too low, clients may roam too frequently, causing unnecessary handovers and potential disruptions. Fast roaming techniques such as 802.11r (Fast BSS Transition) or 802.11k (Neighbor Reports) aim to expedite the roaming process by pre-authenticating clients with neighboring APs or providing information about nearby APs. Misconfigurations or mismatches in fast roaming settings between APs or clients can lead to authentication failures, disruptions, or inconsistencies in roaming behavior.

Cram Quiz

1. Although many routes can be built dynamically, it is often necessary to add the first routes when installing or replacing a router. Which of the following commands can you use on most Cisco routers to do this from the command line?

   ○ **A. ip route**
   ○ **B. add route**
   ○ **C. first route**
   ○ **D. route change**

2. Which of the following best describes the function of the default gateway?

   ○ **A.** It converts hostnames to IP addresses.
   ○ **B.** It converts IP addresses to hostnames.
   ○ **C.** It enables systems to communicate with systems on a remote network.
   ○ **D.** It enables systems to communicate with routers.

3. Consider the following figure. Which of the following statements is true?

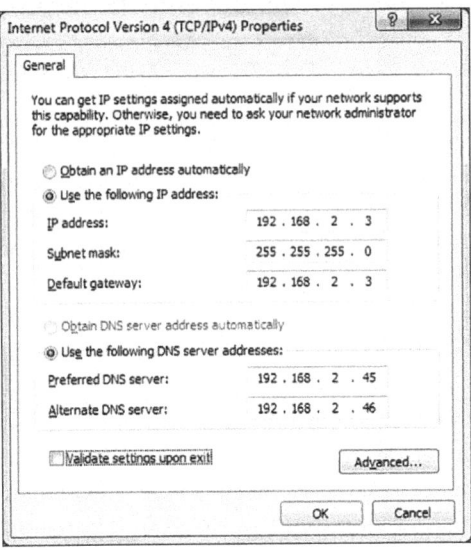

- ○ **A.** The system cannot access the local network.
- ○ **B.** The system cannot access remote networks.
- ○ **C.** The system cannot have hostname resolution.
- ○ **D.** The system has the wrong subnet mask.

4. Which of the following bits of IP information are mandatory to join the network? (Choose two.)
- ○ **A.** Subnet mask
- ○ **B.** IP address
- ○ **C.** DNS address
- ○ **D.** Default gateway

5. All of a sudden, some devices on your local network are not receiving their normal IP addresses, and some of them are now using APIPA addresses in the 169.254 range. What has most likely occurred?
- ○ **A.** Low optical link budget
- ○ **B.** NTP time issue
- ○ **C.** Broadcast storm
- ○ **D.** DHCP scope exhaustion

6. You purchase a new wireless AP that uses no security by default. You change the security settings to use 128-bit encryption. How must the client systems be configured?

- ○ **A.** All client systems must be set to 128-bit encryption.
- ○ **B.** The client system inherits security settings from the AP.
- ○ **C.** Wireless security does not support 128-bit encryption.
- ○ **D.** The client wireless security settings must be set to autodetect.

7. You experience connectivity problems with your SOHO network. What can you change in an attempt to solve this problem?

- ○ **A.** Shorten the SSID.
- ○ **B.** Remove all encryption.
- ○ **C.** Lower the transfer rate.
- ○ **D.** Raise the transfer rate.

8. Which of the following is a web page that is first launched when a user is connecting through a network that usually requires some type of interaction before the user is allowed access to other network resources or Internet sites?

- ○ **A.** EIRP
- ○ **B.** RSSI
- ○ **C.** Captive portal
- ○ **D.** SSID

Cram Quiz Answers

1. **A.** Although many routes can be built dynamically, it is often necessary to add the first routes when installing or replacing a router. You can use the **ip route** command on most Cisco routers to do this from the command line.

2. **C.** The default gateway enables the system to communicate with systems on a remote network, without the need for explicit routes to be defined. The default gateway can be assigned automatically using a DHCP server or can be input manually.

3. **B.** The IP addresses of the client system and the default gateway are the same. This error probably occurred when the IP address information was input. In this configuration, the client system would likely access the local network and resources but not remote networks because the gateway address to remote networks is wrong. The DNS, IP, and subnet mask settings are correct.

4. **A and B.** Configuring a client requires at least the IP address and a subnet mask. The other values are all optional, but network functionality may be limited without them.

5. **D.** The DHCP scope is the pool of possible IP addresses a DHCP server can issue. If that pool becomes exhausted and not enough addresses are available for the devices needing to connect, devices will not be given the values they need (many will then resort to using APIPA addresses in the 169.254 range).

6. **A.** On a wireless connection between an AP and the client, each system must be configured to use the same wireless security settings. In this case, they must both be configured to use 128-bit encryption.

7. **C.** If you experience connectivity problems between wireless devices, try using the lower transfer rate in a fixed mode to achieve a more stable connection. For example, you can manually choose the wireless transfer rate. The higher the transfer rate, the shorter the connection distance.

8. **C.** A captive portal is a web page that is first launched when a user is connecting through a network that usually requires some type of interaction before the user is allowed access to other network resources or Internet sites. Effective Isotropic Radiated Power (EIRP) is used to measure the combination of the power emitted by the transmitter and the ability of the antenna to direct that power in a given direction. Received Signal Strength Indication (RSSI) is an indicator of the power level being received by the receiving host after any antenna or cable loss. The greater the RSSI value, the stronger the signal. The SSID number used on the client system must match the one used on the AP. You might need to change it if you are switching a laptop or other wireless device between different WLANs.

Troubleshooting Tools

▶ **5.5 Given a scenario, use the appropriate tool or protocol to solve networking issues.**

CramSaver

If you can correctly answer these questions before going through this section, save time by skimming the ExamAlerts in this section and then completing the Cram Quiz at the end of the section.

1. What cross-platform tool is used for network performance measurement/tuning and can produce standardized performance measurements for any network?

2. What TCP/IP command can be used to troubleshoot DNS problems?

3. What is the Linux, macOS, and UNIX equivalent of the **ipconfig** command?

4. What utility is part of the TCP/IP suite and has the function of resolving IP addresses to MAC addresses?

Answers

1. The **iperf** tool is used for network performance measurement/tuning and can produce standardized performance measurements for any network.

2. The **nslookup** command is a TCP/IP diagnostic tool used to troubleshoot DNS problems. On Linux, UNIX, and macOS systems, you can also use the **dig** command for the same purpose.

3. The **ifconfig** command is the Linux, macOS, and UNIX equivalent of the **ipconfig** command.

4. The function of **arp** is to resolve IP addresses to MAC addresses.

ExamAlert

Remember that this objective begins with "Given a scenario." This means that you may receive a drag-and-drop, matching, or "live OS" scenario where you have to click through to complete a specific objective-based task.

A large part of network administration involves having the right tools for the job and knowing when and how to use them. Selecting the correct tool for a networking job sounds like an easy task, but network administrators can choose from a mind-boggling number of tools and utilities.

Given the diverse range of tools and utilities available, it is unlikely that you will encounter all the tools available—or even all those discussed in this chapter. For the Network+ exam, you are required to have general knowledge of the tools available and what they are designed to do.

Toner

A *toner* is a hardware tool used to trace and identify network cables within a structured cabling system and is also known as a cable tracer or cable toner. The toner typically consists of two components: a tone generator and a probe. The tone generator is a handheld device typically connected to one end of the cable being tested, and it transmits an electrical signal (usually in the form of an audible tone or a digital signal) onto the conductor of a network cable. The probe is a handheld receiver that is used to detect the signal transmitted by the tone generator. It is equipped with a speaker or headphones to audibly indicate when the signal is detected. The probe is moved along the length of the cable, and when it detects the transmitted signal, it produces an audible tone or visual indicator to help locate the cable.

By using a toner, network technicians can quickly and accurately identify the physical location of network cables, locate faults or breaks in the cable, and verify cable connections. This capability is particularly useful when dealing with large network installations or complex cabling environments where cables may be concealed behind walls, above ceilings, or within cable trays. Toning allows technicians to trace cables from one end to another, facilitating cable management, troubleshooting, and maintenance tasks within the network infrastructure.

Cable Tester

A *cable tester* is a hardware tool used for network troubleshooting and validation of network cabling installations. It is designed to verify the integrity and performance of various types of network cables, such as Ethernet cables (e.g., Cat 5e, Cat 6, Cat 6a), coaxial cables, and fiber-optic cables. Cable testers help identify faults, wiring errors, and connectivity issues within the cabling infrastructure, ensuring reliable network connectivity and optimal performance.

Taps

Taps are hardware devices used to monitor network traffic nonintrusively. Taps are passive devices that are inserted into a network link to capture and monitor data packets as they pass through the network. Unlike network switches or routers, which actively forward and process data packets, taps operate transparently and do not affect the flow of traffic on the network. Taps are commonly used for network analysis, performance monitoring, security monitoring, and troubleshooting.

Since taps capture the entire data packets passing through the network link—including the payload, headers, and metadata—they can be used for comprehensive analysis of network traffic for troubleshooting, forensic analysis, and security investigations. Taps ensure zero packet loss by copying the traffic from the network link without interfering with the original data flow. This ensures that monitoring tools receive all network packets for analysis, even during high-traffic conditions. They support monitoring of both directions of traffic (ingress and egress) on the network link simultaneously. This allows monitoring tools to analyze traffic flows in both directions for comprehensive visibility into network activity.

Some taps feature multiple output ports, allowing the monitored traffic to be sent to multiple monitoring tools simultaneously. This enables parallel analysis by multiple monitoring tools or redundant monitoring for high availability. They typically use passive splitting technology to replicate network traffic to monitoring devices without introducing additional points of failure. Taps are compatible with various network media and speeds, including Ethernet, fiber optic, copper, and different data rates (e.g., 1 Gbps, 10 Gbps, 40 Gbps, 100 Gbps). They support different types of network connections, such as copper Ethernet (RJ-45) and fiber optic (LC, SC, and so on).

Visual Fault Locator

A *visual fault locator (VFL)* is a handheld hardware tool used for network troubleshooting, particularly in fiber-optic cabling systems. It emits a visible laser light that helps technicians identify breaks, bends, or other faults in fiber-optic cables. Visual fault locators are portable, easy to use, and effective for quickly identifying issues in fiber-optic cables, making them an essential tool for network technicians, installers, and maintenance personnel working with fiber-optic infrastructure.

Wi-Fi Analyzer

As the name implies, a *Wi-Fi analyzer* is used to identify Wi-Fi problems. It can be helpful in finding the ideal place for locating an access point or the ideal channel to use. Many software-based Wi-Fi analyzers are available, and some use the word *scanner* in place of *analyzer*. One good feature of most wireless survey tools is that they can be used to create heat maps showing the quantity and quality of wireless network coverage in areas. They can also allow you to see access points (including rogues) and security settings. They can be used to help you design and deploy an efficient network, and they can also be used (by you or others) to find weaknesses in your existing network.

Before we explore some other common troubleshooting tools, Table 10.6 highlights the five just discussed that CompTIA wants you to know for the exam.

TABLE 10.6 **Common Hardware Troubleshooting Tools**

	Toner	Cable Tester	Tap	Visual Fault Locator (VFL)	Wi-Fi Analyzer
Description	Hardware tool that sends a signal down a cable to identify its location and ensure proper connectivity	Hardware tool that tests the continuity and integrity of network cables by checking for shorts, opens, miswires, and cable length	Passive network monitoring device that allows for nonintrusive monitoring of network traffic	Hardware tool that emits a visible laser light to identify breaks, bends, or faults in fiber-optic cables	Software tool that analyzes Wi-Fi networks, including signal strength, channel interference, and network congestion
Common Use Cases	Identifying cable runs and endpoints, tracing cables in large installations	Verifying cable integrity during installation, troubleshooting connectivity issues	Capturing network traffic for analysis, monitoring network performance and security	Locating breaks or faults in fiber-optic cables, verifying fiber continuity	Optimizing Wi-Fi network performance, diagnosing connectivity issues
Advantages	Provides quick identification of cable endpoints, can trace cables through walls and ceilings	Provides detailed diagnostic information, identifies specific cable faults	Is nonintrusive, does not disrupt network traffic, allows for real-time monitoring and analysis	Provides visual indication of fiber faults, helps pinpoint exact location of fiber issues	Provides comprehensive analysis of Wi-Fi networks, helps identify sources of interference

	Toner	Cable Tester	Tap	Visual Fault Locator (VFL)	Wi-Fi Analyzer
Disadvantages	Is limited to testing copper cables, does not provide detailed diagnostic information	Requires physical access to both ends of the cable, may not detect intermittent or complex issues	Is limited to monitoring specific network segments, may not capture all traffic depending on deployment	Is limited to testing fiber-optic cables, does not provide detailed diagnostic information	Requires compatible Wi-Fi hardware and software, may not be effective in highly congested or complex environments

> **ExamAlert**
>
> Be sure that you know the tools listed in Table 10.6.

Protocol Analyzer

Protocol analyzers are used to do just that—analyze network protocols such as TCP, UDP, HTTPS, and FTP. Protocol analyzers can be hardware or software based. In use, protocol analyzers help diagnose computer networking problems, alert you to unused protocols, identify unwanted or malicious network traffic, and help isolate network traffic-related problems.

Like packet sniffers, protocol analyzers capture the communication stream between systems. But unlike the sniffer, the protocol analyzer captures more than network traffic; it reads and decodes the traffic. Decoding enables administrators to view the network communication in English. From this communication, administrators can get a better idea of the traffic that is flowing on the network. As soon as unwanted or damaged traffic is spotted, analyzers make it easy to isolate and repair. For example, if there is a problem with specific TCP/IP communication, such as a broadcast storm, the analyzer can find the source of the TCP/IP problem and isolate the system causing the storm. Protocol analyzers also provide many real-time trend statistics that help you justify to management the purchase of new hardware.

You can use protocol analyzers for two key reasons:

▶ **Identify protocol patterns:** By creating a historical baseline of analysis, administrators can spot trends in protocol errors. That way, when a protocol error occurs, it can be researched in the documentation to see if that error has occurred before and what was done to fix it.

▶ **Decode information:** By capturing and decoding network traffic, administrators can see what exactly is going on with the network at a protocol level. This process helps find protocol errors as well as potential intruders.

Caution

Protocol analyzers enable administrators to examine the bandwidth that a particular protocol is using.

Speed Tester

Two types of websites that can be invaluable when it comes to networking are *speed test sites* and *looking-glass sites*. Speed test sites, as the name implies, are *bandwidth speed testers* that report the speed of the connection that you have to them and can be helpful in determining if you are getting the rate your ISP has promised.

Looking-glass sites are servers running *looking-glass (LG)* software that allows you to see routing information. The servers act as a read-only portal giving information about the backbone connection. Most of these servers will show **ping** information, trace (**tracert/traceroute**) information, and Border Gateway Protocol (BGP) information.

ExamAlert

Think of a looking-glass site as a graphical interface to routing-related information.

Port Scanner

Port scanners are software-based security utilities designed to search a network host for open ports on a TCP/IP-based network. As a refresher, in a TCP/IP-based network, a system can be accessed through one of 65,535 available port numbers. Each network service is associated with a particular port.

Note

Chapter 3, "Network Addressing, Routing, and Switching," includes a list of some of the most common TCP/IP suite protocols and their port assignments.

Many of the thousands of ports are closed by default; however, many others, depending on the OS, are open by default. These are the ports that can cause trouble. Like packet sniffers, port scanners can be used by both administrators and hackers. Hackers use port scanners to try to find an open port that they can use to access a system. Port scanners are easily obtained on the Internet either for free or for a modest cost. After it is installed, the scanner probes a computer system running TCP/IP, looking for a UDP or TCP port that is open and listening.

When a port scanner is used, several port states may be reported:

▶ **Open/listening:** The host sent a reply indicating that a service is listening on the port. There was a response from the port.

▶ **Closed or denied or not listening:** No process is listening on that port. Access to this port will likely be denied.

▶ **Filtered or blocked:** There was no reply from the host, meaning that the port is not listening or the port is secured and filtered.

Note

Sometimes, an Internet service provider (ISP) takes the initiative and blocks specific traffic entering its network before the traffic reaches the ISP's customers, or after the traffic leaves the customers and before it exits the network. This is done to protect customers from well-known attacks.

Because others can potentially review the status of ports, it is critical that administrators know which ports are open and potentially vulnerable. As mentioned, many tools and utilities are available for this purpose. The quickest way to get an overview of the ports used by the system and their status is to issue the **netstat -a** command from the command line. The following is a sample of the output from the **netstat -a** command and active connections for a computer system:

```
Proto    Local Address          Foreign Address      State

TCP      0.0.0.0:135            mike-PC:0            LISTENING

TCP      0.0.0.0:10114          mike-PC:0            LISTENING

TCP      0.0.0.0:10115          mike-PC:0            LISTENING

TCP      0.0.0.0:20523          mike-PC:0            LISTENING

TCP      0.0.0.0:20943          mike-PC:0            LISTENING

TCP      0.0.0.0:49152          mike-PC:0            LISTENING

TCP      0.0.0.0:49153          mike-PC:0            LISTENING

TCP      0.0.0.0:49154          mike-PC:0            LISTENING

TCP      0.0.0.0:49155          mike-PC:0            LISTENING

TCP      0.0.0.0:49156          mike-PC:0            LISTENING
```

TCP	0.0.0.0:49157	mike-PC:0	LISTENING
TCP	127.0.0.1:5354	mike-PC:0	LISTENING
TCP	127.0.0.1:27015	mike-PC:0	LISTENING
TCP	127.0.0.1:27015	mike-PC:49187	ESTABLISHED
TCP	127.0.0.1:49187	mike-PC:27015	ESTABLISHED
TCP	192.168.0.100:49190	206.18.166.15:http	CLOSED
TCP	192.168.1.66:139	mike-PC:0	LISTENING
TCP	[::]:135	mike-PC:0	LISTENING
TCP	[::]:445	mike-PC:0	LISTENING
TCP	[::]:2869	mike-PC:0	LISTENING
TCP	[::]:5357	mike-PC:0	LISTENING
TCP	[::]:10115	mike-PC:0	LISTENING
TCP	[::]:20523	mike-PC:0	LISTENING
TCP	[::]:49152	mike-PC:0	LISTENING
TCP	[::]:49153	mike-PC:0	LISTENING
TCP	[::]:49154	mike-PC:0	LISTENING
TCP	[::]:49155	mike-PC:0	LISTENING
TCP	[::]:49156	mike-PC:0	LISTENING
TCP	[::]:49157	mike-PC:0	LISTENING
UDP	0.0.0.0:123	*:*	
UDP	0.0.0.0:500	*:*	
UDP	0.0.0.0:3702	*:*	
UDP	0.0.0.0:3702	*:*	

As you can see from the output, the system has many listening ports. Not all these suggest that a risk exists, but the output does let you know that there are many listening ports and that they might be vulnerable. To test for actual vulnerability, you use a port scanner. A number of free online scanners and scanning services are available. Although a network administrator might use these free online tools out of curiosity, for better security testing, you should use a quality scanner. Nmap, discussed later in this chapter, is one of the most popular open-source port scanning tools available and it provides a number of different port scanning techniques for different scenarios.

Caution

Administrators use the detailed information revealed from a port scan to ensure network security. Port scans identify closed, open, and listening ports. However, port scanners also can be used by people who want to compromise security by finding open and unguarded ports.

LLDP and CDP

Link Layer Discovery Protocol (LLDP) and *Cisco Discovery Protocol (CDP)* are both network protocols used for device discovery and neighbor identification in Ethernet networks. They operate at the data link layer (Layer 2) of the OSI model and enable network devices to exchange information about their capabilities, configurations, and connected neighbors.

LLDP is an open standard protocol defined by the Institute of Electrical and Electronics Engineers (IEEE) as 802.1AB. It is supported by various networking vendors and is interoperable across different vendor devices. Network administrators can use LLDP to discover and map the network topology, identify neighboring devices, and troubleshoot connectivity issues by examining LLDP neighbor information. LLDP packets are sent periodically by network devices and are multicast to a well-known destination address (01-80-C2-00-00-0E). Devices that support LLDP can receive and process these packets to discover neighboring devices and their attributes.

CDP, on the other hand, is a proprietary network protocol developed by Cisco Systems for device discovery and neighbor identification within Cisco network environments. It operates similarly to LLDP but is specific to Cisco devices and is not interoperable with non-Cisco devices. CDP packets contain information such as device type, model number, software version, IP address, and connected interface. Network administrators can use CDP to gather information about neighboring Cisco devices, verify device connectivity, and troubleshoot network issues within Cisco environments. CDP packets are sent periodically by Cisco devices and are multicast to a specific destination address (01-00-0C-CC-CC-CC). Non-Cisco devices do not understand CDP packets and will ignore them.

ExamAlert

LLDP and CDP can provide valuable information about the network topology, device connectivity, and neighboring devices, helping administrators diagnose and resolve connectivity issues. By examining LLDP or CDP neighbor information, administrators can identify network devices, verify connections, detect potential misconfigurations, and troubleshoot problems related to device discovery, link status, VLAN assignments, and physical connectivity. Additionally, LLDP and CDP can be useful for inventory management, network documentation, and maintaining accurate network maps. However, it's important to note that while LLDP is an open standard supported by various vendors, CDP is specific to Cisco devices and may not be available or compatible with non-Cisco equipment.

NetFlow Analyzer

NetFlow is a proprietary Cisco protocol used for network flow analysis. A *NetFlow analyzer* is used to collect IP network traffic as it enters or exits an interface and can identify such values as the source and destination of traffic, class of service, and the causes of congestion. RFC 3954 (the NetFlow standard) does not specify a specific listening port. The most common port used by NetFlow is port 2055 (UDP), but other ports can also be used (such as 9555, 9995, 9025, and 9026).

TFTP Server

A Trivial File Transfer Protocol (TFTP) can be used to send files between servers and is finding new use today in applications uploading HTML pages on the HTTPS server. The TFTP server uses port 69 by default.

Terminal Emulator

A *terminal emulator* is any software program that emulates a computer terminal. Most often, the terminal being emulated is that of a command-line window, though it need not be (if it is graphical, it is usually called *a terminal window* instead of a terminal emulator), allowing the running of command-line utilities. Those utilities can be running on the local machine or remotely through the use of SSH or a similar utility.

IP Scanner

An *IP scanner* is any tool that can scan for IP addresses and related information. An administrator can use it to scan available ports, discover devices, and get detailed hardware and software information on workstations and servers to manage inventory.

> **ExamAlert**
>
> Remember all of the software tools mentioned in the preceding sections. You will likely be given a scenario where you need to use the appropriate tool.

Command-Line Tools

For anyone working with TCP/IP networks, troubleshooting connectivity is something that must be done. This section describes the tools used in the troubleshooting process and identifies scenarios in which they can be used.

You can use many utilities when troubleshooting TCP/IP. Although the utilities available vary from platform to platform, the functionality between platforms is quite similar. Table 10.7 lists the TCP/IP troubleshooting tools covered on the Network+ exam, along with their purpose.

TABLE 10.7 **Common TCP/IP Troubleshooting Tools and Their Purposes**

Tool	Description
tracert/ traceroute	Tracks the path a packet takes as it travels across a network. **tracert** is used on Windows systems; **traceroute** is used on UNIX, Linux, and macOS systems.
tracert -6 traceroute6 traceroute -6	Performs the same function as **tracert/traceroute**, but using the IPv6 protocol in place of IPv4.
ping	Tests connectivity between two devices on a network with IPv4.
ping6/ping -6	Tests connectivity between two devices on a network using the IPv6 protocol in place of IPv4.
hostname	Displays the name assigned to the host.
arp	Enables you to view and work with the IP address to MAC address resolution cache.
arp ping	Uses ARP to test connectivity between systems rather than the Internet Control Message Protocol (ICMP), as done with a regular ping.
netstat	Enables you to view the current TCP/IP connections on a system.
telnet	Allows remote access to a host. Because it is not secure, its usage is usually discouraged in favor of newer, more secure options such as SSH.
ipconfig	Enables you to view and renew a TCP/IP configuration on a Windows system.
ifconfig	Enables you to view a TCP/IP configuration on a UNIX, Linux, or macOS system.
ip	Enables you to configure network interfaces on Linux. Older Linux distributions used the **ifconfig** command, which operates similarly.
nslookup	Performs manual DNS lookups. **nslookup** can be used on Windows, UNIX, macOS, and Linux systems.
dig	Used on UNIX, Linux, and macOS systems, the Domain Information Groper collects data about Domain Name Servers. The **dig** command is helpful for troubleshooting DNS problems but is also used to display DNS information.
tcpdump	Acts as a Linux-based packet analyzer.
route	Enables you to view and configure the routes in the routing table.
nmap	Acts as a popular vulnerability scanner.

The following sections look in more detail at these utilities and the output they produce.

> **Note**
>
> Many of the utilities discussed in this chapter have a help facility that you can access by typing the command followed by **/?** or **-?**. On a Windows system, for example, you can get help on the **netstat** utility by typing **netstat /?**. Sometimes, using a utility with an invalid switch also brings up the help screen.

> **ExamAlert**
>
> Be prepared to identify what software tool to use in a given scenario. Remember, you might be able to use more than one tool. You will be expected to pick the best one for the situation described.

You will be asked to identify the output from a command, and you should be able to interpret the information provided by the command. In a performance-based question, you may be asked to enter the appropriate command for a given scenario.

The Trace Route Utility (tracert/traceroute)

The trace route utility does exactly what its name implies: it traces the route between two hosts. It does this by using ICMP echo packets to report information at every step in the journey. Each of the common network operating systems provides a trace route utility, but the name of the command and the output vary slightly on each. However, for the purposes of the Network+ exam, you should not concern yourself with the minor differences in the output format. Table 10.8 shows the **traceroute** command syntax used in various operating systems.

> **Note**
>
> The phrase *trace route utility* is used in this section to refer generically to the various route-tracing applications available on common operating systems. In a live environment, you should become familiar with the version of the tool used on the operating systems you are working with.

TABLE 10.8 **Trace Route Utility Commands**

Operating System	Trace Route Command Syntax
Windows systems	tracert *IP address*
	tracert -6 *IP address*
Linux/UNIX/macOS	traceroute *IP address*
	traceroute6 *IP address*
	traceroute -6 *IP address*

ExamAlert

Be prepared to identify the IP tracing command syntax used with various operating systems for the exam. Review Table 10.8 for this information.

The **traceroute** command provides a lot of useful information, including the IP address of every router connection it passes through and, in many cases, the name of the router (although this depends on the router's configuration). **traceroute** also reports the length, in milliseconds, of the round trip the packet made from the source location to the router and back. This information can help identify where network bottlenecks or breakdowns might be. The following is an example of a successful **tracert** command on a Windows system:

```
C:\> tracert 24.7.70.37
Tracing route to c1-p4.sttlwa1.home.net [24.7.70.37]
 over a maximum of 30 hops:
 1 30 ms 20 ms 20 ms 24.67.184.1
 2 20 ms 20 ms 30 ms rd1ht-ge3-0.ok.shawcable.net
 [24.67.224.7]
 3 50 ms 30 ms 30 ms rc1wh-atm0-2-1.vc.shawcable.net
 [204.209.214.193]
 4 50 ms 30 ms 30 ms rc2wh-pos15-0.vc.shawcable.net
 [204.209.214.90]
 5 30 ms 40 ms 30 ms rc2wt-pos2-0.wa.shawcable.net
 [66.163.76.37]
 6 30 ms 40 ms 30 ms c1-pos6-3.sttlwa1.home.net [24.7.70.37]
Trace complete.
```

Similar to the other common operating systems covered on the Network+ exam, the **tracert** display on a Windows-based system includes several columns of information. The first column represents the hop number. You may recall that *hop* is the term used to describe a step in the path a packet takes as it crosses the network. The next three columns indicate the round-trip time, in milliseconds, that a packet takes in its attempts to reach the destination. The last column is the hostname and the IP address of the responding device.

However, not all trace route attempts are successful. The following is the output from a **tracert** command on a Windows system that does not manage to get to the remote host:

```
C:\> tracert comptia.org
Tracing route to comptia.org [216.119.103.72]
over a maximum of 30 hops:
 1 27 ms 28 ms 14 ms 24.67.179.1
 2 55 ms 13 ms 14 ms rd1ht-ge3-0.ok.shawcable.net
[24.67.224.7]
 3 27 ms 27 ms 28 ms rc1wh-atm0-2-1.shawcable.net
[204.209.214.19]
 4 28 ms 41 ms 27 ms rc1wt-pos2-0.wa.shawcable.net
[66.163.76.65]
 5 28 ms 41 ms 27 ms rc2wt-pos1-0.wa.shawcable.net
[66.163.68.2]
 6 41 ms 55 ms 41 ms c1-pos6-3.sttlwa1.home.net
[24.7.70.37]
 7 54 ms 42 ms 27 ms home-gw.st6wa.ip.att.net
[192.205.32.249]
 8 * * * Request timed out.
 9 * * * Request timed out.
10 * * * Request timed out.
11 * * * Request timed out.
12 * * * Request timed out.
13 * * * Request timed out.
14 * * * Request timed out.
15 * * * Request timed out.
```

In this example, the trace route request gets to only the seventh hop, at which point it fails. This failure indicates that the problem lies on the far side of the device in step 7 or on the near side of the device in step 8. In other words, the device at step 7 is functioning but might not make the next hop. The cause of the problem could be a range of things, such as an error in the routing table or a faulty connection. Alternatively, the seventh device might be operating at 100 percent, but device 8 might not be functioning at all. In any case, you can isolate the problem to just one or two devices.

Note

In some cases, the owner of a router might configure it to not return ICMP traffic like that generated by **ping** or **traceroute**. If this is the case, the **ping** or **traceroute** will fail just as if the router did not exist or was not operating.

ExamAlert

Although we have used the Windows **tracert** command to provide sample output in these sections, the output from **traceroute** on a UNIX, Linux, or macOS system is extremely similar.

The trace route utility can also help you isolate a heavily congested network. In the following example, the trace route packets fail in the midst of the **tracert** from a Windows system, but subsequently they continue. This behavior can be an indicator of network congestion:

```
C:\> tracert comptia.org
Tracing route to comptia.org [216.119.103.72]over a maximum of 30 hops:
 1 96 ms 96 ms 55 ms 24.67.179.1
 2 14 ms 13 ms 28 ms rd1ht-ge3-0.ok.shawcable.net
[24.67.224.7]
 3 28 ms 27 ms 41 ms rc1wh-atm0-2-1.shawcable.net
[204.209.214.19]
 4 28 ms 41 ms 27 ms rc1wt-pos2-0.wa.shawcable.net
[66.163.76.65]
 5 41 ms 27 ms 27 ms rc2wt-pos1-0.wa.shawcable.net
[66.163.68.2]
 6 55 ms 41 ms 27 ms c1-pos6-3.sttlwa1.home.net [24.7.70.37]
 7 54 ms 42 ms 27 ms home-gw.st6wa.ip.att.net
[192.205.32.249]
 8 55 ms 41 ms 28 ms gbr3-p40.st6wa.ip.att.net
[12.123.44.130]
```

```
9 * * * Request timed out.

10 * * * Request timed out.

11 * * * Request timed out.

12 * * * Request timed out.

13 69 ms 68 ms 69 ms gbr2-p20.sd2ca.ip.att.net
[12.122.11.254]

14 55 ms 68 ms 69 ms gbr1-p60.sd2ca.ip.att.net
[12.122.1.109]

15 82 ms 69 ms 82 ms gbr1-p30.phmaz.ip.att.net
[12.122.2.142]

16 68 ms 69 ms 82 ms gar2-p360.phmaz.ip.att.net
[12.123.142.45]

17 110 ms 96 ms 96 ms 12.125.99.70

18 124 ms 96 ms 96 ms light.crystaltech.com [216.119.107.1]

19 82 ms 96 ms 96 ms 216.119.103.72

Trace complete.
```

Generally, trace route utilities enable you to identify the location of a problem in the connectivity between two devices. After you determine this location, you might need to use a utility such as **ping** to continue troubleshooting. In many cases, as in the examples provided in this chapter, the routers might be on a network such as the Internet and therefore not within your control. In that case, you can do little except inform your ISP of the problem.

When you're dealing with IPv6, the same tools exist but are followed with -6; so **tracert** becomes **tracert -6** and **traceroute** becomes **traceroute -6**.

ping

Most network administrators are familiar with the **ping** utility and are likely to use it on an almost daily basis. The basic function of the **ping** command is to test the connectivity between the two devices on a network. All the command is designed to do is determine whether the two computers can see each other and to notify you of how long the round trip takes to complete.

Although **ping** is most often used on its own, a number of switches can be used to assist in the troubleshooting process. Table 10.9 shows some of the commonly used switches with **ping** on a Windows system.

TABLE 10.9 **ping Command Switches**

Option	Description
ping -t	Pings a device on the network until stopped
ping -a	Resolves addresses to hostnames
ping -n count	Specifies the number of echo requests to send
ping -r count	Records the route for count hops
ping -s count	Sets the time stamp for count hops
ping -w timeout	Sets the timeout in milliseconds to wait for each reply
ping -6 or **ping6**	Pings a device on the network using IPv6 instead of IPv4

ExamAlert

You will likely be asked about **ping**, its switches, and how it can be used in a troubleshooting scenario.

The **ping** command works by sending ICMP echo request messages to another device on the network. If the other device on the network hears the ping request, it automatically responds with an ICMP echo reply. By default, the **ping** command on a Windows-based system sends four data packets; however, if the **-t** switch is used, a continuous stream of ping requests can be sent.

ping is perhaps the most widely used of all network tools; it is primarily used to verify connectivity between two network devices. On a good day, the results from the **ping** command are successful, and the sending device receives a reply from the remote device. Not all ping results are that successful. To use **ping** effectively, you must interpret the results of a failed **ping** command.

The Destination Host Unreachable Message

The "Destination Host Unreachable" error message means that a route to the destination computer system cannot be found. To remedy this problem, you might need to examine the routing information on the local host to confirm that the local host is correctly configured, or you might need to make sure that the default gateway information is correct. The following is an example of a ping failure that gives the "Destination Host Unreachable" message:

```
Pinging 24.67.54.233 with 32 bytes of data:
Destination host unreachable.
Destination host unreachable.
```

```
Destination host unreachable.
Destination host unreachable.
Ping statistics for 24.67.54.233:
 Packets: Sent = 4, Received = 0, Lost = 4 (100% loss),
Approximate round trip times in milli-seconds:
 Minimum = 0ms, Maximum = 0ms, Average = 0ms
```

The Request Timed Out Message

The "Request Timed Out" error message is common when you use the **ping** command. Essentially, this error message indicates that your host did not receive the ping message back from the destination device within the designated time period. Assuming that the network connectivity is okay on your system, this message typically indicates that the destination device is not connected to the network, is powered off, or is not correctly configured. It could also mean that some intermediate device is not operating correctly. In some rare cases, it can also indicate that the network has so much congestion that timely delivery of the ping message could not be completed. It might also mean that the ping is being sent to an invalid IP address or that the system is not on the same network as the remote host, and an intermediary device is not correctly configured. In any of these cases, the failed ping should initiate a troubleshooting process that might involve other tools, manual inspection, and possibly reconfiguration. The following example shows the output from a ping to an invalid IP address:

```
C:\> ping 169.76.54.3
Pinging 169.76.54.3 with 32 bytes of data:
Request timed out.
Request timed out.
Request timed out.
Request timed out.
Ping statistics for 169.76.54.3:
 Packets: Sent = 4, Received = 0, Lost = 4 (100%
Approximate round trip times in milli-seconds:
 Minimum = 0ms, Maximum = 0ms, Average = 0ms
```

During the ping request, you might receive some replies from the remote host that are intermixed with "Request Timed Out" errors. This is often the result of a congested network. An example follows; notice that this example, which was run on a Windows system, uses the **-t** switch to generate continuous pings:

```
C:\> ping -t 24.67.184.65
Pinging 24.67.184.65 with 32 bytes of data:
Reply from 24.67.184.65: bytes=32 time=55ms TTL=127
Reply from 24.67.184.65: bytes=32 time=54ms TTL=127
Reply from 24.67.184.65: bytes=32 time=27ms TTL=127
Request timed out.
Request timed out.
Request timed out.
Reply from 24.67.184.65: bytes=32 time=69ms TTL=127
Reply from 24.67.184.65: bytes=32 time=28ms TTL=127
Reply from 24.67.184.65: bytes=32 time=28ms TTL=127
Reply from 24.67.184.65: bytes=32 time=68ms TTL=127
Reply from 24.67.184.65: bytes=32 time=41ms TTL=127
Ping statistics for 24.67.184.65:
 Packets: Sent = 11, Received = 8, Lost = 3 (27% loss),
Approximate round trip times in milli-seconds:
 Minimum = 27ms, Maximum = 69ms, Average = 33ms
```

In this example, three packets were lost. If this command continued on your network, you would need to troubleshoot to find out why packets were dropped.

The Unknown Host Message

The "Unknown Host" error message is generated when the hostname of the destination computer cannot be resolved. This error usually occurs when you ping an incorrect hostname, as shown in the following example, or try to use **ping** with a hostname when hostname resolution (via DNS or a HOSTS text file) is not configured:

```
C:\> ping www.comptia.ca
Unknown host www.comptia.ca
```

If the ping fails, you need to verify that the ping is sent to the correct remote host. If it is, and if name resolution is configured, you have to dig a little more to find the problem. This error might indicate a problem with the name resolution process, and you might need to verify that the DNS or WINS server is available. Other commands, such as **nslookup** or **dig**, can help in this process.

The Expired TTL Message

The time to live (TTL) is a key consideration in understanding the **ping** command. The function of the TTL is to prevent circular routing, which occurs when a ping request keeps looping through a series of hosts. The TTL counts each hop along the way toward its destination device. Each time it counts one hop, the hop is subtracted from the TTL. If the TTL reaches 0, it has expired, and you get a message like the following:

```
Reply from 24.67.180.1: TTL expired in transit
```

If the TTL is exceeded with **ping**, you might have a routing problem on the network. You can modify the TTL for **ping** on a Windows system by using the **ping -i** command.

Troubleshooting with ping

Although **ping** does not completely isolate problems, you can use it to help identify where a problem lies. When troubleshooting with **ping**, follow these steps:

1. Ping the IP address of your local loopback using the command **ping 127.0.0.1**. If this command is successful, you know that the TCP/IP protocol suite is installed correctly on your system and is functioning. If you cannot ping the local loopback adapter, TCP/IP might need to be reloaded or reconfigured on the machine you are using.

> **ExamAlert**
>
> The loopback is a special function within the TCP/IP protocol stack that is supplied for troubleshooting purposes. The Class A IP address 127.*x.x.x* is reserved for the IPv4 loopback. Although convention dictates that you use 127.0.0.1, you can use any address in the 127.*x.x.x* range, except for the network number itself (127.0.0.0) and the broadcast address (127.255.255.255). You can also ping by using the default hostname for the local system, which is called localhost (for example, **ping localhost**). The same function can be performed in IPv6 by using the address ::1.

2. Ping the assigned IP address of your local network interface card (NIC). If the ping is successful, you know that your NIC is functioning on the network and has TCP/IP correctly installed. If you cannot ping the local

NIC, TCP/IP might not be correctly bound to the NIC, or the NIC drivers might be improperly installed.

3. Ping the IP address of another known good system on your local network. By doing so, you can determine whether the computer you are using can see other computers on the network. If you can ping other devices on your local network, you have network connectivity.

   If you cannot ping other devices on your local network, but you could ping the IP address of your system, you might not be connected to the network correctly.

4. After you confirm that you have network connectivity for the local network, you can verify connectivity to a remote network by sending a ping to the IP address of the default gateway.

5. If you can ping the default gateway, you can verify remote connectivity by sending a ping to the IP address of a system on a remote network.

ExamAlert

You might be asked to relate the correct procedure for using **ping** for a connectivity problem. A performance-based question may ask you to implement the **ping** command to test for connectivity.

Using just the **ping** command in these steps, you can confirm network connectivity on not only the local network but also on a remote network. The whole process requires as much time as it takes to enter the command, and you can do it all from a single location.

If you are an optimistic person, you can perform step 5 first. If that works, all the other steps will also work, saving you the need to test them. If your step 5 trial fails, you can go to step 1 and start the troubleshooting process from the beginning.

Note

All but one of the **ping** examples used in this section show the **ping** command using the IP address of the remote host. It is also possible to ping the Domain Name Service (DNS) name of the remote host (for example, **ping www.comptia.org**, **ping server1**). However, you can do this only if your network uses a DNS server. On a Windows-based network, you can also **ping** by using the Network Basic Input/Output System (NetBIOS) computer name.

When you're dealing with IPv6, the same tools exist but are followed with 6 or -6; so **ping** becomes **ping6** or **ping -6**.

hostname

Sometimes all you really want to know is the hostname assigned to a particular host. When that is the case, you can use the **hostname** command to provide that information. It reports back the character string that refers to the name of the host the command was entered on. The following example illustrates this command in action:

```
C:\> hostname
EADULANEY7040
```

ARP

Address Resolution Protocol (ARP) is used to resolve IP addresses to MAC addresses. This is significant because on a network, devices find each other using the IP address, but communication between devices requires the MAC address.

> **ExamAlert**
>
> Remember that the function of ARP is to resolve IP addresses to Layer 2 or MAC addresses.

When a computer wants to send data to another computer on the network, it must know the MAC address (physical address) of the destination system. To discover this information, ARP sends out a discovery packet to obtain the MAC address. When the destination computer is found, it sends its MAC address to the sending computer. The ARP-resolved MAC addresses are stored temporarily on a computer system in the ARP cache. Inside this ARP cache is a list of matching MAC and IP addresses. This ARP cache is checked before a discovery packet is sent to the network to determine whether there is an existing entry.

Entries in the ARP cache are periodically flushed so that the cache does not fill up with unused entries. The following code shows an example of the **arp** command with the output from a Windows system:

```
C:\> arp -a
Interface: 24.67.179.22 on Interface 0x3
  Internet Address Physical Address Type
  24.67.179.1 00-00-77-93-d8-3d dynamic
```

As you might notice, the type is listed as dynamic. Entries in the ARP cache can be added statically or dynamically. Static entries are added manually and do not expire. The dynamic entries are added automatically when the system accesses another on the network.

As with other command-line utilities, several switches are available for the **arp** command. Table 10.10 shows the available switches for Windows-based systems.

TABLE 10.10 **arp Switches**

Switch	Description
-a or **-g**	Displays both the IP and MAC addresses and whether they are dynamic or static entries
inet_addr	Specifies a specific Internet address
-N if_addr	Displays the ARP entries for a specified network interface
eth_addr	Specifies a MAC address
if_addr	Specifies an Internet address
-d	Deletes an entry from the ARP cache
-s	Adds a static permanent address to the ARP cache

arp ping

Earlier in this chapter we talked about the **ping** command and how it is used to test connectivity between devices on a network. Using the **ping** command is often an administrator's first step to test connectivity between network devices. If the ping fails, it is assumed that the device you are pinging is offline. But this may not always be the case.

Most companies now use firewalls or other security measures that may block Internet Control Message Protocol (ICMP) requests. This means that a ping request will not work. Blocking ICMP is a security measure; if would-be hackers cannot hit the target, they may not attack the host.

> **ExamAlert**
>
> One type of attack is called an *ICMP flood attack* (also known as a *ping attack*). The attacker sends continuous ping packets to a server or network system, eventually tying up that system's resources, making it unable to respond to requests from other systems.

If ICMP is blocked, you have still another option to test connectivity with a device on the network: the **arp ping**. As mentioned, the ARP utility is used to resolve IP addresses to MAC addresses. The **arp ping** utility does not use the ICMP protocol to test connectivity like **ping** does; rather, it uses the ARP protocol. However, ARP is not routable, and the **arp ping** cannot be routed to work over separate networks. The **arp ping** works only on the local subnet.

Just like with a regular **ping**, an **arp ping** specifies an IP address; however, instead of returning regular **ping** results, the **arp ping** responds with the MAC address and name of the computer system. So, when a regular **ping** using ICMP fails to locate a system, the **arp ping** uses a different method to find the system. With **arp ping**, you can directly ping a MAC address. From this, you can determine whether duplicate IP addresses are used and, as mentioned, determine whether a system is responding.

arp ping is not built in to Windows, but you can download a number of programs that allow you to ping using ARP. Linux, however, has an **arp ping** utility ready to use.

> **ExamAlert**
>
> **arp ping** is not routable and can be used only on the local network.

The netstat Command

The **netstat** command displays the protocol statistics and current TCP/IP connections on the local system. Used without any switches, the **netstat** command shows the active connections for all outbound TCP/IP connections. In addition, several switches are available that change the type of information **netstat** displays. Table 10.11 shows the various switches available for the **netstat** utility.

TABLE 10.11 **netstat Switches**

Switch	Description
-a	Displays the current connections and listening ports
-e	Displays Ethernet statistics
-n	Lists addresses and port numbers in numeric form
-p	Shows connections for the specified protocol
-r	Shows the routing table
-s	Lists per-protocol statistics
interval	Specifies how long to wait before redisplaying statistics

ExamAlert

You can use the **netstat** and **route print** commands to show the routing table on a local or remote system.

The **netstat** utility is used to show the port activity for both TCP and UDP connections, showing the inbound and outbound connections. When used without switches, the **netstat** utility has four information headings:

▶ **Proto:** Lists the protocol being used, either UDP or TCP

▶ **Local address:** Specifies the local address and port being used

▶ **Foreign address:** Identifies the destination address and port being used

▶ **State:** Specifies whether the connection is established

In its default use, the **netstat** command shows outbound connections that have been established by TCP. The following shows sample output from a **netstat** command without using any switches:

```
C:\> netstat
Active Connections
  Proto Local Address Foreign Address State
  TCP laptop:2848 MEDIASERVICES1:1755 ESTABLISHED
  TCP laptop:1833 www.dollarhost.com:80 ESTABLISHED
  TCP laptop:2858 194.70.58.241:80 ESTABLISHED
  TCP laptop:2860 194.70.58.241:80 ESTABLISHED
  TCP laptop:2354 www.dollarhost.com:80 ESTABLISHED
  TCP laptop:2361 www.dollarhost.com:80 ESTABLISHED
  TCP laptop:1114 www.dollarhost.com:80 ESTABLISHED
  TCP laptop:1959 www.dollarhost.com:80 ESTABLISHED
  TCP laptop:1960 www.dollarhost.com:80 ESTABLISHED
  TCP laptop:1963 www.dollarhost.com:80 ESTABLISHED
  TCP laptop:2870 localhost:8431 TIME_WAIT
  TCP laptop:8431 localhost:2862 TIME_WAIT
  TCP laptop:8431 localhost:2863 TIME_WAIT
  TCP laptop:8431 localhost:2867 TIME_WAIT
  TCP laptop:8431 localhost:2872 TIME_WAIT
```

As with any other command-line utility, the **netstat** utility has a number of switches. The following sections briefly explain the switches and give sample output from each.

netstat -e

The **netstat -e** command shows the activity for the NIC and displays the number of packets that have been both sent and received. Here's an example:

```
C:\WINDOWS\Desktop> netstat -e
Interface Statistics
 Received Sent
Bytes 17412385 40237510
Unicast packets 79129 85055
Non-unicast packets 693 254
Discards 0 0
Errors 0 0
Unknown protocols 306
```

As you can see, the **netstat -e** command shows more than just the packets that have been sent and received:

▶ **Bytes:** The number of bytes that the NIC has sent or received since the computer was turned on.

▶ **Unicast packets:** Packets sent and received directly by this interface.

▶ **Nonunicast packets:** Broadcast or multicast packets that the NIC picked up.

▶ **Discards:** The number of packets rejected by the NIC, perhaps because they were damaged.

▶ **Errors:** The errors that occurred during either the sending or receiving process. As you would expect, this column should be a low number. If it is not, this could indicate a problem with the NIC.

▶ **Unknown protocols:** The number of packets that the system could not recognize.

netstat -a

The **netstat -a** command displays statistics for both Transmission Control Protocol (TCP) and User Datagram Protocol (UDP). Here is an example of the **netstat -a** command:

```
C:\WINDOWS\Desktop> netstat -a
Active Connections
 Proto Local Address Foreign Address State
 TCP laptop:1027 LAPTOP:0 LISTENING
 TCP laptop:1030 LAPTOP:0 LISTENING
 TCP laptop:1035 LAPTOP:0 LISTENING
 TCP laptop:50000 LAPTOP:0 LISTENING
 TCP laptop:5000 LAPTOP:0 LISTENING
 TCP laptop:1035 msgr-ns41.msgr.hotmail.com:1863 ESTABLISHED
 TCP laptop:nbsession LAPTOP:0 LISTENING
 TCP laptop:1027 localhost:50000 ESTABLISHED
 TCP laptop:50000 localhost:1027 ESTABLISHED
 UDP laptop:1900 *:*
 UDP laptop:nbname *:*
 UDP laptop:nbdatagram *:*
 UDP laptop:1547 *:*
 UDP laptop:1038 *:*
 UDP laptop:1828 *:*
 UDP laptop:3366 *:*
```

As you can see, the output includes four columns, which show the protocol, the local address, the foreign address, and the port's state. The TCP connections show the local and foreign destination addresses and the connection's current state. UDP, however, is a little different. It does not list a state status because, as mentioned throughout this book, UDP is a connectionless protocol and does not establish connections. The following list further explains the information provided by the **netstat -a** command:

▶ **Proto:** The protocol used by the connection.

▶ **Local address:** The IP address of the local computer system and the port number it is using. If the entry in the local address field is an asterisk (*), the port has not yet been established.

▶ **Foreign address:** The IP address of a remote computer system and the associated port. When a port has not been established, as with the UDP connections, *:* appears in the column.

▶ **State:** The current state of the TCP connection. Possible states include established, listening, closed, and waiting.

netstat -r

The **netstat -r** command is often used to view a system's routing table. A system uses a routing table to determine routing information for TCP/IP traffic. The following is an example of the **netstat -r** command from a Windows system:

```
C:\WINDOWS\Desktop> netstat -r

Route table

=====================================================================
=====================================================================
Active Routes:
Network Destination Netmask Gateway Interface Metric
 0.0.0.0 0.0.0.0 24.67.179.1 24.67.179.22 1
 24.67.179.0 255.255.255.0 24.67.179.22 24.67.179.22 1
 24.67.179.22 255.255.255.255 127.0.0.1 127.0.0.1 1
 24.255.255.255 255.255.255.255 24.67.179.22 24.67.179.22 1
 127.0.0.0 255.0.0.0 127.0.0.1 127.0.0.1 1
 224.0.0.0 224.0.0.0 24.67.179.22 24.67.179.22 1
255.255.255.255 255.255.255.255 24.67.179.22 2 1
Default Gateway: 24.67.179.1

=====================================================================
Persistent Routes:
 None
```

Caution

The **netstat -r** command output shows the same information as the output from the **route print** command.

netstat -s

The **netstat -s** command displays a number of statistics related to the TCP/IP protocol suite. Understanding the purpose of every field in the output is beyond the scope of the Network+ exam, but for your reference, sample output from the **netstat -s** command is shown here:

```
C:\> netstat -s
IP Statistics
 Packets Received = 389938
 Received Header Errors - 0
 Received Address Errors = 1876
 Datagrams Forwarded = 498
 Unknown Protocols Received = 0
 Received Packets Discarded = 0
 Received Packets Delivered = 387566
 Output Requests = 397334
 Routing Discards = 0
 Discarded Output Packets = 0
 Output Packet No Route = 916
 Reassembly Required = 0
 Reassembly Successful = 0
 Reassembly Failures = 0
 Datagrams Successfully Fragmented = 0
 Datagrams Failing Fragmentation = 0
 Fragments Created = 0
ICMP Statistics
 Received Sent
 Messages 40641 41111
 Errors 0 0
 Destination Unreachable 223 680
 Time Exceeded 24 0
 Parameter Problems 0 0
 Source Quenches 0 0
 Redirects 0 38
 Echos 20245 20148
 Echo Replies 20149 20245
 Timestamps 0 0
```

```
Timestamp Replies 0 0

Address Masks 0 0

Address Mask Replies 0 0

TCP Statistics

 Active Opens = 13538

 Passive Opens = 23132

 Failed Connection Attempts = 9259

 Reset Connections = 254

 Current Connections = 15

 Segments Received = 330242

 Segments Sent = 326935

 Segments Retransmitted = 18851

UDP Statistics

 Datagrams Received = 20402

 No Ports = 20594

 Receive Errors = 0

 Datagrams Sent = 10217
```

telnet

The **telnet** utility is used for remote access to a host via the Telnet service. This utility was mentioned in previous chapters, and the same caveat that accompanies it there must be given here: because it is an older utility that lacks security features, it is highly recommended that it not be used and that other utilities—such as SSH—which provide the same functionality be used in its place.

ipconfig

The **ipconfig** command is a technician's best friend when it comes to viewing the TCP/IP configuration of a Windows system. Used on its own, the **ipconfig** command shows basic information, such as the name of the local network interface, the IP address, the subnet mask, and the default gateway. Combined with the **/all** switch, it shows a detailed set of information, as shown in the following example:

```
C:\> ipconfig /all

Windows IP Configuration

 Host Name . . . . . . . . . . . : server
```

```
Primary Dns Suffix . . . . . . . :

Node Type . . . . . . . . . . . : Hybrid

IP Routing Enabled. . . . . . . . : No

WINS Proxy Enabled. . . . . . . . : No

DNS Suffix Search List. . . . . . : tampabay.rr.com

Ethernet adapter Local Area Connection:

Connection-specific DNS Suffix . : tampabay.rr.com

Description . . . . . . . . . . . : Broadcom NetLink (TM)
Gigabit Ethernet

Physical Address. . . . . . . . . : 00-25-64-8C-9E-BF

DHCP Enabled. . . . . . . . . . . : Yes

Autoconfiguration Enabled . . . . : Yes

Link-local IPv6 Address . . . . . : fe80::51b9:996e:9fac:7715%10
(Preferred)

IPv4 Address. . . . . . . . . . . : 192.168.1.119(Preferred)

Subnet Mask . . . . . . . . . . . : 255.255.255.0

Lease Obtained. . . . . . . . . . : Thursday, January 23, 2025
6:00:54 AM

Lease Expires . . . . . . . . . . : Friday, January 24, 2025 6:00:54 AM

Default Gateway . . . . . . . . . : 192.168.1.1

DHCP Server . . . . . . . . . . . : 192.168.1.1

DHCPv6 IAID . . . . . . . . . . . : 234890596

DHCPv6 Client DUID. . . . . . . . :
00-01-00-01-13-2A-5B-37-00-25-64-8C-9E-BF

DNS Servers . . . . . . . . . . . : 192.168.1.1

NetBIOS over Tcpip. . . . . . . . : Enabled

Connection-specific DNS Suffix Search List :

tampabay.rr.com
```

As you can imagine, you can use the output from the **ipconfig /all** command in a massive range of troubleshooting scenarios. Table 10.12 lists some of the most common troubleshooting symptoms, along with where to look for clues about solving them in the **ipconfig /all** output.

Note

When looking at **ipconfig** information, you should be sure that all information is present and correct. For example, a missing or incorrect default gateway parameter limits communication to the local segment.

TABLE 10.12 **Common Troubleshooting Symptoms That** ipconfig **Can Help Solve**

Symptom	Field to Check in the Output
The user cannot connect to any other system.	Ensure that the TCP/IP address and subnet mask are correct. If the network uses DHCP, ensure that DHCP is enabled.
The user can connect to another on the same subnet but cannot connect to a remote system.	Ensure the default gateway is configured correctly.
The user is unable to browse the Internet.	Ensure the DNS server parameters are correctly configured.
The user cannot browse across remote subnets.	Ensure the WINS or DNS server parameters are correctly configured, if applicable.

ExamAlert

You should be prepared to identify the output from an **ipconfig** command in relationship to a troubleshooting scenario.

Using the **/all** switch might be the most popular, but there are a few others. They include the switches listed in Table 10.13.

ExamAlert

ipconfig and its associated switches are widely used by network administrators and therefore should be expected to make an appearance on the exam.

TABLE 10.13 **ipconfig Switches**

Switch	Description
?	Displays the **ipconfig** help screen
/all	Displays additional IP configuration information
/release	Releases the IPv4 address of the specified adapter
/release6	Releases the IPv6 address of the specified adapter
/renew	Renews the IPv4 address of a specified adapter
/renew6	Renews the IPv6 address of a specified adapter
/flushdns	Purges the DNS cache
/registerdns	Refreshes the DHCP lease and reregisters the DNS names
/displaydns	Displays the information in the DNS cache

ifconfig

ifconfig performs the same function as **ipconfig**, but on a Linux, UNIX, or macOS system. Because Linux relies more heavily on command-line utilities than Windows, the Linux and UNIX version of **ifconfig** provides much more functionality than **ipconfig**. On a Linux or UNIX system, you can get information about the usage of the **ifconfig** command by using **ifconfig -help**. The following output provides an example of the basic **ifconfig** command run on a Linux system:

```
eth0 Link encap:Ethernet HWaddr 00:60:08:17:63:A0
 inet addr:192.168.1.101 Bcast:192.168.1.255 Mask:255.255.255.0
 UP BROADCAST RUNNING MTU:1500 Metric:1
 RX packets:911 errors:0 dropped:0 overruns:0 frame:0
 TX packets:804 errors:0 dropped:0 overruns:0 carrier:0
 collisions:0 txqueuelen:100
 Interrupt:5 Base address:0xe400
lo Link encap:Local Loopback
 inet addr:127.0.0.1 Mask:255.0.0.0
 UP LOOPBACK RUNNING MTU:3924 Metric:1
 RX packets:18 errors:0 dropped:0 overruns:0 frame:0
 TX packets:18 errors:0 dropped:0 overruns:0 carrier:0
 collisions:0 txqueuelen:0
```

Although the **ifconfig** command displays the IP address, subnet mask, and default gateway information for both the installed network adapter and the local loopback adapter, it does not report DHCP lease information. Instead, you can use the **pump -s** command to view detailed information on the DHCP lease, including the assigned IP address, the address of the DHCP server, and the time remaining on the lease. You can also use the **pump** command to release and renew IP addresses assigned via DHCP and to view DNS server information.

> **Note**
>
> On newer Linux implementations, **ip** has replaced **ifconfig** for most functions.

nslookup

The **nslookup** utility is used to troubleshoot DNS-related problems. Using **nslookup**, you can, for example, run manual name resolution queries against DNS servers, get information about your system's DNS configuration, or specify what kind of DNS record should be resolved.

When **nslookup** is started, it displays the current hostname and the IP address of the locally configured DNS server. You then see a command prompt that enables you to specify further queries. This is known as *interactive* mode. Table 10.14 lists the commands you can enter in interactive mode.

TABLE 10.14 **nslookup Switches**

Switch	Description
all	Prints options, as well as current server and host information
[no]debug	Prints debugging information
[no]d2	Prints exhaustive debugging information
[no]defname	Appends the domain name to each query
[no]recurse	Asks for a recursive answer to the query
[no]search	Uses the domain search list
[no]vc	Always uses a virtual circuit
domain=NAME	Sets the default domain name to **NAME**
srchlist=N1 [/N2/.../N6]	Sets the domain to **N1** and the search list to **N1**, **N2**, and so on
root=NAME	Sets the root server to **NAME**
retry=X	Sets the number of retries to X

Switch	Description
timeout=X	Sets the initial timeout interval to X seconds
type=X	Sets the query type (for example, **A**, **ANY**, **CNAME**, **MX**, **NS**, **PTR**, **SOA**, or **SRV**)
querytype=X	Same as **type**
class=X	Sets the query class (for example, **IN** [Internet], **ANY**)
[no]msxfr	Uses Microsoft fast zone transfer
ixfrver=X	Sets the current version to use in an IXFR transfer request
server NAME	Sets the default server to **NAME**, using the current default server
exit	Exits the program

Instead of using interactive mode, you can execute **nslookup** requests directly at the command prompt. The following sample shows the output from the **nslookup** command when a domain name is specified to be resolved:

```
C:\> nslookup comptia.org
Server: nsc1.ht.ok.shawcable.net
Address: 64.59.168.13
Non-authoritative answer:
Name: comptia.org
Address: 208.252.144.4
```

As you can see from the output, **nslookup** shows the hostname and IP address of the DNS server against which the resolution was performed, along with the hostname and IP address of the resolved host.

dig

The **dig** command is used on a Linux, UNIX, or macOS system to perform manual DNS lookups. **dig** performs the same basic task as **nslookup**, but with one major distinction: the **dig** command does not have an interactive mode and instead uses only command-line switches to customize results.

dig generally is considered a more powerful tool than **nslookup**, but in the course of a typical network administrator's day, the minor limitations of **nslookup** are unlikely to be too much of a factor. Instead, **dig** is often the tool of choice for DNS information and troubleshooting on UNIX, Linux, or macOS systems. Like **nslookup**, **dig** can be used to perform simple name resolution requests. The output from this process is shown in the following listing:

```
; <<>> DiG 8.2 <<>> examcram.com
;; res options: init recurs defnam dnsrch
;; got answer:
;; ->>HEADER<<- opcode: QUERY, status: NOERROR, id: 4
;; flags: qr rd ra; QUERY: 1, ANSWER: 1, AUTHORITY: 2, ADDITIONAL: 0
;; QUERY SECTION:
;; examcram.com, type = A, class = IN
;; ANSWER SECTION:
examcram.com. 7h33m IN A 63.240.93.157
;; AUTHORITY SECTION:
examcram.com. 7h33m IN NS usrxdns1.pearsontc.com.
examcram.com. 7h33m IN NS oldtxdns2.pearsontc.com.
;; Total query time: 78 msec
;; FROM: localhost.localdomain to SERVER: default - 209.53.4.130
;; WHEN: Sat Oct 16 20:21:24 2018
;; MSG SIZE sent: 30 rcvd: 103
```

As you can see, **dig** provides a number of pieces of information in the basic output—more so than **nslookup**. Network administrators can gain information from three key areas of the output: **ANSWER SECTION**, **AUTHORITY SECTION**, and the last four lines of the output.

The **ANSWER SECTION** of the output provides the name of the domain or host being resolved, along with its IP address. The *A* in the results line indicates the record type that is being resolved.

The **AUTHORITY SECTION** provides information on the authoritative DNS servers for the domain against which the resolution request was performed. This information can be useful in determining whether the correct DNS servers are considered authoritative for a domain.

The last four lines of the output show how long the name resolution request took to process and the IP address of the DNS server that performed the resolution. It also shows the date and time of the request, as well as the size of the packets sent and received.

The tcpdump Command

The **tcpdump** command is used on Linux/UNIX systems to print the contents of network packets. It can read packets from a network interface card or from a previously created saved packet file and write packets to either standard output or a file.

The route Utility

The **route** utility is an often-used and very handy tool. With the **route** command, you display and modify the routing table on your Windows and Linux systems. Figure 10.4 shows the output from a **route print** command on a Windows system.

```
C:\>route print
===========================================================================
Interface List
 10...00 25 64 8c 9e bf ......Broadcom NetLink (TM) Gigabit Ethernet
  1...........................Software Loopback Interface 1
 14...00 00 00 00 00 00 00 e0 Teredo Tunneling Pseudo-Interface
 15...00 00 00 00 00 00 00 e0 Microsoft ISATAP Adapter #2
===========================================================================

IPv4 Route Table
===========================================================================
Active Routes:
Network Destination        Netmask          Gateway       Interface  Metric
          0.0.0.0          0.0.0.0      192.168.1.1   192.168.1.119     10
        127.0.0.0        255.0.0.0         On-link        127.0.0.1    306
        127.0.0.1  255.255.255.255         On-link        127.0.0.1    306
  127.255.255.255  255.255.255.255         On-link        127.0.0.1    306
      192.168.1.0    255.255.255.0         On-link    192.168.1.119    266
    192.168.1.119  255.255.255.255         On-link    192.168.1.119    266
    192.168.1.255  255.255.255.255         On-link    192.168.1.119    266
        224.0.0.0        240.0.0.0         On-link        127.0.0.1    306
        224.0.0.0        240.0.0.0         On-link    192.168.1.119    266
  255.255.255.255  255.255.255.255         On-link        127.0.0.1    306
  255.255.255.255  255.255.255.255         On-link    192.168.1.119    266
===========================================================================
Persistent Routes:
  None

IPv6 Route Table
===========================================================================
Active Routes:
 If Metric Network Destination      Gateway
 14     58 ::/0                    On-link
  1    306 ::1/128                 On-link
 14     58 2001::/32              On-link
 14    306 2001:0:9d38:6ab8:2001:1013:3f57:fe88/128
                                   On-link
 10    266 fe80::/64              On-link
 14    306 fe80::/64              On-link
 14    306 fe80::2001:1013:3f57:fe88/128
                                   On-link
 10    266 fe80::51b9:996e:9fac:7715/128
                                   On-link
  1    306 ff00::/8               On-link
 14    306 ff00::/8               On-link
 10    266 ff00::/8               On-link
===========================================================================
Persistent Routes:
  None
```

FIGURE 10.4 **The Output from a route print Command on a Windows System**

> **Note**
>
> The discussion here focuses on the Windows **route** command, but other operating systems have equivalent commands. On a Linux system, for example, the command is also **route**, but the usage and switches are different.

In addition to displaying the routing table, the Windows version of the **route** command has a number of other switches, as detailed in Table 10.15. For complete information about all the switches available with the **route** command on a Windows system, type **route** at the command line. To see a list of the **route** command switches on a Linux system, use the command **route -help**.

TABLE 10.15 **Switches for the route Command in Windows**

Switch	Description
add	Enables you to add a static route to the routing table.
delete	Enables you to remove a route from the routing table.
change	Enables you to modify an existing route.
-p	Makes the route permanent when used with the **add** command. If the **-p** switch is not used when a route is added, the route is lost upon reboot.
print	Enables you to view the system's routing table.
-f	Removes all gateway entries from the routing table.

nmap

The **nmap** utility is a free download for Windows or Linux used to scan ports on machines. Those scans can show what services are running as well as information about the target machine's operating system. The utility can be used to scan a range of IP addresses or just a single IP address.

Basic Networking Device Commands

When you're working with routers, one of the most useful troubleshooting command-line tools is **show**. This Cisco-based utility takes a plethora of options after it and can be used to view almost any variable or value. Seven of the most popular options are shown in Table 10.16.

TABLE 10.16 **Seven Popular Options for the show Command**

Option	Description
mac-address-table	Displays the MAC address table maintained by the device (usually a switch). This table contains information about MAC addresses learned by the device on its switch ports, along with associated information such as VLAN membership and port numbers.
route	Displays the routing table.
interface	Displays statistics for the interface (or interfaces with **interfaces**) configured on the router or access server.
config	Displays the current system configuration.
arp	Displays the ARP cache maintained by the router. The ARP cache contains mappings of IP addresses to MAC addresses learned through ARP requests and responses.
vlan	Displays information about VLANs configured on a switch or router in Cisco IOS-based networking environments including their names, VLAN IDs, and associated interfaces.
power	Displays information about the power supply units and power usage on a switch or router in Cisco IOS-based networking environments including their capacities, voltages, currents, and power consumption.

ExamAlert

Know the options for the **show** command shown in Table 10.16 for the Network+ exam.

Cram Quiz

1. What command can you issue from the command line to view the status of the system's ports?

   ○ A. netstat -p

   ○ B. netstat -o

   ○ C. netstat -a

   ○ D. netstat -y

2. Which of the following tools can you use to perform manual DNS lookups on a Linux system? (Choose two.)

   ○ A. dig

   ○ B. nslookup

   ○ C. tracert

   ○ D. dnslookup

3. Which of the following commands generates a "Request Timed Out" error message?

   ○ **A. ping**

   ○ **B. netstat**

   ○ **C. ipconfig**

   ○ **D. ip**

4. Which of the following commands would you use to add a static entry to the ARP table of a Windows system?

   ○ **A. arp -a *IP Address MAC Address***

   ○ **B. arp -s *MAC Address IP Address***

   ○ **C. arp -s *IP Address MAC Address***

   ○ **D. arp -i *IP Address MAC Address***

5. Which command created the following output?

   ```
   Server: nen.bx.ttfc.net
   Address: 209.55.4.155
   Name: examcram.com
   Address: 63.240.93.157
   ```

   ○ **A. ip**

   ○ **B. ipconfig**

   ○ **C. tracert**

   ○ **D. nslookup**

6. Which command displays statistics for all interfaces configured on the router or access server?

   ○ **A. show ip route**

   ○ **B. show config**

   ○ **C. show interfaces**

   ○ **D. show me state**

7. Which of the following are Layer 2 network protocols used for device discovery and neighbor identification in Ethernet networks? (Choose two.)

   ○ **A. NetFlow**

   ○ **B. LLDP**

   ○ **C. CDP**

   ○ **D. IP scanner**

8. Which of the following is like a packet sniffer and can capture the communication stream between systems? But unlike a sniffer, it can capture more than network traffic; it reads and decodes the traffic, which enables the administrator to view the network communication in English.

   ○ **A.** VFL

   ○ **B.** Cable tester

   ○ **C.** Toner

   ○ **D.** Protocol analyzer

9. Which of the following commands uses ARP to test connectivity between systems rather than Internet Control Message Protocol (ICMP)?

   ○ **A. ping**

   ○ **B. arp ping**

   ○ **C. ping6**

   ○ **D. netstat**

Cram Quiz Answers

1. **C.** You can quickly determine the status of common ports by issuing the **netstat -a** command from the command line. This command output lists the ports used by the system and whether they are open and listening.

2. **A and B.** Both the **dig** and **nslookup** commands can be used to perform manual DNS lookups on a Linux system. You cannot perform a manual lookup with the **tracert** command. There is no such command as **dnslookup**.

3. **A.** The **ping** command generates a "Request Timed Out" error when it cannot receive a reply from the destination system. None of the other commands listed produce this output.

4. **C.** The **arp -s *IP Address MAC Address*** command would correctly add a static entry to the ARP table. None of the other answers are valid ARP switches.

5. **D.** The output was produced by the **nslookup** command. The other commands listed produce different output.

6. **C.** The **show interfaces** command displays statistics for all interfaces configured on the router or access server. **show ip route** displays the routing table. **show config** displays the current system configuration. "Show Me State" is Missouri's unofficial nickname, which appears on its license plates.

7. **B and C.** Link Layer Discovery Protocol (LLDP) and Cisco Discovery Protocol (CDP) are both network protocols used for device discovery and neighbor identification in Ethernet networks. They operate at the data link layer (Layer 2) of the OSI model and enable network devices to exchange information about their capabilities, configurations, and connected neighbors. NetFlow is a proprietary Cisco protocol used for network flow analysis. A NetFlow analyzer is used to collect IP network traffic as it enters or exits an interface and can identify such values as the

source and destination of traffic, class of service, and the causes of congestion. An IP scanner is any tool that can scan for IP addresses and related information. An administrator can use it to scan available ports, discover devices, and get detailed hardware and software information on workstations and servers to manage inventory.

8. **D.** Like a packet sniffer, a protocol analyzer captures the communication stream between systems. But unlike the sniffer, the protocol analyzer captures more than network traffic; it reads and decodes the traffic. Decoding enables administrators to view the network communication in English. From this communication, administrators can get a better idea of the traffic that is flowing on the network. A visual fault locator (VFL) is a handheld hardware tool used for network troubleshooting, particularly in fiber-optic cabling systems. It emits a visible laser light that helps technicians identify breaks, bends, or other faults in fiber-optic cables. A cable tester is a hardware tool used for network troubleshooting and validation of network cabling installations. It is designed to verify the integrity and performance of various types of network cables, such as Ethernet cables (e.g., Cat 5e, Cat 6, Cat 6a), coaxial cables, and fiber-optic cables. A toner is a hardware tool used to trace and identify network cables within a structured cabling system and is also known as a cable tracer or cable toner. The toner typically consists of two components: a tone generator and a probe. The tone generator is a handheld device typically connected to one end of the cable being tested, and it transmits an electrical signal (usually in the form of an audible tone or a digital signal) onto the conductor of a network cable. The probe is a handheld receiver that is used to detect the signal transmitted by the tone generator. It is equipped with a speaker or headphones to audibly indicate when the signal is detected.

9. **B. arp ping** uses ARP to test connectivity between systems rather than Internet Control Message Protocol (ICMP), as done with a regular **ping**. **ping6** tests connectivity between two devices on a network using the IPv6 protocol in place of IPv4. **netstat** enables you to view the current TCP/IP connections on a system.

What's Next?

Congratulations! You finished the reading and are now familiar with all the objectives on the Network+ exam. You are now ready for the practice exams that are posted online to accompany this book. There are two multiple-choice question exams to help you determine how prepared you are for the actual exam and which topics you need to review further.

Network+ Cram Sheet

This Cram Sheet contains the distilled key facts about the CompTIA Network+ exam. Review this information as the last thing you do before you enter the testing center, paying special attention to those areas in which you think you need the most review. You can transfer any of these facts from your head onto a blank sheet of paper immediately before you begin the exam.

Networking Concepts

▶ As data is passed up or down through the OSI model structure, headers are added (going down) or removed (going up) at each layer—a process called encapsulation (when added) or decapsulation (when removed).

TABLE 1 **Summary of the OSI Model**

OSI Layer	Description
Application (Layer 7)	Provides access to the network for applications and certain end-user functions. Displays incoming information and prepares outgoing information for network access. Proxy servers operate at this layer. Load balancers can operate at this layer. Some firewalls (application-level) can operate at this layer.
Presentation (Layer 6)	Converts data from the application layer into a format that can be sent over the network. Converts data from the session layer into a format that the application layer can understand. Encrypts and decrypts data. Provides compression and decompression functionality.
Session (Layer 5)	Synchronizes the data exchange between applications on separate devices. Handles error detection and notification to the peer layer on the other device. Load balancers and firewalls (transport layer) can operate at this layer.
Transport (Layer 4)	Establishes, maintains, and breaks connections between two devices. Determines the ordering and priorities of data. Performs error checking and verification and handles retransmissions if necessary.
Network (Layer 3)	Provides mechanisms for the routing of data between devices across single or multiple network segments. Handles the discovery of destination systems and addressing. Routers, Layer 3 switches, IDS/IPS, and firewalls (network layer) can operate at this layer.

OSI Layer	Description
Data link (Layer 2)	Has two distinct sublayers: Link Layer Control (LLC) and Media Access Control (MAC). Performs error detection and handling for the transmitted signals. Defines the method by which the medium is accessed. Defines hardware addressing through the MAC sublayer. NICs, access points (APs), bridges, and Layer 2 switches operate at this layer.
Physical (Layer 1)	Defines the network's physical structure. Defines voltage/signal rates and the physical connection methods. Defines the physical topology. Hubs and repeaters operate at this layer.

▶ A local-area network (LAN) is a data network that is restricted to a single geographic location and typically encompasses a relatively small area, such as an office building or school. The function of the LAN is to interconnect workstation computers and devices for the purpose of sharing files and resources.

▶ A wide-area network (WAN) is a network that spans more than one geographic location, often connecting separated LANs. WANs are slower than LANs and often require additional and costly hardware such as routers, dedicated leased lines, and complicated implementation procedures.

TABLE 2 **Network Devices Summary**

Device	Description	Key Points
Router	Connects networks	A router uses the software-configured network address to make forwarding decisions.
Switch	Connects devices on a twisted-pair network	A switch forwards data to its destination by using the MAC address embedded in each packet. It only forwards data to nodes that need to receive it.
Firewall	Provides controlled data access between networks	Firewalls can be hardware or software based. They are an essential part of a network's security strategy.
IDS/IPS	Detects and prevents intrusions	An IDS monitors the network and attempts to detect/alert suspicious activity. An IPS can actually prevent intrusion attempts.
Load balancer	Distributes network load	Load balancing increases redundancy and performance by distributing the load to multiple servers.
Proxy	Acts on behalf of another	Proxy servers typically are part of a firewall system. They have become so integrated with firewalls that the distinction between the two can sometimes be lost. A proxy server typically provides an increased level of security, caching, NAT, and administrative control.

Device	Description	Key Points
Network-attached storage (NAS)	Allows multiple users and client devices to access data simultaneously	NAS devices use various protocols for file sharing and access, including NFS, SMB, TCP/IP, and FTP.
Storage-area network (SAN)	Used to connect storage devices to servers	Unlike NAS, which provides file-level access, SANs provide block-level access to storage volumes.
Access point	Used to create a wireless LAN and to extend a wired network	This device uses the wireless infrastructure network mode to provide a connection point between WLANs and a wired Ethernet LAN.
Wireless LAN controller	Used with branch/remote office deployments for wireless authentication	When an AP boots, it authenticates with a controller before it can start working as an AP.

▶ The National Institute of Standards and Technology (NIST) defines three cloud computing service models: Software as a Service (SaaS), Platform as a Service (PaaS), and Infrastructure as a Service (IaaS).

▶ NIST defines three possible cloud delivery models: private, public, and hybrid.

▶ In the context of computer networking, multitenancy often relates to cloud computing and virtualization environments, where a single physical infrastructure is shared among multiple users or organizations.

▶ Elasticity is the ability of a network infrastructure or service to dynamically adapt and scale resources, such as bandwidth, computing capacity, or storage, in response to changing demand or workload requirements.

▶ Virtual private cloud (VPC) technology allows users to create isolated and logically segmented private networks within a public cloud environment via an endpoint that allows the VPC to be connected with other services without the need for additional technologies.

▶ Infrastructure as Code (IaC) automation plays a crucial role in managing upgrades and dynamic inventories by providing automated and repeatable processes for updating infrastructure configurations and managing inventory changes. IaC automation typically leverages playbooks, templates, and reusable tasks to streamline the provisioning and management of infrastructure resources. Playbooks are high-level definitions of infrastructure configurations and workflows written in a declarative or imperative language. Playbooks provide a structured and reusable format for defining infrastructure configurations, making it easier to manage and automate complex deployment processes.

▶ AWS Direct Connect is a cloud service that links your network directly to Amazon Web Services (AWS) to deliver consistent, low-latency performance.

▶ Network security groups serve as a virtual firewall within cloud computing environments to control and regulate network traffic to and from resources like virtual machines, network interfaces, and subnets.

▶ Network functions virtualization (NFV) is a network architecture in which virtualization technologies are used to connect/create communication services. NFV allows for the virtualization of network functions such as routers, firewalls, and switches, resulting in increased flexibility and scalability.

TABLE 3   **Port Assignments for Commonly Used Protocols**

Protocol	Port Assignment
FTP	20, 21
SSH	22
SFTP	22
Telnet	23
SMTP	25
DNS	53
DHCP	67, 68
TFTP	69
HTTP	80
POP3	110
NTP	123
SNMP	161, 162
LDAP	389
HTTPS	443
SMB	445
Syslog	514
SMTPS	587
LDAPS	636
SQL server	1433
RDP	3389
SIP	5060, 5061

▶ A software-defined wide-area network (SD-WAN) is an extension of software-defined networking (SDN), which is commonly used in telecom and datacenters, on a large scale. The concept is to take many of the principles that make cloud computing so attractive and make them accessible at the WAN level.

▶ A storage-area network (SAN) consists of just what the name implies: networked/shared storage devices. With clustered storage, you can use multiple devices to increase performance. SANs are subsets of LANs and offer block-level data storage that appears within the operating systems of the connected devices as locally attached devices.

▶ A content delivery network (CDN) is a geographically distributed network of servers and datacenters designed to efficiently deliver web content, such as web pages, images, videos, and other multimedia files, to users based on their geographic location.

▶ A topology refers to a network's physical and logical layout. A network's physical topology refers to the actual layout of the computer cables and other network devices. A network's logical topology refers to the way in which the network appears to the devices that use it.

▶ Wireless networks typically are implemented using one of two wireless topologies: infrastructure (managed, wireless topology) or ad hoc (unmanaged, wireless topology).

▶ The term *hybrid topology* also can refer to the combination of wireless and wired networks but often just refers to the combination of physical networks.

▶ Unicast traffic involves communication between a single sender and a single receiver, multicast traffic is sent from one sender to multiple specific recipients, anycast traffic targets the nearest available server from a group to which a request is made, while broadcast traffic is sent from one sender to all devices in the network.

▶ F-type connectors are used with coaxial cable, most commonly to connect cable modems and TVs. F-type connectors are screw-type connectors.

▶ MPO, ST, SC, and LC connectors are associated with fiber cabling. ST connectors offer a twist-type attachment, and SC, LC, and MPO connectors are push-on. You can choose to purchase ones that are either angled physical contact (APC) or ultra-physical contact (UPC).

▶ RJ-45 connectors are used with UTP cable and are associated with networking applications.

▶ North-south traffic flows between the internal network and external networks, such as the Internet, while east-west traffic flows between devices or servers within the internal network.

▶ Plenum-rated cables are used to run cabling through walls or ceilings.

▶ A MAC address is a 6-byte hexadecimal address that allows a device to be uniquely identified on the network. A MAC address combines numbers and the letters *A* to *F*. An example of a MAC address is 00:D0:59:09:07:51.

▶ A Class A TCP/IP address uses only the first octet to represent the network portion, a Class B address uses two octets, and a Class C address uses three octets.

▶ Class A addresses span from 1 to 126, with a default subnet mask of 255.0.0.0.

▶ Class B addresses span from 128 to 191, with a default subnet mask of 255.255.0.0.

▶ Class C addresses span from 192 to 223, with a default subnet mask of 255.255.255.0.

▶ The 127 network ID is reserved for the IPv4 local loopback.

TABLE 4   **IPv4 Private Address Ranges**

Class	Address Range	Default Subnet Mask
A	10.0.0.0 to 10.255.255.255	255.0.0.0
B	172.16.0.0 to 172.31.255.255	255.255.0.0
C	192.168.0.0 to 192.168.255.255	255.255.255.0

▶ Subnetting is a process in which parts of the host ID portion of an IP address are used to create more network IDs.

▶ Automatic Private IP Addressing (APIPA) is a system used on Windows to automatically self-assign an IP address in the 169.x.x.x range in the absence of a DHCP server.

▶ In addressing terms, the CIDR value is expressed after the address, using a slash. So, the address 192.168.2.1/24 means that the node's IP address is 192.168.2.1, and the subnet mask is 255.255.255.0.

▶ IPv6 networks use Stateless Address Auto Configuration (SLAAC) to assign IP addresses. With SLAAC, devices send the router a request for the network prefix, and the device then uses the prefix along with its own MAC address to create an IP address.

▶ RFC 1918 addresses are IP addresses that an enterprise can assign to internal hosts without requiring coordination with an Internet registry.

▶ NAT64 is a translation mechanism used to facilitate communication between IPv6-only and IPv4-only devices by mapping IPv6 addresses to IPv4 addresses and vice versa.

TABLE 5 **Comparing IPv4 and IPv6**

Address Feature	IPv4 Address	IPv6 Address
Loopback address	127.0.0.1	0:0:0:0:0:0:0:1 (::1)
Network-wide addresses	IPv4 public address ranges	Global unicast IPv6 addresses
Private network addresses	10.0.0.0 172.16.0.0 192.168.0.0	Site-local address ranges (FEC0::)
Autoconfigured addresses	IPv4 automatic private IP addressing (169.254.0.0)	Link-local addresses of FE80:: prefix

▶ In a star/hub-and-spoke configuration, all devices on the network connect to a central device, and this central device creates a single point of failure on the network.

▶ The mesh topology requires each computer on the network to be individually connected to every other device. This configuration provides maximum reliability and redundancy for the network.

▶ A wireless infrastructure network uses a centralized device known as a wireless access point (WAP). Ad hoc wireless topologies are a peer-to-peer configuration and do not use a wireless access point.

▶ A three-tiered architecture separates the user interface, the functional logic, and the data storage/access as independent modules/platforms.

▶ Fibre Channel is widely used for high-speed fiber networking and has become common in enterprise SANs.

▶ A virtual switch (vSwitch) works the same as a physical switch but allows multiple switches to exist on the same host, saving the implementation of additional hardware.

▶ A virtual firewall (VF) is either a network firewall service or an appliance running entirely within the virtualized environment. Regardless of which implementation, a virtual firewall serves the same purpose as a physical one: packet filtering and monitoring. The firewall can also run in a guest OS VM.

▶ In a virtual environment, shared storage can be done on storage-area network (SAN), network-attached storage (NAS), and so on, but the virtual machine sees only a "physical disk." With clustered storage, you can use multiple devices to increase performance.

TABLE 6 **802.11 Wireless Standards**

IEEE Standard	Frequency/ Medium	Speed	Topology	Transmission Range	Access Method
802.11	2.4 GHz RF	1 to 2 Mbps	Ad hoc/ infrastructure	20 feet indoors	CSMA/CA
802.11a	5 GHz	Up to 54 Mbps	Ad hoc/ infrastructure	25 to 75 feet indoors; range can be affected by building materials	CSMA/CA
802.11b	2.4 GHz	Up to 11 Mbps	Ad hoc/ infrastructure	Up to 150 feet indoors; range can be affected by building materials	CSMA/CA
802.11g	2.4 GHz	Up to 54 Mbps	Ad hoc/ infrastructure	Up to 150 feet indoors; range can be affected by building materials	CSMA/CA
802.11n (Wi-Fi 4)	2.4 GHz/5 GHz	Up to 600 Mbps	Ad hoc/ infrastructure	175+ feet indoors; range can be affected by building materials	CSMA/CA
802.11ac (Wi-Fi 5)	5 GHz	Up to 1.3 Gbps	Ad hoc/ infrastructure	115+ feet indoors; range can be affected by building materials	CSMA/CA
802.11ax (Wi-Fi 6)	2.4 GHz/5 GHz	Up to 14 Gbps	Ad hoc/ infrastructure	~230 feet indoors; range can be affected by building materials	CSMA/CA

▶ Nonoverlapping channels are a set of wireless frequency channels that do not interfere with each other. For example, in the 2.4 GHz band, channels 1, 6, and 11 are commonly considered nonoverlapping channels for Wi-Fi networks.

▶ Quality of Service (QoS) allows administrators to predict bandwidth use, monitor that use, and control it to ensure that bandwidth is available to applications that need it.

▶ An intrusion detection system (IDS) can detect malware or other dangerous traffic that may pass undetected by the firewall. Most IDSs can detect potentially dangerous content by its signature.

▶ An intrusion prevention system (IPS) is a network device that continually scans the network, looking for inappropriate activity. It can shut down or prevent any potential threats.

▶ A virtual private network (VPN) extends a LAN by establishing a remote connection, a connection tunnel, using a public network such as the Internet. Common VPN implementations include site to site/host to site/host to host.

▶ VPNs are created and managed by using protocols such as Point-to-Point Tunneling Protocol (PPTP) and Layer 2 Tunneling Protocol (L2TP), which build on the functionality of PPP. This makes it possible to create dedicated point-to-point tunnels through a public network such as the Internet. Currently, the most common methods for creating secure VPNs include IP Security (IPsec) and Secure Sockets Layer/Transport Layer Security (SSL/TLS).

▶ 802.3 defines the Carrier Sense Multiple Access with Collision Detection (CSMA/CD) media access method used in Ethernet networks. This is the most popular networking standard used today.

Network Implementation

▶ Network Address Translation (NAT) translates private network addresses into public network addresses.

▶ Port Address Translation (PAT) is a variation on NAT in which all systems on the LAN are translated into the same IP address but with different port number assignment.

▶ A router that uses a link-state protocol differs from a router that uses a distance-vector protocol because it builds a map of the entire network and then holds that map in memory. Link-state protocols include Open Shortest Path First (OSPF).

▶ Hops are the means by which distance-vector routing protocols determine the shortest way to reach a given destination. Each router constitutes one hop, so if a router is four hops away from another router, there are three routers, or hops, between itself and the destination.

▶ Routing Information Protocol version 2 (RIPv2) is a distance-vector routing protocol used for TCP/IP.

▶ The **route add** command adds a static route to the routing table. The **route add** command with the **-p** switch makes the static route persistent.

▶ Distance-vector routing protocols operate by having each router send updates about all the other routers it knows about to the routers directly connected to it. Protocols include RIP, RIPv2, and EIGRP.

▶ When you want the best of both worlds, a hybrid protocol can be the answer. A popular hybrid protocol is the Border Gateway Protocol (BGP).

▶ Default gateways are the means by which a device can access hosts on other networks for which it does not have a specifically configured route.

▶ Half-duplex mode enables each device to both transmit and receive, but only one of these processes can occur at a time.

▶ Full-duplex mode enables devices to receive and transmit simultaneously.

▶ An antenna's strength is its gain value.

TABLE 7 **Comparing Omnidirectional and Unidirectional Antennas**

Characteristic	Omnidirectional	Unidirectional	Advantage/Disadvantage
Wireless area coverage	General coverage area	Focused coverage area.	Omnidirectional allows 360-degree coverage, giving it a wide coverage area. Unidirectional provides a targeted path for signals to travel.
Wireless transmission range	Limited	Long point-to-point range.	Omnidirectional antennas provide a 360-degree coverage pattern and, as a result, far less range. Unidirectional antennas focus the wireless transmission; this focus enables greater range.
Wireless coverage shaping	Restricted	The unidirectional wireless range can be increased and decreased.	Omnidirectional antennas are limited to their circular pattern range. Unidirectional antennas can be adjusted to define a specific pattern, wider or more focused.

▶ Multiuser multiple input, multiple output (MU-MIMO) is an enhancement over the original MIMO technology. It allows antennas to be spread over a multitude of independent access points.

▶ Virtual LANs (VLANs) are used for network segmentation. 802.1Q is the Institute of Electrical and Electronics Engineers (IEEE) specification developed to ensure interoperability of VLAN technologies from the various vendors. A voice VLAN is a type of VLAN configuration commonly used in VoIP (voice over Internet Protocol) networks to ensure the efficient and secure transmission of voice traffic.

▶ A native VLAN is a default VLAN to which untagged traffic is assigned on a trunk port. Trunk ports are used to carry traffic for multiple VLANs between switches or between a switch and a router. When traffic enters a trunk port without a VLAN tag (i.e., untagged traffic), it is automatically placed into the native VLAN. The native VLAN is used to carry control traffic, such as management and Spanning Tree Protocol (STP) traffic, and can also be used to carry user data.

▶ VLAN trunking is the application of trunking to the virtual LAN—now common with routers, firewalls, VMware hosts, and wireless access points. VLAN trunking provides a simple and cheap way to offer a nearly unlimited number of virtual network connections. The requirements are only that the switch, the network adapter, and the OS drivers all support VLANs.

▶ The VLAN Trunking Protocol (VTP) is a proprietary protocol from Cisco.

▶ IEEE 802.11 wireless systems communicate with each other using radio frequency signals in the band between 2.4 GHz and 2.5 GHz or 5.0 GHz. Of those in the 2.4 to 2.5 range, neighboring channels are 5 MHz apart. Applying two channels that allow the maximum channel separation decreases the amount of channel crosstalk and provides a noticeable performance increase over networks with minimal channel separation.

▶ Type C fire extinguishers are used for electrical fires.

▶ The major drawback to gas-based fire suppression systems is that they require sealed environments to operate.

▶ Temperature monitors keep track of the temperature in wiring closets and server rooms.

▶ Humidity control prevents the buildup of static electricity in the environment. If the humidity drops much below 50 percent, electronic components are extremely vulnerable to damage from electrostatic shock.

Network Operations

▶ Documentation should include diagrams of the physical and logical network design. The physical topology refers to how a network is physically constructed, or how it looks.

▶ A baseline/golden configuration is a baseline configuration version used in configuration management as an ideal configuration against which configurations from similar devices can be compared.

▶ A DHCP relay is nothing more than an agent on the router that acts as a go-between for clients and the server. One level above DHCP relay is *IP helper*. These two terms are often used as synonyms, but they are not; a better way to think of it is with IP helper being a superset of DHCP relay.

TABLE 8 **Standard Business Documents**

Document	Description
SLA (service-level agreement)	An agreement between a customer and provider detailing the level of service to be provided on a regular basis and in the event of problems
MOU (memorandum of understanding)	An agreement (bilateral or multilateral) between parties defining terms and conditions of an agreement
NDA (nondisclosure agreement)	A document agreeing that information shared will not be shared further with other parties
AUP (acceptable use policy)	A plan that describes how the employees in an organization can use company systems and resources: both software and hardware

▶ Domain Name Service (DNS) resolves hostnames to IP addresses. DNS record types include A, MX, AAAA, CNAME, and PTR. Dynamic DNS (DDNS) automatically updates DNS information, often in real time.

▶ In a network that does not use Dynamic Host Configuration Protocol (DHCP), you need to watch for duplicate IP addresses that prevent a user from logging on to the network.

▶ DNS over HTTP (DoH) encrypts DNS queries and responses within HTTP traffic, while DNS over TLS (DoT) encrypts DNS traffic using the Transport Layer Security (TLS) protocol, providing privacy and security enhancements for domain name resolution.

▶ Network Time Protocol (NTP) synchronizes time across a network with moderate accuracy, Precision Time Protocol (PTP) provides highly accurate time synchronization for devices in a local-area network, and Network Time Security (NTS) enhances NTP by securing time synchronization exchanges against attacks and tampering.

▶ In-band network device management is local management (the most common method), and out-of-band management is done remotely.

▶ A system's security log contains events related to security incidents, such as successful and unsuccessful logon attempts and failed resource access. An application log contains information logged by applications that run on a particular system rather than the operating system itself. System logs record information about components or drivers in the system.

▶ Syslog is a standard for message logging. It is available on most network devices (such as routers, switches, and firewalls), as well as printers and UNIX/Linux-based systems. Over a network, a syslog server listens for and then logs data messages coming from the syslog client.

▶ An SNMP management system is a computer running a special piece of software called a network management system (NMS). SNMP uses databases of information called Management Information Bases (MIBs) to define what parameters are accessible, which of the parameters are read-only, and which can be set. MIBs are available for thousands of devices and services, covering every imaginable need.

▶ SNMPv2c does not provide robust security features out-of-the-box; it relies primarily on community strings for authentication, which are essentially plaintext passwords used to control access to SNMP-enabled devices. This makes SNMPv2c vulnerable to security threats, such as eavesdropping. SNMPv3 is the latest and most secure version of SNMP and allows for the use of strong cryptographic algorithms, such as HMAC-SHA, HMAC-MD5, and AES, to ensure data integrity and confidentiality.

▶ Fault tolerance is the capability to withstand a fault (failure) without losing data. This can be accomplished through the use of RAID, backups, and similar technologies. Popular fault-tolerant RAID implementations include RAID 1, RAID 5, and RAID 10 (1+0).

▶ The MTBF is the measurement of the anticipated or predicted incidence of failure of a system or component between inherent failures, whereas the MTTR is the measurement of how long it takes to repair a system or component after a failure occurs. The RTO is the maximum amount of time that a process or service is allowed to be down and the consequences still to be considered acceptable. The RPO is the maximum time in which transactions could be lost from a major incident.

▶ Within a few hours, a hot recovery site can become a fully functioning element of an organization. A cold recovery site is a site that can be up and operational in a relatively short amount of time, such as a day or two. Provision of services, such as telephone lines and power, is taken care of, and the basic office furniture might be in place. A warm site typically has computers but is not configured ready to go. This means that data might need to be upgraded, or other manual interventions might need to be performed before the network is again operational.

▶ In a full backup, all data is backed up. Full backups do not use the archive bit, but they do clear it.

▶ Incremental backups back up all data that has changed since the last full or incremental backup. They use and clear the archive bit.

▶ Security information and event management (SIEM) products provide notifications and real-time analysis of security alerts and can help you head off problems quickly.

▶ Tabletop exercises are scenario-based discussions or simulations conducted in a classroom or meeting room setting, where key stakeholders participate in guided discussions and decision-making exercises to evaluate their response to hypothetical disaster scenarios.

▶ Validation tests, also known as technical or operational tests, involve the actual execution of recovery procedures and the testing of DR infrastructure, systems, and processes in a controlled environment.

Network Security

▶ Authentication refers to the mechanisms used to verify the identity of the computer or user attempting to access a particular resource. This includes passwords and biometrics.

▶ Authorization is the method used to determine whether an authenticated user has access to a particular resource. This is commonly determined through group association; for example, a particular group may have a specific level of security clearance.

▶ Accounting refers to the tracking mechanisms used to keep a record of events on a system.

▶ User authentication methods include multifactor authentication (MFA), two-factor authentication (2FA), and single sign-on (SSO). The factors used in authentication systems or methods are based on one or more of these five factors:

  ▶ Something you know, such as a password or PIN

  ▶ Something you have, such as a smartcard, token, or identification device

  ▶ Something you are, such as your fingerprints or retinal pattern (often called biometrics)

  ▶ Somewhere you are (based on geolocation)

  ▶ Something you do, such as an action you must take to complete authentication

▶ The Internet of Things (IoT) refers to interconnected devices and systems aimed at enhancing daily life and consumer experiences, while the Industrial Internet of Things (IIoT) focuses on interconnected industrial devices and systems for improving efficiency, productivity, and automation in sectors like manufacturing, energy, and transportation.

▶ The IEEE standard 802.1X defines port-based security for wireless network access control.

▶ Network access control (NAC) is a method to restrict access to the network based on identity or posture. A posture assessment is any evaluation of a system's security based on settings and applications found.

▶ A public key infrastructure (PKI) is a collection of software, standards, and policies that are combined to allow users from the Internet or other unsecured public networks to securely exchange data.

▶ A public key is a nonsecret key that forms half of a cryptographic key pair that is used with a public key algorithm. The public key is freely given to all potential receivers.

▶ A private key is the secret half of a cryptographic key pair that is used with a public key algorithm. The private part of the public key cryptography system is never transmitted over a network.

▶ A certificate is a digitally signed statement that associates the credentials of a public key to the identity of the person, device, or service that holds the corresponding private key. Certificate authorities (CAs) issue and manage certificates. They validate the identity of a network device or user requesting data.

▶ A certificate revocation list (CRL) is list of certificates that were revoked before they reached the certificate expiration date.

▶ Physical security controls include access control vestibules (previously known as mantraps), video monitoring, proximity readers/key fobs, keypad/cipher locks, biometrics, and security guards.

▶ A honeypot is a computer that has been designated as a target for computer attacks. A honeynet is an entire network set up to monitor attacks from outsiders.

▶ Terminal Access Controller Access Control System Plus (TACACS+) is a security protocol designed to provide centralized validation of users who are attempting to gain access to a router or network access server (NAS).

▶ Remote Authentication Dial-In User Service (RADIUS) is a security standard that uses a client/server model to authenticate remote network users.

▶ Both RADIUS and TACACS+ provide authentication, authorization, and accounting services. One notable difference between TACACS+ and RADIUS is that TACACS+ relies on the connection-oriented TCP, whereas RADIUS uses the connectionless UDP.

▶ A screened subnet (previously known as a demilitarized zone or DMZ) is part of a network on which you place servers that must be accessible by sources both outside and inside your network.

▶ An access control list (ACL) typically refers to specific access permissions assigned to an object or device on the network. For example, using Media Access Control (MAC) address filtering, wireless routers can be configured to restrict who can and cannot access the router based on the MAC address.

▶ When a port is blocked, you disable the capability for traffic to pass through that port, thereby filtering that traffic.

▶ To create secure data transmissions, IPsec uses two separate protocols: Authentication Header (AH) and Encapsulating Security Payload (ESP).

▶ In role-based access control (RBAC), access decisions are determined by the roles that individual users have within the organization.

▶ Risk management involves recognizing and acknowledging that risks exist and then determining what to do about them.

▶ Social engineering involves tricking people into revealing their password or some form of security information. It might include trying to get users to send passwords or other information over email, following someone closely into a secured area (known as tailgating), walking in with them (known as piggybacking), or looking over someone's shoulder at their screen (known as shoulder surfing).

▶ A rogue DHCP server added to a network has the potential to issue an address to a client, isolating it on an unauthorized network where its data can be captured.

▶ A rogue access point describes a situation in which a wireless access point has been placed on a network without the administrator's knowledge.

▶ Often users receive a variety of emails offering products, services, information, or opportunities. Unsolicited email of this type is called phishing.

▶ With DNS poisoning, the DNS server is given information about a name server that it thinks is legitimate when it isn't.

▶ Spoofing is a technique in which the real source of a transmission, file, or email is concealed or replaced with a fake source.

▶ Zero trust architecture emphasizes a granular approach to access control, where access decisions are made based on contextual factors such as user identity, device posture, location, and behavior. Implementing fine-grained access controls requires a comprehensive understanding of the organization's network, applications, and user requirements.

▶ A zero-day vulnerability is a newly discovered vulnerability for which a patch or fix has not yet been issued.

Network Troubleshooting

TABLE 9 **Network Troubleshooting Methodology**

Steps	Actions	Considerations
Identify the problem.	Gather information.	
	Question users.	
	Identify symptoms.	
	Determine whether anything has changed.	
	Duplicate the problem, if possible.	
	Approach multiple problems individually.	
Establish a theory of probable cause.	Question the obvious.	
	Consider multiple approaches.	Top-to-bottom/ bottom-to-top OSI model
		Divide and conquer
Test the theory to determine the cause.	If the theory is confirmed, determine next steps to resolve the problem.	
	If the theory is not confirmed, reestablish a new theory or escalate.	
Establish a plan of action to resolve the problem and identify potential effects.		
Implement the solution or escalate as necessary.		
Verify full system functionality and implement preventive measures if applicable.		
Document findings, actions, outcomes, and lessons learned throughout the process.		

▶ Latency is one of the biggest problems with satellite access. Latency is the time lapse between sending or requesting information and the time it takes to return. Satellite communication experiences high latency due to the distance it has to travel as well as weather conditions. Although latency is not restricted solely to satellites, it is one of the easiest forms of transmission to associate with it. In reality, latency can occur with almost any form of transmission.

▶ Jitter is closely tied to latency but differs in the length of the delay between received packets. While the sender continues to transmit packets in a continuous stream and space them evenly apart, the delay between packets received varies instead of remaining constant. This delay can be caused by network congestion, improper queuing, or configuration errors.

▶ Giants are packets that are discarded because they exceed the medium's maximum packet size. They are frames received that are larger than 1,518 bytes.

▶ Runts are frames that did not meet the minimum frame size requirement of 64 bytes. They are typically caused by collisions. When an interface receives an incoming frame smaller than 64 bytes, the frame is considered a runt.

▶ The **netstat -a** command can be used on a Windows-based system to see the status of ports. It is used to view both inbound and outbound TCP/IP network connections.

▶ You can ping the local loopback adapter by using the command **ping 127.0.0.1**. If this command is successful, you know that the TCP/IP suite is installed correctly on your system and is functioning.

▶ In Windows, the **tracert** command reports how long it takes to reach each router in the path. It's a useful tool for isolating bottlenecks in a network. The **traceroute** command performs the same task on UNIX and Linux systems.

▶ Address Resolution Protocol (ARP) is the part of the TCP/IP suite whose function is to resolve IP addresses to MAC addresses.

▶ **ipconfig** shows the IP configuration information for all NICs installed in a system.

▶ **ipconfig /all** is used to display detailed TCP/IP configuration information.

▶ **ipconfig /renew** is used on Windows operating systems to renew the system's DHCP information.

▶ When looking for client connectivity problems using **ipconfig**, you should ensure that the gateway is set correctly.

▶ The **ifconfig** or **ip** command is the Linux equivalent of the **ipconfig** command.

▶ The **nslookup** command is a TCP/IP diagnostic tool used to troubleshoot DNS problems.

▶ The weakening of data signals as they traverse the media is called attenuation.

▶ Toner probes are used to locate cables hidden in floors, ceilings, or walls and to track cables from the patch panel to their destination.

▶ A Wi-Fi analyzer is used to identify Wi-Fi problems. It can be helpful in finding the ideal place for locating an access point or the ideal channel to use.

▶ Protocol analyzers can be hardware or software based. Their primary function is to analyze network protocols such as Transmission Control Protocol (TCP), User Datagram Protocol (UDP), Hypertext Transfer Protocol Secure (HTTPS), File Transfer Protocol (FTP), and more.

▶ The **dig** command is used on a Linux, UNIX, or macOS system to perform manual DNS lookups.

▶ The **tcpdump** command is used on Linux/UNIX systems to print the contents of network packets.

▶ The **iperf** tool is used for active measurements of the maximum achievable bandwidth and used for network tuning.

▶ An IP scanner is any tool that can scan for IP addresses and related information. An administrator can use it to scan available ports, discover devices, and get detailed hardware and software information on workstations and servers to manage inventory.

▶ With the **route** command, you display and modify the routing table on your Windows and Linux systems.

▶ When you're working with routers, one of the most useful troubleshooting command-line tools is **show**. The **show** options include **show interfaces**, **show config**, and **show ip route**.

▶ **nmap** is an open-source tool used for network exploration and security auditing. It utilizes various scanning techniques, such as TCP SYN scan, UDP scan, and ICMP echo request (ping) scan, to gather information about devices, open ports, and running services on the network.

Index

Numerics

A

D

X-Y-Z

To receive your 10% off Exam Voucher, register your product at:

www.pearsonitcertification.com/register

and follow the instructions.